SOLID STATE PHYSICS

VOLUME 60

Founding Editors

FREDERICK SEITZ

DAVID TURNBULL

SOLID STATE PHYSICS

Advances in
Research and Applications

Editors

HENRY EHRENREICH

FRANS SPAEPEN

Division of Engineering and Applied Sciences
Harvard University
Cambridge, Massachusetts

VOLUME 60

AMSTERDAM • BOSTON • HEIDELBERG • LONDON • NEW YORK • OXFORD
PARIS • SAN DIEGO • SAN FRANCISCO • SINGAPORE • SYDNEY • TOKYO

Academic Press is an imprint of Elsevier

Academic Press is an imprint of Elsevier
84 Theobald's Road, London WC1X 8RR, UK
Radarweg 29, PO Box 211, 1000 AE, Amsterdam, The Netherlands
30 Corporate Drive, Suite 400, Burlington, MA 01803, USA
525 B Street, Suite 1900, San Diego, California 92101-4495, USA

This book is printed on acid-free paper. ∞

Copyright © 2006, Elsevier Inc. All rights reserved

No part of this publication may be reproduced, stored in a retrieval system, or transmitted in any form or by any means electronic, mechanical, photocopying, recording or otherwise, without the prior written permission of the publisher.

Permissions may be sought directly from Elsevier's Science and Technology Rights Department in Oxford, UK: phone: (+44) (0) 1865 843830; fax: (+44) (0) 1865 853333; e-mail: permissions@elsevier.co.uk. You may also complete your request on-line via the Elsevier homepage (http://www.elsevier.com), by selecting 'Customer Support' and then 'Obtaining Permissions'

ISBN-13: 978-0-12-607760-5
ISBN-10: 0-12-607760-6
ISSN: 0081-1947

For information on all Academic Press publications
visit our web site at http://books.elsevier.com

Printed and bound in USA

05 06 07 08 09 10   10 9 8 7 6 5 4 3 2 1

Contents

CONTRIBUTORS TO VOLUME 60 vii
PREFACE ix

Strongly Correlated Electrons
P. Fulde, P. Thalmeier and G. Zwicknagl

I.	Introduction	2
II.	Special Features of Strong Correlations	8
III.	Kondo Lattice Systems	14
IV.	Quantum Phase Transitions	30
V.	Partial Localization	48
VI.	Charge Ordering	66
VII.	Geometrically Frustrated Lattices	122
VIII.	High-Energy Excitations	147
IX.	Summary and Outlook	176
	Acknowledgment	179

Defect-Induced Dynamic Pattern Formation in Metals and Alloys
Y. Bréchet and C. Hutchinson

I.	Introduction	182
II.	Free Energy Changes, Driving Forces and Energy Input	188
III.	Chemical Patterning: Interface-Mediated Transformations: Eutectoid Decomposition and Discontinuous Precipitation	192
IV.	Structural Defect Patterning: Grain Growth, Recovery and Recrystallization	233
V.	Structural Defect Patterning: Irradiation and Plastic Deformation	248
VI.	Spatio-Temporal Patterning in Plasticity: The Portevin–Le Chatelier Effect	273
VII.	Concluding Remarks	285

AUTHOR INDEX 289
SUBJECT INDEX 303

Contributors to Volume 60

Numbers in parentheses indicate the pages on which the authors' contributions begin.

YVES BRÉCHET (181) *Laboratoire de Thermodynamique et Physico-Chimie Métallurgiques, Institut National Polytechnique de Grenoble, 38402, St. Martin D'Hères, France*

PETER FULDE (1) *Max-Planck-Institut für Physik komplexer Systeme, 01187 Dresden, Germany*

CHRISTOPHER HUTCHINSON (181) *Department of Materials Engineering, Monash University, Clayton, 3168, Vic, Australia*

PETER THALMEIER (1) *Max-Planck-Institut für Chemische Physik fester Stoffe, 01187 Dresden, Germany*

GERTRUD ZWICKNAGL (1) *Institut für Mathematische Physik, Technische Universität Braunschweig, 38106 Braunschweig, Germany*

Preface

The article by Fulde, Thalmeier and Zwicknagl traces many of the recent developments in the field of strongly correlated many electron systems. It is very useful both as a reference and a pedagogical exposition since it places these developments into a historical context beginning with early developments in the electron theory of solids. Together with its extensive references and its formal elucidation of important theoretical subjects, the article supplies an excellent entry point into the field which is closely coupled to the classic works on the subject. Many early important ideas, such as Hückel and Heitler–London theory, the Wigner lattice, metal-insulator transitions originated in the thirties and were extensively developed in the sixties when it came to be realized that transition and rare earth metals with their characteristic d and f electrons respectively should be viewed as at least moderately strongly correlated systems requiring more sophisticated theoretical treatments than those supplied by the simplest band theories. The development of the relevant theories are associated with Hubbard and Mott among many others. They introduced model Hamiltonians, which were amenable to easily performed calculation whose results exhibited metal-insulator transitions for appropriate choices of the inter- and intra-atomic physical parameters. The Hubbard Hamiltonian, for example, motivated by lattices containing transition metal constituents has remained, in the authors' words one of the "working horses of many studies of strongly correlated" electron systems. Even though early conventional applications are limited to d electrons, many of the generic effects of strongly correlated electrons are captured by its solutions.

    A principal impetus for the development of the field is associated with the discovery of high temperature superconductivity in copper-oxide based perovskites in 1986. Indeed, but for the strong electron correlations in hole-doped superconductors like $La_{2-x}Ba_xCuO_4$ would be metallic instead of insulating antiferromagnetic. It is the richness of the periodic table resulting from the presence of inner shell electrons in the heavier elements and their associated correlations that is responsible for these otherwise unexpected physical effects.

    Although superconducting phenomena play an important role in the article, we stress that the principal topic concerns the more general behavior of strongly correlated electrons in solids. The article presents a detailed description of Kondo lattice systems and its associated heavy fermions, which preceded the discovery of high temperature superconductivity. It relates the results to a wide variety of experimental observations in real systems wherever possible. Readers having special interests in given heavy fermion materials will thus be well rewarded by browsing its pages and detailed figures of, for example, Fermi surfaces, photoemission results and phase diagrams.

The article begins with a detailed exposition of the physical features that characterize strong electron correlations, and presents a detailed description of the central feature associated with strong correlations leading to the presence of energy scales low compared to the typical metallic Fermi energy possibly resulting in deviations from Fermi-liquid behavior.

This introduction is followed by a detailed description of Kondo lattice systems which deal with magnetic impurities associated with spins embedded in a metal and interacting via exchange scattering. The understanding of Ce-based heavy Fermi systems present particular challenges. The authors describe theoretical techniques for describing such systems that combine *ab initio* approaches with a phenomenologically based Landau theory. A so-called renormalized band scheme is used for calculating realistic quasi-particle bands of materials grown and observed in the laboratory.

Quantum phase transitions in localized and itinerant magnetic systems characterize many systems of direct interest here. Quantum criticality in the Kondo lattice and scaling theory close to quantum critical points are discussed in considerable detail. Charge ordering, which began with the notion of crystallization first introduced in the 30s by Wigner was subsequently observed in 2D electron systems. Originating from a variety of effects, charge ordering can occur in many types of systems. For example, Yb_4As_3, which is discussed in considerable detail from both experimental and theoretical points of view in the present exposition, exhibits charge ordering associated with $4f$ holes. It illustrates the formation of heavy quasiparticles caused by spin chains without involvement of the Kondo effect.

Other subjects discussed here include partial localization found in some actinide-based heavy fermion compounds in which experiments are used to infer the co-existence of delocalized and localized $5f$ electrons, superconductivity mediated by intra-atomic excitations, geometrically frustrated lattices such as metallic spinels, fractional charges resulting from strong correlations, and high energy correlation induced excitations, for example, the appearance of shadow bands and satellites associated with nickel.

Because of its significant reliance on and comparison with experiment, its frequent use of phenomenologically based theory used to calculate quasi-particle band structures and Fermi surfaces, this article is to some extent less focussed on fundamental microscopic theoretical aspects than on techniques leading to optimal physical insight permitting ready comparison with observations relevant to important materials systems. Clearly both types of expositions are needed.

The second article in this volume, by Bréchet and Hutchinson, concerns pattern formation in metals and alloys. Spontaneous pattern formation is the development of a regularity, either in the spatial distribution of the material in a system or in its development in time, of a lower symmetry than that of its cause. These phenomena have been of considerable interest to the non-linear physics community, in particular in fluid dynamics and in chemical reactions. This article deals with pattern formation in the solid state, which is comparatively less studied. Usually, crystal defects, such as dislocations or interfaces, play an important role and the energetic cost required by their formation requires that the system be prepared or maintained far from equilibrium.

Interface-mediated formation of lamellar microstructures from supersaturated solution is discussed as an example of chemical patterning in a system prepared far from equilibrium. Striking examples are the formation of pearlite and discontinuous precipitation. Grain growth, recovery and recrystallization are discussed as examples of defect patterning in a system prepared far from equilibrium. Grain growth is a rich topological subject, several aspects of which have been presented in earlier articles in this Series, such as that by Weaire and McMurry in vol. 50 and by Thompson in vol. 55.

Systems that are maintained far from equilibrium can be called "driven systems". Martin and Bellon have reviewed the chemical aspects of such systems in their article in vol. 50 of the Series. The present article therefore concentrates more on structural patterning in driven systems, such as the formation of arrays of dislocation loops and voids under irradiation or dislocation patterning in deformation.

Finally, plastic deformation provides an interesting example of pattering in both space and time: the Portevin–Le Chatelier effect, which is a stick-slip effect of plastic waves that move through the sample. The authors show that this can be usefully analyzed as a case of self-organized criticality.

HENRY EHRENREICH
FRANS SPAEPEN

SOLID STATE PHYSICS

VOLUME 60

Strongly Correlated Electrons

PETER FULDE

Max-Planck-Institut für Physik komplexer Systeme, 01187 Dresden, Germany

PETER THALMEIER

Max-Planck-Institut für Chemische Physik fester Stoffe, 01187 Dresden, Germany

GERTRUD ZWICKNAGL

Institut für Mathematische Physik, Technische Universität Braunschweig, 38106 Braunschweig, Germany

I. Introduction	2
II. Special Features of Strong Correlations	8
1. Low-Energy Scales: a Signature of Strong Correlations	10
2. Deviations from Fermi-Liquid Behavior	12
III. Kondo Lattice Systems	14
3. Fermi-Liquid State and Heavy Quasiparticles: Renormalized Band Theory	18
4. Heavy Fermions in $CeRu_2Si_2$ and $CeCu_2Si_2$	19
5. Low-Temperature Phase Diagram of $CeCu_2Si_2$	23
IV. Quantum Phase Transitions	30
6. Quantum Phase Transition in Localized and Itinerant Magnets	32
7. Quantum Criticality in the Kondo Lattice	38
8. Scaling Theory Close to the Quantum Critical Point	45
V. Partial Localization	48
9. Heavy Quasiparticles in UPd_2Al_3	51
10. Microscopic Model Calculation	54
11. Superconductivity Mediated by Intra-Atomic Excitations	59
VI. Charge Ordering	66
12. Wigner Crystallization in Homogeneous 2D Electron Systems	69
13. Generalized Wigner Lattice: Yb_4As_3	74
14. Charge Ordering and 1D Spin Excitations in α'-NaV_2O_5	91
15. Reentrant Charge Ordering and Polaron Formation in Double Exchange Bilayer Manganites $La_{2-2x}Sr_{1+2x}Mn_2O_7$	108
VII. Geometrically Frustrated Lattices	122
16. Metallic Spinels: LiV_2O_4 — a Metal with Heavy Quasiparticles	125

ISBN 0-12-607760-6
ISSN 0081-1947/06

© 2006 Elsevier Inc. (USA)
All rights reserved.

17. Structural Transition and Charge Disproportionation: AlV_2O_4 133
18. Fractional Charges Due to Strong Correlations 139
VIII. High-Energy Excitations 147
19. Projection Operators 149
20. The Hubbard Model: Appearance of Shadow Bands 151
21. Marginal Fermi Liquid Behavior and Kink Structure 153
22. Nickel and its Satellite 163
23. Multiplet Effects in $5f$ Systems 168
24. Excitations in Copper-Oxide Planes 171
IX. Summary and Outlook 176
Acknowledgment 179
List of Acronyms 179

I. Introduction

The field of strongly correlated electron systems has been constantly growing for almost three decades. A milestone in its development was the discovery by Andres, Graebner and Ott[1] of heavy-quasiparticle excitations in $CeAl_3$. Additional verve came from the discovery of superconductivity in the related compounds $CeCu_2Si_2$,[2] UBe_{13}[3] and UPt_3.[4] But a real great push for the field was provided by the discovery of high-temperature superconductivity in the copper-oxide based perovskites.[5] Were it not for strong electron correlations La_2CuO_4, one of the key compounds of that class of materials and the basis of the hole doped superconductors $La_{2-x}Ba_xCuO_4$ and $La_{2-x}Sr_xCuO_4$ would be metallic. Instead it is an antiferromagnet which remains insulating even above the Néel temperature where the unit cell is not doubled anymore. Therefore, electron correlations are apparently so strong that the metallic character of the material is suppressed in favor of an insulating state. That electron correlation may induce a metal to insulator transition had been suggested long before the discovery of heavy quasiparticles and high-T_c cuprates. The names of Mott[6] and Hubbard[7] stand for that phenomenon. At their time the interests in the effects of strong correlations resulted from the transition metal oxides and their various phase transitions. It is

---

[1] K. Andres, J. E. Graebner, and H. R. Ott, *Phys. Rev. Lett.* **35**, 1779 (1975).
[2] F. Steglich, J. Aarts, C. D. Bredl, W. Liecke, D. Meschede, W. Franz, and H. Schäfer, *Phys. Rev. Lett.* **43**, 1892 (1979).
[3] H. R. Ott, H. Rudigier, Z. Fisk, and J. L. Smith, *Phys. Rev. Lett.* **50**, 1595 (1983).
[4] G. R. Stewart, Z. Fisk, J. O. Willis, and J. L. Smith, *Phys. Rev. Lett.* **52**, 679 (1984).
[5] J. G. Bednorz and K. A. Müller, *Z. Phys. B* **64**, 189 (1986).
[6] N. F. Mott, *Metal–Insulator Transition*, Taylor and Francis, London (1990), 2nd ed.
[7] J. Hubbard, *Proc. R. Soc. London A* **276**, 238 (1963).

worth recalling that the famous Verwey[8] transition in magnetite Fe_3O_4 falls into the same category. One may even go back to Wigner[9] or Heitler and London[10] who dealt with strongly correlated electrons long before corresponding experiments were available. While Wigner pointed out that electrons may form a lattice when their correlations become sufficiently strong, Heitler and London developed a theory for chemical bonding based on strongly correlated electrons. It is the opposite limit of Hückel's theory[11–13] based on molecular orbitals in which electron correlations are completely neglected. This raises the question of how to quantify the strength of electronic correlations. For example, one would like to know how much more strongly electrons are correlated in $LaCu_2O_4$ than, e.g., in iron or nickel or in transition metal oxides.

The differences between systems with strongly and with weakly correlated electrons may be seen by considering the ground state of the simplest possible example, i.e., of a H_2 molecule in the Heitler–London- and in the molecular orbital limit. The Heitler–London form of the ground-state wavefunction is

$$\psi_{HL}(\mathbf{r}_1, \mathbf{r}_2) = \frac{1}{2}[\phi_1(\mathbf{r}_1)\phi_2(\mathbf{r}_2) + \phi_2(\mathbf{r}_1)\phi_1(\mathbf{r}_2)](\alpha_1\beta_2 - \beta_1\alpha_2) \qquad (1.1)$$

where the single-electron wavefunctions $\phi_{1,2}(\mathbf{r})$ are centered on atoms 1 and 2 of the molecule and α and β denote spinors for up and down spins. In distinction to Eq. (1.1) the molecular-orbital form of the ground-state wavefunction is

$$\psi_{MO}(\mathbf{r}_1, \mathbf{r}_2) = \frac{1}{2^{3/2}}[\phi_1(\mathbf{r}_1)\phi_1(\mathbf{r}_2) + \phi_1(\mathbf{r}_1)\phi_2(\mathbf{r}_2) + \phi_2(\mathbf{r}_1)\phi_1(\mathbf{r}_2)$$
$$+ \phi_2(\mathbf{r}_1)\phi_2(\mathbf{r}_2)](\alpha_1\beta_2 - \beta_1\alpha_2). \qquad (1.2)$$

It is seen that $\psi_{MO}(\mathbf{r}_1, \mathbf{r}_2)$ but not $\psi_{HL}(\mathbf{r}_1, \mathbf{r}_2)$ contains ionic configurations $\phi_1(\mathbf{r}_1)\phi_1(\mathbf{r}_2)$ and $\phi_2(\mathbf{r}_1)\phi_2(\mathbf{r}_2)$. In Eq. (1.2) they have equal weight as the non-ionic configurations. But ionic configurations cost additional Coulomb repulsion energy of the electrons. Therefore they are completely suppressed in the Heitler–London- or strong correlation limit. This demonstrates an important feature of electron correlations, namely a partial suppression of electronic charge fluctuations on an atomic site. The former are called interatomic correlations because charge fluctuations at an atomic site are caused by an overlap of wavefunctions of

---

[8] E. J. W. Verwey and P. W. Haayman, *Physica* **8**, 979 (1941).
[9] E. Wigner, *Phys. Rev.* **46**, 1002 (1934).
[10] W. Heitler and F. London, *Z. Phys.* **44**, 455 (1927).
[11] E. Hückel, *Z. Phys.* **70**, 204 (1931).
[12] E. Hückel, *Z. Phys.* **72**, 310 (1931).
[13] E. Hückel, *Z. Phys.* **76**, 628 (1932).

different atoms. They are favored by a kinetic energy gain due to electron delocalization. Reducing them compared with uncorrelated electrons keeps the Coulomb repulsions small.

In addition to interatomic correlations we must also consider intra-atomic correlations. Consider an atom of a solid in a configuration with a given number of electrons, for example, a C atom in diamond with, e.g., 4 or 5 valence electrons. Those electrons will optimize their on-site Coulomb repulsions by arranging according to Hund's rules and by in-out correlations. Hund's rules ensure that electrons on an atom are optimally distributed over the angular segments of the atom, so that their repulsions are as small as possible. In-out correlations achieve the same by proper radial distribution of the electrons. Intra-atomic correlations are strongest for $4f$ electrons, i.e., for atoms or ions of the lanthanide series. But also in actinides or transition-metals they play a big role. Large overlaps with atomic wavefunctions of the chemical environment will weaken them. This is understandable: before the electrons can fully establish intra-atomic correlations they leave for the neighboring sites by hopping off the site. Interatomic correlations can be strong even when intra-atomic correlations are moderate or weak. Let us make a gedanken experiment and consider a Si crystal with artificially enlarged lattice parameter. The intra-atomic correlations on a Si site are fairly moderate, but the interatomic correlations are becoming strong when the lattice constant is increased, i.e., when the limit of separate atoms is approached. In that case fluctuations in the electron number at a site reduce to zero.

From the above considerations it follows that a suitable measure of the *interatomic* correlation strength is the reduction of electron number fluctuations on a given atom. An independent-electron or Hartree–Fock description implies too large fluctuations. Let $|\psi_0\rangle$ denote the exact ground state of an electronic system and $|\Phi_{\text{SCF}}\rangle$ the corresponding self-consistent field (SCF) or Hartree–Fock (HF) state. The normalized mean-square deviation of the electron number n_i on atom i is given by

$$\Sigma(i) = \frac{\langle \Phi_{\text{SCF}}|(\Delta n_i)^2|\Phi_{\text{SCF}}\rangle - \langle \psi_0|(\Delta n_i)^2|\psi_0\rangle}{\langle \Phi_{\text{SCF}}|(\Delta n_i)^2|\Phi_{\text{SCF}}\rangle} \quad (1.3)$$

where $\Delta n_i = n_i - \bar{n}_i$ and $\bar{n}_i$ denotes the average value. One notices that $0 \leqslant \Sigma(i) \leqslant 1$. When $\Sigma(i) = 0$ the interatomic correlations vanish, i.e., the Coulomb repulsions between the electrons can be treated in mean-field approximation. In a solid atoms or ions with strongly correlated electrons have $\Sigma(i)$ values near unity. One can also define a correlation strength for different bonds instead of atoms. In that case the denominator is modified when heteropolar bonds are considered. Then we must subtract from $\langle \Phi_{\text{SCF}}|(\Delta n_i)^2|\Phi_{\text{SCF}}\rangle$ a term $(\Delta n)_{\text{pc}}^2$. It takes into account that some number fluctuations are required even when the electrons are perfectly correlated in order to ensure a heteropolar charge distribution within the bond. Let α_p denote the bond polarity. It is defined by the difference in the average occupation numbers of the two half-bonds 1 and 2 which form the heteropolar bond, i.e., $\bar{n}_{1(2)} = (1 \pm \alpha_p)$. In that case $(\Delta n)^2_{pc} = \alpha_p(1 - \alpha_p)$. Those considerations apply to a solid as well as to a molecule.

For the H_2 molecule one checks immediately that approximating $|\psi_0\rangle$ by $\psi_{MO}(\mathbf{r}_1, \mathbf{r}_2)$ gives $\Sigma = 0$ while a replacement by $\psi_{HL}(\mathbf{r}_1, \mathbf{r}_2)$ yields $\Sigma = 1$ since $\langle\psi_0|(\Delta n)^2|\psi_0\rangle = 0$ in that case. For a C=C or N=N π bond one finds $\Sigma \approx 0.5$ while for a C–C or N–N σ bond $\Sigma = 0.30$ and 0.35, respectively. Let us consider the ground state of La_2CuO_4 and let $P(d^\nu)$ denote the probability of finding ν $3d$ electrons on a given Cu site. Within the independent electron or Hartree–Fock approximation the average d count is found to be $\bar{n}_d \simeq 9.5$ and the probabilities of different configurations are $P(d^{10}) = 0.56$, $P(d^9) = 0.38$ and $P(d^8) = 0.06$. When correlations are included, i.e., the correlated ground state $|\psi_0\rangle$ is used the average d electron number changes to $\bar{n}_d \simeq 9.3$ and $P(d^{10}) = 0.29$, $P(d^9) = 0.70$ while $P(d^8) = 0.0$. One notices that the d^8 configurations are almost completely suppressed in agreement with photoemission experiments. The fluctuations between the d^9 and d^{10} configurations are fixed by the value of $\bar{n}_d$. A similar analysis for the oxygen atoms reveals that there the $2p^4$ configurations are *not* completely suppressed because the Coulomb integrals are not as large as for Cu. Indeed, these configurations are important for superexchange to occur, which determines the antiferromagnetic coupling between Cu ions. In accordance with the above consideration one finds $\Sigma(Cu) \simeq 0.8$ and $\Sigma(O) \simeq 0.7$.[14] So indeed, correlations are quite strong in La_2CuO_4. On the other hand, they are still smaller than those of $4f$ electrons in a system like $CeAl_3$.

A measure for the strength of intra-atomic correlations is more difficult to define. One way is by finding out to which extent Hund's rule correlations are building up on a given atomic site i. A possible measure for that is the degree of spin alignment at a given atomic site i

$$S_i^2 = \langle\psi_0|\mathbf{S}^2(i)|\psi_0\rangle \tag{1.4}$$

where $\mathbf{S}(i) = \sum_\nu \mathbf{s}_\nu(i)$ and $\mathbf{s}_\nu(i)$ is the spin operator for orbital ν. The quantity S_i^2 should be compared with the values when the SCF ground-state wavefunction $|\Phi_{SCF}\rangle$ is used and when instead the ground state $|\Phi_{loc}\rangle$ in the limit of complete suppression of interatomic charge fluctuations is taken, i.e., for large atomic distances. Therefore we may define

$$\Delta S_i^2 = \frac{\langle\psi_0|\mathbf{S}^2(i)|\psi_0\rangle - \langle\Phi_{SCF}|\mathbf{S}^2(i)|\Phi_{SCF}\rangle}{\langle\Phi_{loc}|\mathbf{S}^2(i)|\Phi_{loc}\rangle - \langle\Phi_{SCF}|\mathbf{S}^2(i)|\Phi_{SCF}\rangle} \tag{1.5}$$

---

[14] A. Oleś, J. Zaanen, and P. Fulde, *Physica B* **148**, 260 (1987).

as a possible measure of the strength of intra atomic correlations. Note that $0 \leqslant \Delta S_i^2 \leqslant 1$. For example, for the transition metals Fe, Co and Ni ΔS_i^2 is approximately 0.5.

Those findings show that the much discussed transition metals are just in the middle between the limits of uncorrelated and strongly correlated electrons. Hund's rule correlations are important in them but relatively large overlaps of atomic wavefunctions on neighboring sites prevent their complete establishment. Starting from the work of Slater[15] and Van Vleck[16] in particular Friedel,[17] Gutzwiller,[18,19] Hubbard[20] and Kanamori[21] have discussed their effects in detail. One of the outcomes of the studies of transition metals is the Hubbard Hamiltonian. It was in fact used independently also by Gutzwiller and in a slightly modified version by Kanamori. This Hamiltonian was extensively treated in various approximations. The multiband Hubbard model has remained until present times the working horse of many studies of strongly correlated electrons.[22,23] The shortcomings of that model are known. For example, it considers d electrons only, i.e., s electrons are neglected. Also it cannot provide for orbital relaxations when electrons hop on or off a site because only one basis function per atomic orbital is used. Nevertheless, it is believed that it covers the most important generic effects of strongly correlated electrons.

The valence electrons which are most strongly correlated are the $4f$ ones because their atomic wavefunction is close to the nucleus and the tendency to delocalize is very small. In fact, in intermetallic rare-earth compounds only f-electrons in Ce or Yb ions show a noticeable degree of itinerancy. The consequence are new low-energy scales which may appear in those compounds and as a result heavy-quasiparticle excitations. Not always do quasiparticles show conventional Fermi liquid behavior which governs the low-temperature thermodynamic properties of many metals. In a number of cases one observes what is called non-Fermi liquid behavior, i.e., quantities like the temperature dependence of the specific heat or of the susceptibility deviate from normal metallic behavior. In particular this holds true near a quantum critical point where apparently no characteristic energy scale is prevailing. Fermi liquid behavior requires that at low temperatures all thermodynamic quantities scale with $k_B T^*$, a characteristic energy which in

---

[15] J. C. Slater, *Phys. Rev.* **49**, 537 and 931 (1936).
[16] J. H. V. Vleck, *Rev. Mod. Phys.* **25**, 220 (1953).
[17] J. Friedel, *The Physics of Metals: 1. Electrons*, Cambridge Univ. Press, Cambridge (1969).
[18] M. C. Gutzwiller, *Phys. Rev. A* **134**, 923 (1964).
[19] M. C. Gutzwiller, *Phys. Rev. A* **137**, 1726 (1965).
[20] J. Hubbard, *Proc. R. Soc. London A* **281**, 401 (1964).
[21] J. Kanamori, *Progr. Theor. Phys.* **30**, 275 (1963).
[22] M. Imada, A. Fujimori, and Y. Tokura, *Rev. Mod. Phys.* **70**, 1039 (1998).
[23] F. Mancini and A. Avella, *Adv. Phys.* **53**, 537 (2004).

strongly correlated electron systems takes the role of the Fermi energy. When such a characteristic scale does not exist deviations from Fermi liquid behavior do occur. In a way it is more astonishing that to good approximation Fermi liquid behavior is observed in a number of strongly correlated electron systems than that it is not. A study of the Hubbard model shows ways for obtaining deviations from standard features of a metal.

One interesting aspect of strong electron correlations is the possible occurrence of charge order. A charge ordered state minimizes the repulsive energy between electrons at the expense of the kinetic energy. Wigner was the first to study this subject by considering a homogeneous electron gas and specifying the conditions under which the formation of an electronic lattice is possible. Chances for charge ordering are larger for inhomogeneous systems, i.e., lattices than for homogeneous ones, the reason being that the kinetic energy gain of electrons due to delocalization may become very much reduced as compared with homogeneous electron systems. A prototype example is Yb_4As_3 where charge order occurs close to room temperature and there are many other cases.

While $4f$ electrons are localized in most cases and are very strongly correlated, $5f$ electrons are more delocalized but still more strongly correlated as, e.g., $3d$ electrons in transition metals. It turns out that in this case a dual picture applies: while $5f$ electrons become itinerant in some of the orbitals they remain localized in others. Such a model explains very well a number of experiments on U compounds.

Heavy quasiparticles have also been observed in LiV_2O_4, a metal with $3d$ electrons. A special feature of that material is that the $3d$ electrons are placed on a pyrochlore or geometrically frustrated lattice. Model calculations show that charge degrees of freedom of strongly correlated electrons in frustrated lattice structures can give rise to new phenomena at special band fillings. There may exist large numbers of low-energy excitations for which Landau's Fermi liquid approach fails and there may be even excitations with fractional electron charges. Although phenomena of this kind have not been observed yet, the theoretical results may stimulate further thinking.

Strongly correlated electrons show in addition to the quasiparticle bands also satellite structures in photoemission experiments. They are contained in the incoherent part of the one-particle Green function. It appears that detailed studies of the incoherent part of Green's function have not been done to the extent they deserve. The reason for their importance is the following. A quasiparticle in a solid can be considered as a bare particle (electron or hole) surrounded by a correlation hole. The whole object, i.e., particle plus correlation hole moves in form of a Bloch wave through the system. The internal degrees of freedom of the correlation hole give rise to excitations which are contained in the incoherent part of Green's function. Therefore it is very instructive to study general features of that incoherent part. Hubbard's upper band can be considered a satellite feature for filling factors $n < 1/2$, i.e., for less than one electron per site. Other examples will be presented.

It is impossible to cover all aspects of strongly correlated electrons in a review of reasonable size. Therefore selections have to be made. Naturally, authors select topics for reviews for which they feel particularly competent. These are usually areas in which they have actively worked. This holds also true here and the selection we made may do injustice to other interesting developments in fields not covered here. So we apologize for an incomplete covering of topics as well as for incomplete lists of contributions of authors to the subjects discussed here.

II. Special Features of Strong Correlations

Metals with strongly correlated electrons exhibit characteristic deviations from the behavior of independent electrons. The latter are reflected in thermodynamic and transport properties as well as in the high energy spectra.

Traditional electron theory of metals proceeds from the electron gas model formulated by Sommerfeld and Bethe.[24] The electrons are described as a system of non-interacting fermions. The eigenstates are formed by filling single-particle levels in a manner consistent with the Pauli principle which permits at most one electron per spin direction to occupy any single-electron level. The ground state of an N electron state is obtained by filling the $N/2$ single particle levels with the lowest energies. It is non-degenerate and characterized by a surface in **k**-space separating the occupied levels from their unoccupied counterparts. The existence of this surface, the Fermi surface, follows directly for a system of independent electrons. But note that the observation of a Fermi surface does not imply that the independent electron approximation is a valid description of a system. The low-temperature properties which are dominated by the low-energy excitations are universal, the detailed character of the system under consideration being reflected in a characteristic energy—the Fermi energy E_F. The energy scale is set by the variation of the single particle levels with wave number **k**. A measure of it is the Fermi velocity v_F. The linear variation with temperature of the specific heat, $C(T) \simeq \gamma T$, and the temperature-independent magnetic susceptibility, $\chi_s(T) \to$ const are also characteristic features of free electrons. Finally, the spectrum for adding or removing a particle in a single-particle level **k**, σ to the ground state $A(\mathbf{k}, \omega)$, exhibits a well-defined peak

$$A(\mathbf{k}, \omega) = \delta(\omega - \epsilon_k) \tag{2.1}$$

---

[24] A. Sommerfeld and H. Bethe, *Elektronentheorie der Metalle*, vol. 24/2 of *Handbuch der Physik*, Springer, Berlin, Heidelberg (1933), 2nd ed.

of weight unity centered at the single-particle energy $\epsilon_\mathbf{k}$. The independent electron model has proven to be very successful in explaining experimentally observed properties of simple metals. That was a surprise for some time since electron–electron repulsions are not weak in any metal and one might therefore expect that they modify strongly the properties of a system of independent electrons. That this is not necessarily the case was shown by Landau.[25–28]

The Landau theory assumes that there exists a one-to-one correspondence between the excitations of the complex interacting electron system and those of independent electron. The former are called quasiparticles and their orbitals and energies $E(\mathbf{k})$ are determined from an effective Hamiltonian. It contains an effective, not necessarily local potential. The many-body aspects are contained in the construction of the effective potential which must be determined specifically for the problem under consideration.

The quasiparticle energies may be altered when the overall configuration is changed. A characteristic feature of interacting Fermi liquids is that the energy dispersion $\tilde{E}_\sigma(\mathbf{k})$ of a quasiparticle depends on how many other quasiparticles are present,

$$\tilde{E}_\sigma(\mathbf{k}) = E(\mathbf{k}) + \sum_{\mathbf{k}'\sigma'} f_{\sigma\sigma'}(\mathbf{k},\mathbf{k}')\delta n'_\sigma(\mathbf{k}'). \tag{2.2}$$

Here $E(\mathbf{k})$ denotes the energy dispersion of a quasiparticle when there are no other quasiparticles around (dilute gas limit). In systems with strong correlations it reflects the electron interactions and hence cannot be calculated from the overlap of single-electron wave functions. Interactions among quasiparticles are characterized by the matrix $f_{\sigma\sigma'}(\mathbf{k},\mathbf{k}')$. The deviations from a step-function-like Fermi distribution $f(E(\mathbf{k}), T=0)$ are given by $\delta n_\sigma(\mathbf{k})$.

The scattering amplitudes $f_{\sigma\sigma'}(\mathbf{k},\mathbf{k}')$ are parameterized and the parameters are adjusted to experiments. Their form is strictly applicable only to a homogeneous translationally invariant electron system. Therefore applying it to an inhomogeneous periodic solid requires some modifications (see, e.g., Ref. [29]) which are usually not discussed. From this point of view Landau's theory is more of a useful theoretical concept rather than a quantitative computational scheme.

The assumed one-to-one correspondence of the excitations implies that the low temperature thermodynamic properties resemble those of independent electrons but with renormalized parameters such as the effective electron or hole

---

[25] L. D. Landau, *Zh. Eksp. Teor. Fiz.* **30**, 1058 (1956).

[26] L. D. Landau, *Zh. Eksp. Teor. Fiz.* **32**, 59 (1957).

[27] L. D. Landau, *Zh. Eksp. Teor. Fiz.* **35**, 97 (1958).

[28] A. A. Abrikosov, L. P. Gorkov, and I. E. Dzyaloshinski, *Methods of Quantum Field Theory in Statistical Physics*, Prentice-Hall, Englewood Cliffs, New York (1963).

[29] P. Fulde, J. Keller, and G. Zwicknagl, in *Solid State Physics*, vol. 41, eds. H. Ehrenreich and D. Turnbull, Academic Press, New York (1988), p. 1.

mass. Also the weight of the peak in the spectral density $A(\mathbf{k}, \omega)$ is modified to $Z \cdot \delta(\omega - E(\mathbf{k}))$ where the renormalization factor $0 < Z \leqslant 1$ describes the weight of the bare electron in the quasiparticle. The latter contains in addition to the bare electron also the correlation hole around it.

An interacting electron system to which Landau's theory applies has also a Fermi surface. Luttinger has proven[30] that in case that perturbation theory is applicable the volume enclosed by the Fermi surface is independent of the electron interactions. Another important property of quasiparticles is that they can be considered as 'rigid' with respect to low-energy and long-wavelength perturbations. That is to say that excitations involving degrees of freedom of the correlation hole show up at high energies only and are neglected as regards low-temperature properties. They are discussed in Section VIII.

When electron correlations are strong the quasiparticle concept is still applicable for a number of substances. In that case the renormalization factor Z may become very small. This results in heavy quasiparticles because the Fermi velocity is reduced by the same factor. Probably in many cases the one-to-one correspondence between the excitations of a strongly correlated electron system and a corresponding system of independent electrons is only approximately fulfilled. But then the low-temperature properties of the system may still look very similar to those of independent electrons with renormalized parameters. For example, the specific heat will still be nearly linear in T at low temperatures etc.

However, we want to stress that from the observation of a specific heat linear in T or a temperature independent spin susceptibility in the low temperature regime one may not conclude that the quasiparticle picture is applicable. In fact, Luttinger liquids in quasi-one-dimensional systems show many properties as quasiparticles do. This is so despite the fact that the key assumption of a one-to-one correspondence of excitations to those of independent electrons is unjustified here. There are strongly correlated systems where the quasiparticle picture seems totally inappropriate. This is outlined in the following Section II.2 and discussed in more detail in various sections of this chapter.

1. LOW-ENERGY SCALES: A SIGNATURE OF STRONG CORRELATIONS

As mentioned above, the characteristic energy scale of a free electron gas is the Fermi energy E_F or, alternatively the Fermi temperature T_F. A typical value for E_F is 5 eV corresponding to a T_F of $5 \cdot 10^4$ K. A special feature of strongly correlated electrons is that they introduce new low-energy scales. It is customary to associate a temperature T^* with them. In metals with heavy quasiparticles, i.e., with very strong electron correlations T^* ranges from a few Kelvin to a few

---

[30] J. M. Luttinger and J. C. Ward, *Phys. Rev.* **118**, 1417 (1960).

hundred Kelvin. As correlations become weaker T^* increases until it is no longer justified to speak of a separate low-energy scale. The microscopic origin of the low-energy scales can be quite different. A widely recognized case is the Kondo effect. Here T^* is the Kondo temperature, i.e., it is given by the binding energy of the spins of the conduction electrons to local spins. Local spins imply incomplete inner shells of an atom or ion. The fact that they remain partially filled only, when surrounded by conduction electrons is due to strong correlations. Any conduction electron which tries to enter the incomplete inner shell is expelled by strong on-site Coulomb repulsion. An example are Ce^{3+} ions immersed in a sea of (generally) weakly correlated conduction electron. Due to a weak hybridization the number of $4f$ electrons is nearly one. It forms a singlet with the conduction electrons. The aforementioned $CeAl_3$ falls into that category. Breaking those singlets results in low-energy excitations and fixes the low-energy scale T^*. The low-energy excitations make it plausible that there will be a large low-temperature specific heat. To explain heavy quasiparticles the singlet-triplet excitations on different sites must lock together and form coherent Bloch-like excitations. That takes place at a somewhat lower energy scale T_{coh}. One expects that T_{coh} is of order of T^* but no detailed theory for a relation between the two temperatures is available. There is also no theory existing which tells us that the coherent excitations are in one-to-one correspondence to excitations of (nearly) free electrons. Nevertheless this assumption has worked remarkably well.

The origin of a low T^* is quite different in the strongly correlated semimetal Yb_4As_3. Here the Coulomb repulsion of the $4f$ holes in neighboring Yb ions leads to charge order in the form of well separated chains of Yb^{3+} ions. Spin excitations in those chains by which light mobile $4p$ holes of As are scattered, lead to low temperature properties which resemble very much those of other systems with heavy quasiparticles.[31,32] Despite of this the system is not really a heavy Landau Fermi liquid any more as is explained in the next subsection and discussed in more detail in Section VI.13.

A third mechanism is found to be responsible for a low energy scale T^* in U compounds like UPd_2Al_3 or UPt_3. Strong intra-atomic or Hund's rule correlations lead here to pronounced anisotropies of the effective hybridization of different $5f$ orbitals. As a result some of the $5f$ electrons remain localized while others delocalize. The crystalline environment lifts degeneracies of the localized electrons on a low energy scale. The delocalized or itinerant electrons couple to the excitations of the local system and in this way generate a low T^*.[33] Heavy quasiparticles may also appear near a quantum critical point like in YMn_2.[34]

---

[31] P. Fulde, B. Schmidt, and P. Thalmeier, *Europhys. Lett.* **31**, 323 (1995).
[32] M. Kohgi, K. Iwasa, J.-M. Mignot, A. Ochiai, and T. Suzuki, *Phys. Rev. B* **56**, R11388 (1997).
[33] G. Zwicknagl and P. Fulde, *J. Phys.: Condens. Matter* **15**, S1911 (2003).
[34] C. Pinettes and C. Lacroix, *J. Phys. Cond. Mat.* **6**, 10093 (1994).

Finally, in Ce doped Nd_2CuO_4 a fourth origin of a low T^* is observed. Here it is essentially a fluctuating internal molecular field originating from the Nd ions which causes a low energy scale in the strongly correlated d-electron system of the Cu–O planes to which is couples.[35,36]

It seems obvious that there will be other physical processes identified in the future resulting in low-energy scales of strongly correlated electron systems.

2. DEVIATIONS FROM FERMI-LIQUID BEHAVIOR

There is no obvious reason why strongly correlated metallic electron systems should be Fermi liquids. But as pointed out above a large number of them behave very nearly like ordinary metals, i.e., Fermi liquids with renormalized parameters like the effective mass. Even in these cases, high energy excitations show characteristic satellite structures which reflect strong correlations in partially filled inner shells. This topic is discussed in Section VIII. However, there are also numerous examples where the Fermi liquid concept for low-energy excitations is not applicable.

One much discussed item is the separation of charge and spin degrees of freedom and moreover the appearance of fractional charges. Separate spin and charge excitations occur always when electron correlations are so strong that the electrons remain localized. In that case the coupling of spins on different sites leads to magnetic excitations with energies of order J, the intersite coupling constant. In contrast, charge excitations from the partially filled inner shells as observed, e.g., by photoelectron spectroscopy have much higher energies. But this kind of spin-charge separation is trivial and does not require further consideration. It is well known that in one dimension (1D) spin and charge degrees of freedom lead to different kinds of excitations even when the correlations are weak (Luttinger liquid). For a review see, e.g., Ref. [37]. Spin-charge separation is also found for kink excitations (solitons) in polyacethylene.[38] Those excitations exist even within the independent electron approximation, but require inclusion of lattice degrees of freedom. Doped polyacethylene can have also excitations with fractional charges, again within the one-electron picture but requiring lattice (chain) deformations.[39] In 2D electron correlations, e.g., in semiconducting inversion layers may become strong when a magnetic field is applied perpendicular to the

---

[35] T. Brugger, T. Schreiner, G. Roth, P. Adelmann, and G. Czjzek, *Phys. Rev. Lett.* **71**, 2481 (1993).
[36] P. Fulde, V. Zevin, and G. Zwicknagl, *Z. Phys. B* **92**, 133 (1993).
[37] A. O. Gogolin, A. A. Nersesyan, and A. N. Tsvelik, *Bosonization and Strongly Correlated Systems*, Cambridge University Press, Cambridge (1998).
[38] W. P. Su, J. R. Schrieffer, and A. J. Heeger, *Phys. Rev. Lett.* **42**, 1698 (1979).
[39] W. P. Su and J. R. Schrieffer, *Phys. Rev. Lett.* **46**, 738 (1981).

plane. The kinetic energy of the electrons is strongly reduced in a high field and therefore the Coulomb repulsions become dominant. This results in the fractional quantum Hall effect (FQHE) and quasiparticles with fractional charges.[40] Thus in two dimensions electron correlations are essential for the appearance of fractional charges. The same holds true for 3D systems. There it turns out that excitations with fractional charges may exist in certain geometrically frustrated lattice structures like the pyrochlore lattice.[41,42] There is also spin-charge separation. A Fermi liquid description is inapplicable here. This intriguing possibility is discussed in Section VII.

Another interesting case of a breakdown of Landau's Fermi liquid description is found in Yb_4As_3. This system is metallic in a high temperature phase and semimetallic in the low temperature phase.[32,43] The change is related to a partial electronic charge order in form of well separated Yb^{3+} chains with an effective spin 1/2 per site. It is well known that a Heisenberg spin chain has a specific heat of the form $C = \gamma T$ like a metal. It is due to spinons which obey Fermi statistics. The reader should note that in one dimension one can convert fermions into bosons and vice versa.[44] The coefficient γ is large here because of a weak coupling of the spins in a chain and therefore the specific heat resembles that of heavy quasiparticles.[31] But the charge carriers, which are mainly $4f$ holes in the high temperature phase consist of a small number of As $4p$ holes in the low temperature phase.[45] Therefore one may speak of spin-charge separation and a breakdown of the conventional Fermi liquid picture. The one-to-one correspondence between the excitations in the low temperature phase and those of an independent electron system is no longer given. Nevertheless the system shows many properties of an ordinary metal with heavy quasiparticles at low temperature. A detailed discussion of that interesting material is found in Section VI.

Another form of deviation from classical Fermi liquid behavior is found in the cuprates perovskite structures. In the underdoped regime many of their physical properties show marginal Fermi liquid behavior.[46,47] This implies that they can be described by assuming a frequency dependence of the electron self-energy $\Sigma(\omega)$ for small values of ω of the form

$$\text{Re}\,\Sigma(\omega) \sim \omega \ln \omega, \quad \text{Im}\,\Sigma(\omega) \sim |\omega| \qquad (2.3)$$

---

[40] R. B. Laughlin, *Phys. Rev. Lett.* **50**, 1395 (1983).
[41] P. Fulde, *Adv. Physics* **51**, 909 (2002).
[42] E. Runge and P. Fulde, *Phys. Rev. B* **70**, 245113 (2004).
[43] A. Ochiai, T. Suzuki, and T. Kasuya, *J. Phys. Soc. Jpn.* **59**, 4129 (1990).
[44] A. Luther and I. Peschel, *Phys. Rev. B* **12**, 3908 (1975).
[45] V. N. Antonov, A. N. Yaresko, A. Y. Perlov, P. Thalmeier, P. Fulde, P. M. Oppeneer, and H. Eschrig, *Phys. Rev. B* **58**, 9752 (1998).
[46] C. M. Varma, P. B. Littlewood, S. Schmitt-Rink, E. Abrahams, and A. Ruckenstein, *Phys. Rev. Lett.* **63**, 1996 (1989).
[47] P. B. Littlewood and C. M. Varma, *J. Appl. Phys.* **69**, 4979 (1991).

instead of the Fermi-liquid form

$$\operatorname{Re} \Sigma(\omega) \sim \omega, \quad \operatorname{Im} \Sigma(\omega) \sim \omega^2. \tag{2.4}$$

The latter would be required for the one-to-one correspondence of the excitations. The relations (2.3) hold only for $T < \omega$. Otherwise T replaces ω. They yield, e.g., a resistivity $\rho(T) \sim T$ as is observed in a number of the strongly correlated cuprates. The microscopic origin of marginal Fermi liquid behavior in the presence of strong electron correlations has been an open problem. It is also unclear down to which small ω (or T) values the relations (2.3) must hold in order to explain the relevant experiments. It is shown in Section VIII that marginal Fermi liquid behavior is obtained for a certain parameter range of the Hubbard Hamiltonian on a square lattice near half filling when the one-site Coulomb repulsions dominate.

Last but not least, non-Fermi liquid behavior is also found near a quantum critical point (QCP). It is a point in parameter space at which the system would undergo a phase transition at $T = 0$, if we were able to reach the limit of zero temperature. In that case quantum fluctuations instead of thermal fluctuations determine the critical behavior of the system. It is intuitively obvious that near a QCP the conventional Fermi liquid description breaks down since the self-energy is no longer expected to be of the form (2.4). Instead, quantum fluctuations down to arbitrary low wave numbers will modify this form. The scattering length of electrons diverges at a QCP while it must remain finite for a Fermi liquid. It should be emphasized that those features do not require strong electron correlations but appear also at QCPs of weakly correlated systems. An example of the latter case is the theory of Moriya[48] (see also Ref. [49]) for the resistivity near a QCP of a weak ferromagnet. Quantum critical points are discussed in Section IV.

III. Kondo Lattice Systems

The Kondo Hamiltonian describes magnetic impurities with free spins embedded in a metal and interacting with metal electrons via exchange scattering. The key ingredient is an antiferromagnetic interaction term

$$H_{\text{int}} = J \mathbf{s}(0) \cdot \mathbf{S}, \quad J > 0, \tag{3.1}$$

where $\mathbf{S}$ and $\mathbf{s}(0)$ are the $S = 1/2$ impurity spin and the conduction electron spin density at the impurity site which is taken here to be the origin. The model explains the characteristic Kondo behavior in dilute magnetic alloys which is determined by the phenomena of asymptotic freedom and confinement. They give

---

[48] T. Moriya and A. Kawabata, *J. Phys. Soc. Jpn.* **34**, 639 (1973).
[49] K. K. Murata and S. Doniach, *Phys. Rev. Lett.* **29**, 285 (1972).

rise to anomalies in the variation with temperature of equilibrium and transport properties and the "quenching" of the magnetic moment at low temperatures.

The presence of a highly complex many-body ground state is highlighted by the breakdown of conventional perturbation theory which starts from free electrons and magnetic moments. The divergence of the conduction electron scattering matrix sets the low-energy characteristic scale $k_B T_K$ where T_K is usually referred to as the Kondo temperature. Microscopically it arises because the local degeneracy associated with the magnetic ion is removed through the exchange coupling between the conduction electrons and the impurity spin. The coupling leads to the formation of a singlet ground state and low-energy excitations which can be described in terms of a local Fermi liquid. In close analogy to confinement the local quasiparticles are composite objects formed by conduction electrons and magnetic degrees of freedom.

The problem of magnetic impurities is well understood theoretically. There is a wide variety of techniques available which allow for an accurate description of the impurity contributions to physical properties. For detailed discussion, we refer to[50-52] and references therein.

Challenging problems are posed by work on concentrated systems, in particular on Ce-based compounds with heavy quasiparticles (heavy-fermion systems). At first glance these systems share many properties with dilute magnetic alloys. Those materials differ from ordinary metals in that there exists a characteristic temperature scale $T^* \simeq 10$–100 K, that is much smaller than the usual Fermi temperatures in ordinary metals, on which the electronic behavior of the compounds changes drastically. In the high-temperature regime for $T \gg T^*$ the systems behave like ordinary magnetic rare-earth systems which have itinerant conduction electrons with conventional masses and well-localized f-electrons. This picture is derived from the temperature dependence of the specific heat which exhibits pronounced Schottky anomalies corresponding to crystalline electric field (CEF) excitations. In addition, the magnetic susceptibility is Curie–Weiss-like reflecting the magnetic moment of the partially filled f-shell in a CEF. The low-temperature behavior, however, observed for $T \ll T^*$ is highly unusual and rather surprising: The specific heat varies approximately linearly with temperature (that is $C = \gamma T + \cdots$), and the magnetic susceptibility, χ_s, approaches a Pauli-like form, becoming almost independent of temperature. Values of the coefficients γ are of the order of J/mol K^2 and consequently two to three orders of magnitude larger than those of ordinary metals

---

[50] A. C. Hewson, *The Kondo Problem to Heavy Fermions*, Cambridge University Press (1993).

[51] Y. Kuramoto and Y. Kitaoka, *Dynamics of Heavy Electrons*, vol. 105 of *International Series of Monographs in Physics*, Clarendon Press, Oxford (2000).

[52] D. L. Cox and A. Zawadowski, *Exotic Kondo Effects in Metals: Magnetic Ions in a Crystalline Electric Field and Tunnelling Centres*, Taylor and Francis (1999).

which are of the order of $\frac{\pi^2}{2} N k_B \frac{1}{T_F} \simeq$ mJ/mol K^2. In addition, the magnetic susceptibility χ_s is enhanced by a factor of comparable magnitude. A recent survey of the experimental properties can be found in Refs. [53,54] and references therein.

The similarities in the behavior of Ce-based heavy-fermion systems to that of dilute magnetic alloys have led to the assumption that these systems are "Kondo lattices" where the observed anomalous behavior can be explained in terms of periodically repeated resonant Kondo scattering. This ansatz provides a microscopic model for the formation of a singlet ground state and the appearance of a small energy scale characterizing the low-energy excitations. The Kondo picture has been confirmed by the observation of the Kondo resonance which forms at low temperatures.[55]

In contrast to the impurity case the Kondo model cannot be solved for a periodic lattice of magnetic ions. A major difficulty is the competition between the formation of (local) Kondo singlets and the lifting of degeneracies by long-range magnetic order. In the high-temperature regime the moments of the Ce $4f$-shells are coupled by the RKKY interaction which can favor parallel as well as antiparallel orientation of the moments at neighboring sites. Model calculations for two Kondo impurities[56–58] showed that antiferromagnetic correlations between the magnetic sites weaken the Kondo singlet formation reducing the characteristic energy scale kT^* to rather small values. Consequences for an extended lattice will be discussed in the subsequent section on Quantum Phase Transitions.

The general difficulties in understanding the low temperature properties of Kondo lattices seem partially due to the lack of an adequate common "language" for the two regimes where local singlet formation is dominating on one side and where the magnetic intersite interactions dominate on the other. Such a language is usually provided by a mean-field theory which maps the complex quantum problem onto an appropriate classical model. In the case of the Kondo lattice separate mean-field descriptions exist for the two regimes which cannot be reconciled in a straightforward way to provide a unified approach.

In the present section we focus on the heavy Fermi liquid regime. The corresponding mean-field theory was described, e.g., in Ref. [29] and references

---

[53] P. Thalmeier and G. Zwicknagl, in *Handbook of the Physics and Chemistry of Rare Earth*, vol. 34, eds. K. A. Gschneider, Jr., J.-C. Bünzli, and V. Pecharsky, Elsevier, Amsterdam (2005), p. 135.

[54] P. Thalmeier, G. Zwicknagl, O. Stockert, G. Sparn, and F. Steglich, in *Frontiers in Superconducting Materials*, Springer, Berlin (2005), p. 109.

[55] F. Reinert, D. Ehm, S. Schmidt, G. Nicolay, S. Hüfner, J. Kroha, O. Trovarelli, and C. Geibel, *Phys. Rev. Lett.* **87**, 106401 (2001).

[56] C. Jayprakash, H.-R. Krishna-murthy, and J. W. Wilkins, *Phys. Rev. Lett.* **47**, 737 (1981).

[57] C. Jayprakash, H.-R. Krishna-murthy, and J. W. Wilkins, *J. Appl. Phys.* **53**, 2142 (1982).

[58] B. Jones and C. M. Varma, *J. Magn. Magn. Mater.* **63 & 64**, 251 (1987).

therein. The majority of recent microscopic studies of the Kondo lattice adopted the Dynamical Mean Field Theory which—by construction—explicitly neglects the subtle magnetic intersite correlations. It accounts for the complex local dynamics in terms of a local self-energy which has to be determined self-consistently. The applications include model calculations and a study of the γ-α-transition in Ce.[59] A major restriction on the general validity is imposed by the fact that the $4f$ valence has to be kept fixed at unity, i.e., $n_f = 1$, throughout the calculation.

The novel feature observed in stoichiometric Ce-compounds is the formation of narrow coherent bands of low-energy excitations. They give rise to the temperature dependence of the electrical resistivity which approximately follows $\rho(T) = \rho_0 + AT^2$. While these findings unambiguously show that the low-energy excitations are heavy quasiparticles involving the f degrees of freedom, they nevertheless do not provide conclusive information on how the latter have to be incorporated into a Fermi liquid description. A characteristic property of a Fermi liquid is the existence of a Fermi surface whose volume is determined by the number of itinerant fermions. It is rather obvious that at high temperatures the f electrons should be excluded from the Fermi surface due to their apparent localized character. The latter, however, is lost at low temperatures. The conjecture that the f-degrees of freedom have to be treated as itinerant fermions and, consequently, have to be included in the Fermi surface was met with great scepticism.[60] This hypothesis implies that the strong local correlations in Kondo lattices lead to an observable many-body effect, i.e., a change with temperature of the volume of the Fermi surface. At high temperatures, the f-degrees of freedom appear as localized magnetic moments, and the Fermi surface contains only the itinerant conduction electrons. At low temperatures, however, the f degrees of freedom are now tied into itinerant fermionic quasiparticle excitations and accordingly, have to be included in the Fermi volume following Luttinger's theorem. Consequently the Fermi surface is strongly modified. This scenario[61,62] was confirmed experimentally by measurements of the de Haas–van Alphen (dHvA) effect[63–65] and recent photoemission studies.[66,67]

---

[59] A. K. McMahan, K. Held, and R. T. Scalettar, *Phys. Rev. B* **67**, 075108 (2003).
[60] T. Kasuya, *Prog. Theor. Phys. Supplement* **108**, 1 (1992).
[61] G. Zwicknagl, *Physica Scripta T* **49**, 34 (1993).
[62] P. Fulde, in *Narrow Band Phenomena: Infuence of Electrons with both Band and Localized Character*, eds. J. C. Fuggle, G. A. Sawatzky, and J. Allen, Plenum (1988).
[63] G. G. Lonzarich, *J. Magn. Magn. Mat.* **76–77**, 1 (1988).
[64] H. Aoki, S. Uji, A. K. Albessard, and Y. Onuki, *Phys. Rev. Lett.* **71**, 2110 (1993).
[65] F. S. Tautz, S. R. Julian, G. J. McMullen, and G. G. Lonzarich, *Physica B* **206–207**, 29 (1995).
[66] J. D. Denlinger, G.-H. Gweon, J. W. Allen, C. G. Olson, Y. Daliachaouch, B.-W. Lee, M. B. Maple, Z. Fisk, P. C. Canfield, and P. E. Armstrong, *Physica B* **281 & 282**, 716 (2000).
[67] J. D. Denlinger, G.-H. Gweon, J. W. Allen, C. G. Olson, M. B. Maple, J. L. Sarrao, P. E. Armstrong, Z. Fisk, and H. Yamagami, *J. Electron Spectrosc. Relat. Phenom.* **117 & 118**, 347 (2001).

The present section is mainly devoted to the theoretical description of the Fermi liquid state at low temperatures. We briefly introduce the renormalized band scheme which has been devised for calculating realistic quasiparticle bands of real materials. This is achieved by combining ab initio electronic structure methods and phenomenological concepts in the spirit of Landau theory. Concerning the applications we shall not elaborate on the results for the Fermi surface and anisotropic effective masses in $CeRu_2Si_2$ for which we refer to.[61,68] We rather present recent results concerning the instabilities of the Fermi liquid state in $CeCu_2Si_2$.

We would like to emphasize the predictive power of the renormalized band method. In both cases mentioned above (Fermi surface of $CeRu_2Si_2$ and SDW instability in $CeCu_2Si_2$) the effects were first calculated theoretically and later confirmed experimentally—sometimes with a delay of up to several years.

3. Fermi-Liquid State and Heavy Quasiparticles: Renormalized Band Theory

The energy dispersion $E(\mathbf{k})$ of a dilute gas of noninteracting quasiparticles is parameterized by the Fermi wave vector $\mathbf{k}_F$ and the Fermi velocity $\mathbf{v}_F$

$$E(\mathbf{k}) = \mathbf{v}_F(\hat{\mathbf{k}}) \cdot (\mathbf{k} - \mathbf{k}_F) \qquad (3.2)$$

where $\hat{\mathbf{k}}$ denotes the direction on the Fermi surface. The key idea of the renormalized band method is to determine the quasiparticle states by computing the band structure for a given effective potential. Coherence effects which result from the periodicity of the lattice are then automatically accounted for. The quantities to be parameterized are the effective potentials which include the many-body effects. The parameterization of the quasiparticles is supplemented by information from conventional band structure calculations as they are performed for "ordinary" metals with weakly correlated electrons. The periodic potential leads to multiple-scattering processes involving scattering off the individual centers as well as to propagation between the centers which mainly depends on the lattice structure and is therefore determined by geometry. The characteristic properties of a given material enter through the information about single-center scattering which can be expressed in terms of a properly chosen set of phase shifts $\{\eta_\nu^i(E)\}$ specifying the change in phase of a wave incident on site i with energy E and symmetry ν with respect to the scattering center. Within the scattering formulation of the band structure problem the values of the phase shifts at the Fermi energy $\{\eta_\nu^i(E_F)\}$ together with their derivatives $\{(d\eta_\nu^i/dE)_{E_F}\}$ determine the Fermi wave vectors $\mathbf{k}_F$ and the Fermi velocity $\mathbf{v}_F$.

---

[68] G. Zwicknagl, *Adv. Phys.* **41**, 203 (1992).

A detailed description of the renormalized band method is given in Ref. [68]. The first step is a standard LDA band-structure calculation by means of which the effective single-particle potentials are self-consistently generated. The calculation starts, like any other ab initio calculation, from atomic potentials and structure information. In this step, no adjustable parameters are introduced. The effective potentials and hence the phase shifts of the conduction states are determined from first principles to the same level as in the case of "ordinary" metals. The f-phase shifts at the lanthanide sites, on the other hand, are described by a resonance type expression

$$\tilde{\eta}_f \simeq \arctan \frac{\tilde{\Delta}_f}{E - \tilde{\epsilon}_f} \qquad (3.3)$$

which renormalizes the effective quasiparticle mass. One of the two remaining free parameters $\tilde{\epsilon}_f$ and $\tilde{\Delta}_f$ is eliminated by imposing the condition that the charge distribution is not significantly altered as compared to the LDA calculation by introducing the renormalization. The renormalized band method devised to calculate the quasiparticles in heavy-fermion compounds thus is essentially a one-parameter theory. We mention that spin-orbit and CEF splittings can be accounted for in a straightforward manner.[68]

4. HEAVY FERMIONS IN CERU$_2$SI$_2$ AND CECU$_2$SI$_2$

The archetype heavy fermion superconductor CeCu$_2$Si$_2$ as well as CeRu$_2$Si$_2$ crystallize in the tetragonal ThCr$_2$Si$_2$ structure. The unit cell is shown in Figure III.1.

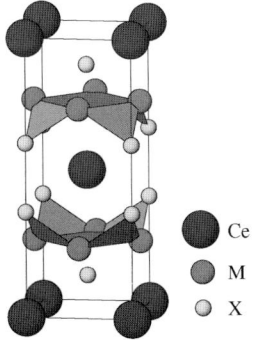

FIG. III.1. Conventional unit cell of the ThCr$_2$Si$_2$ and CeM$_2$X$_2$ structure where M = Cu, Ni, Ru, Rh, Pd, Au, ... and X = Si, Ge.

To study the electronic structure, we compare the results of two different models, i.e., treating the Ce $4f$ degrees of freedom as localized (atomic like) states and as delocalized yet strongly renormalized electrons. The first procedure provides a good quantitative description of the properties at elevated temperatures, high excitation energies, and above the metamagnetic transition. The latter ansatz yields a model for the Fermi liquid state.

The low-temperature behavior of $CeRu_2Si_2$ is well described by a paramagnetic Fermi liquid with weak residual interactions. The relevant low-energy excitations are heavy quasiparticles as inferred from the linear specific heat coefficient $\gamma \simeq 350$ mJ/mol K^2.[69] In the local moment regime, the Fermi surface is determined exclusively by the conduction states. The strongly renormalized Fermi liquid state, on the other hand, is described by the renormalized band method using $\tilde{\Delta}_f \simeq 10$ K in Eq. (3.3) for the intrinsic width of the quasiparticle band. The value is consistent with inelastic neutron data[70] as well as thermopower and specific heat data.[69] CEF effects are accounted for by adopting a Γ_7 ground state. The details of the calculation are described in Ref. [68].

The renormalized band scheme gives the correct Fermi surface topology for $CeRu_2Si_2$ and thus consistently explains the measured dHvA data.[61,68,71] The character of quasiparticles in $CeRu_2Si_2$ varies quite strongly over the Fermi surface. The validity of the Fermi liquid picture is concluded from a comparison of the effective masses on Fermi surface sheets with large f contribution. From the large linear specific heat the renormalized band scheme deduces a characteristic energy $kT^* \simeq 10$ K and predicts heavy masses of order $m^*/m \simeq 100$. This value was confirmed by experiments[72–74] where the ψ orbit with $m^*/m \simeq 120$ was observed. The corresponding Fermi surface cross-section is in agreement with estimates from the renormalized band theory. This proofs that the heavy quasiparticles exhaust the low-energy excitations associated with the f-states.

The change in volume of the Fermi surface when going from $T \ll T^*$ to $T \gg T^*$ is observed by comparing the Fermi surface of $CeRu_2Si_2$ to that of its ferromagnetic isostructural counterpart $CeRu_2Ge_2$ where the f-states are clearly localized. In a series of beautiful experiments[75] it was demonstrated that the Fermi

---

[69] F. Steglich, U. Rauchschwalbe, U. Gottwick, H. M. Mayer, G. Sparn, N. Grewe, and U. Poppe, *J. Appl. Phys.* **57**, 3054 (1985).

[70] L. P. Regnault, W. A. C. Erkelens, J. Rossat-Mignot, P. Lejay, and J. Flouquet, *Phys. Rev. B* **38**, 4481 (1988).

[71] G. Zwicknagl, E. Runge, and N. E. Christensen, *Physica B* **163**, 97 (1990).

[72] A. K. Albessard, T. Ebihara, I. Umehara, K. Satoh, Y. Onuki, H. Aoki, S. Uji, and T. Shimizu, *Physica B* **186–188**, 147 (1993).

[73] H. Aoki, S. Uji, A. K. Albessard, and Y. Onuki, *Phys. Rev. Lett.* **71**, 2110 (1993).

[74] F. S. Tautz, S. R. Julian, G. J. McMullen, and G. G. Lonzarich, *Physica B* **206–207**, 29 (1995).

[75] C. A. King and G. G. Lonzarich, *Physica B* **171**, 161 (1991).

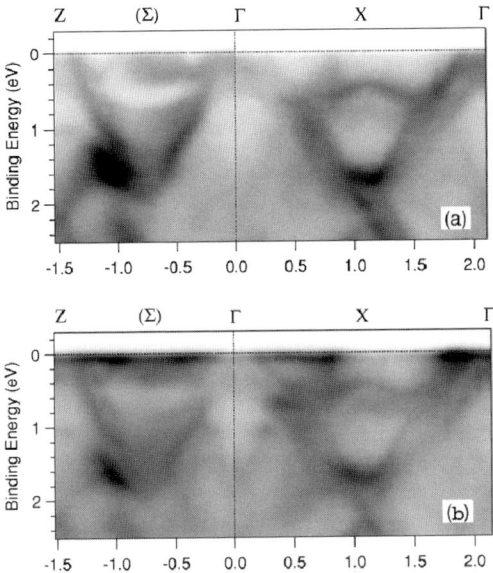

FIG. III.2. Photoemission results for (a) LaRu$_2$Si$_2$ in comparison to (b) CeRu$_2$Si$_2$ at $T = 25$ K, i.e., above the Kondo temperature $T^* = 15$ K of that system. Band structures are very similar for both compounds. (After Ref. [67].)

surfaces of these two compounds are rather similar. However, the enclosed Fermi volume is smaller in the case of CeRu$_2$Ge$_2$, the difference being roughly one electron per unit cell. More direct evidence is provided by recent photoemission experiments (see Figure III.2). Denlinger et al.[67] have shown that at temperatures around 25 K, the Fermi surface of CeRu$_2$Si$_2$, is that of its counterpart LaRu$_2$Si$_2$ which has no f electrons.

At this point the general question arises how the formation of heavy quasiparticles is reflected in the angular resolved photoelectron spectroscopy (ARPES) data. A major difficulty stems from the fact that photoemission experiments probe the occupied part of the spectrum. The most dramatic changes, however, are expected to occur in the empty part. In Figure III.3 we compare the dispersion of the heavy quasiparticle bands at low temperatures to their light high-temperature counterparts. In the occupied part the main difference is a bending close to the Fermi energy which changes the volume of the Fermi surface. The characteristic bending was recently observed in CeCoIn$_5$.[76]

---

[76] A. Koitzsch, Priv. comm.

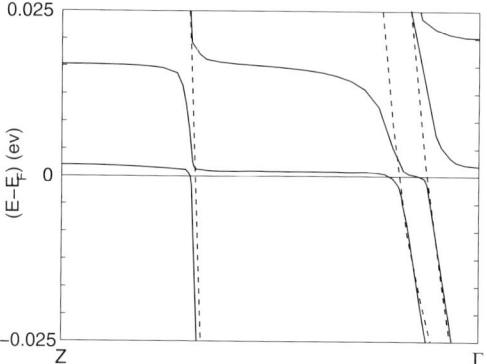

FIG. III.3. Band dispersion for $CeCu_2Si_2$ along Z–Γ for low temperatures $T \ll T^*$ (full lines) and high temperatures (dashed lines). The formation of the heavy quasiparticles leads to a characteristic bending in the occupied part of the spectrum.

Let us now turn to the heavy fermion superconductor $CeCu_2Si_2$ which exhibits a highly complex phase diagram at low temperatures discussed in Section III.5. It results from an extreme sensitivity of the physical properties with respect to variations of the stoichiometry and external magnetic fields.

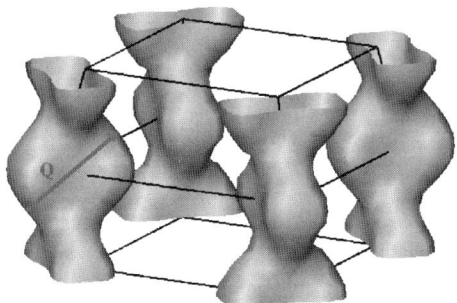

FIG. III.4. $CeCu_2Si_2$: Main Fermi surface sheet of heavy quasiparticles ($m^*/m \simeq 500$) calculated with the renormalized band method. It consists of modulated columns which are oriented parallel to the tetragonal axis. The calculations adopt the CEF scheme of Ref. [77] consisting of a singlet ground state separated from an excited quartet by a CEF splitting $\delta \simeq 330$ K. Therefore $\delta \gg T^* \simeq 10$ K (obtained from the γ-value). The nesting vector $\mathbf{Q} = (0.23, 0.23, 0.52)$ connects flat parts ("nesting") of the Fermi surface. (After Refs. [78] and [79].)

---

[77] E. A. Goremychkin and R. Osborn, *Phys. Rev. B* **47**, 14280 (1993).

[78] G. Zwicknagl and U. Pulst, *Physica B* **186**, 895 (1993).

[79] O. Stockert, E. Faulhaber, G. Zwicknagl, N. Stuesser, H. Jeevan, T. Cichorek, R. Loewenhaupt, C. Geibel, and F. Steglich, *Phys. Rev. Lett.* **92**, 136401 (2004).

To calculate the quasiparticle bands in $CeCu_2Si_2$ by means of the renormalized band method, we adopt the doublet-quartet CEF scheme suggested in Ref. [77]. The ground state is separated from the excited quartet by $\delta \simeq 330$ K.

The results for the Fermi surface[78,80] can be summarized as follows: We find two separate sheets of the Fermi surface for heavy and light quasiparticles. The light quasiparticles have effective masses of the order of $m^*/m \simeq 5$. They can be considered as weakly renormalized conduction electrons. Of particular interest are heavy quasiparticles of effective masses $m^*/m \simeq 500$ which are found on a separate sheet. This surface whose shape (see Figures III.4 and III.8) is rather different from the corresponding LDA surface mainly consists of columns parallel to the tetragonal axis and of small pockets. The topology of the Fermi surface suggests that the strongly correlated Fermi liquid state should become unstable at sufficiently low temperatures. Firstly, it exhibits pronounced nesting features which may eventually lead to the formation of a ground state with a spin-density modulation. This will be discussed in detail below. Secondly, the topology of this surface depends rather sensitively on the position of the Fermi energy. The band filling and hence the f-valence are critical quantities. Reducing the f-occupancy from the initial value of $n_f \simeq 0.95$ by approximately 2% leads to changes in the topology as shown in Refs. [78,80]. As a result, the quasiparticle density of states (DOS) exhibits rather pronounced structures in the immediate vicinity of the Fermi energy which, in turn, can induce instabilities.[81]

5. Low-Temperature Phase Diagram of $CeCu_2Si_2$

The phase diagram of $CeCu_2Si_2$ contains three different phases: the A and the B phase and a superconducting phase. In some samples the superconducting phase expels the A phase while in other samples the two phases may coexist. While the A phase has been identified as a spin-density wave phase as discussed below, the character of the B phase has remained unknown. The instability may result from a reconstruction of the Fermi surface.[78] Much effort has been devoted to the characterization of the A phase which originally had the appearance of a 'hidden order' phase. However, later a spin-density wave character was first inferred from resistivity results[82] and was supported by specific heat and high-resolution magnetization measurements.[83] The transition temperature T_A is suppressed by increasing

---

[80] U. Pulst, *Ph.D. thesis*, TH Darmstadt (1993).
[81] M. I. Kaganov and I. M. Lifshits, *Sov. Phys. Usp.* **22**, 904 (1979).
[82] P. Gegenwart, C. Langhammer, C. Geibel, R. Helfrich, M. Lang, G. Sparn, F. Steglich, R. Horn, L. Donnevert, A. Link, et al., *Phys. Rev. Lett.* **81**, 1501 (1998).
[83] F. Steglich, N. Sato, T. Tayama, T. Lühmann, C. Langhammer, P. Gegenwart, P. Hinze, C. Geibel, M. Lang, G. Sparn, et al., *Physica C* **341–348**, 691 (2000).

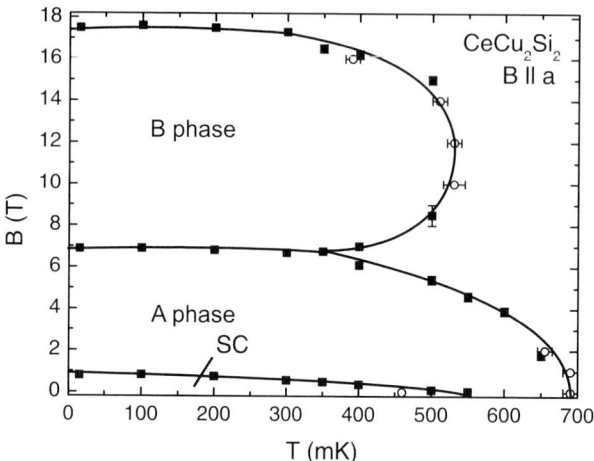

FIG. III.5. B–T phase diagram of $CeCu_2Si_2$ for $B \parallel a$. Original version from Ref. [84], completed version from Ref. [85]. In this sample the SDW A-phase is expelled from the superconducting (SC) region (no coexistence). The nature of the B-phase is yet undetermined.

the $4f$-conduction electron hybridization and eventually vanishes. This can be achieved by applying hydrostatic pressure or choosing a few percent excess of Cu. The ordered moments are expected to be rather small.

The important question in this context is: What is the origin of the antiferromagnetic correlations, showing up in the A phase? How do they arise in the heavy fermion state and finally, how do they affect the heavy quasiparticles? The key to the answers comes from the Fermi surface of $CeCu_2Si_2$ and its nesting properties. As shown in Figure III.4 there are parallel portions which are connected by a wave vector close to $(1/4, 1/4, 1/2)$. As a consequence, the static susceptibility $\chi(\mathbf{q})$ exhibits a maximum for momentum transfer $\mathbf{q}$ close to the nesting vector (Figure III.6).

Recent neutron scattering experiments[79] (Figure III.6) for the stoichiometric compound ($x = 0$) show a spin-density wave (SDW) which forms below $T_N \simeq 0.7$ K. The experimental propagation vector $\mathbf{Q}$ is close to $(0.22, 0.22, 0.55)$ and the ordered moment amounts to $\mu \simeq 0.1\mu_B$. These findings show that the SDW in $CeCu_2Si_2$ arises out of the renormalized Fermi liquid state. The transition is driven by the nesting properties of the heavy quasiparticles.

---

[84] G. Bruls, D. Weber, B. Wolf, P. Thalmeier, B. Lüthi, A. de Visser, and A. Menovsky, *Phys. Rev. Lett.* **65**, 2294 (1990).
[85] F. Weickert, Private comm. (2003).

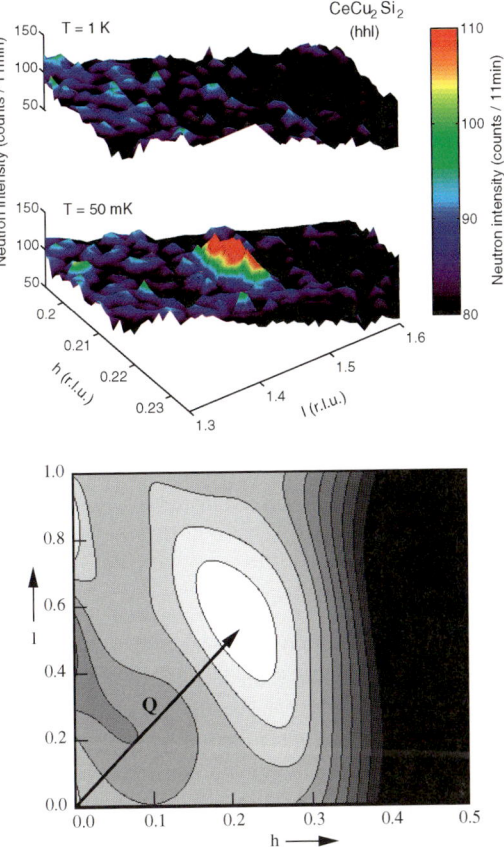

FIG. III.6. Upper panel: Neutron diffraction intensity in CeCu$_2$Si$_2$ at temperature above and below the A-phase transition temperature T_A. Incommensurate peak is at $\mathbf{Q} = (0.22, 0.22, 0.55)$. (After Ref. [79].) Lower panel: Nesting of heavy FS columns (Figure III.4) leads to a peak in the static susceptibility $\chi(\mathbf{q})$ at $\mathbf{q} = \mathbf{Q}$. Intensity map of $\chi(\mathbf{q})$ (value increasing from dark to bright) in the reciprocal (h, h, l)-plane as calculated for the renormalized bands at $T = 100$ mK. The *experimental* $\mathbf{Q}$ at 50 mK from the left panel shows perfect agreement with the calculated maximum position of $\chi(\mathbf{q})$.

Having classified the nature of the A phase we next turn to the question how the latter competes with superconductivity. Itinerant electron antiferromagnetism as realized in the A phase and superconductivity both form in the system of the heavy quasiparticles. Their interplay therefore depends sensitively on the geometric properties of the paramagnetic Fermi surface and the symmetries of the ordered phases. This can be seen from realistic model calculations investigating the

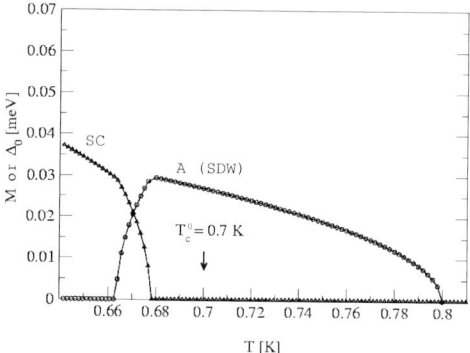

FIG. III.7. Temperature dependence of the superconducting (SC) and magnetic A (SDW) order parameters in a AS crystal. The sublattice magnetization $M = h_0(T)$ and the amplitude Δ_0 of the SC order parameter are calculated from Eq. (3.23) using realistic quasiparticles and the experimentally determined propagation vectors for the A phase. A SC order parameter $\Delta(\mathbf{k}) = \Delta_0(\cos k_x a - \cos k_y a)$ can form in the A phase. Below the superconducting transition temperature $T < T_c < T_N^{(1)}$ the two order parameters coexist and compete. Itinerant SDW antiferromagnetism is expelled at a temperature $T_N^{(2)} < T_c$. (After Ref. [86,87].)

variation with temperature of the two order parameters in crystals where the two ordering phenomena coexist in some temperature range ("AS-type" crystals).[86,87] Figure III.7 summarizes the results for an unconventional superconducting state with Γ_3 (d-wave) symmetry of the gap function, i.e., $\Delta(\mathbf{k}) \sim (\cos k_x a - \cos k_y a)$.

The theory leading to the results displayed in Figure III.7 starts from the model Hamiltonian

$$H = H_0 + H_{\text{int}} \quad (3.4)$$

where the free quasiparticles are described by

$$H_0 = \sum_{\mathbf{k}\sigma} E(\mathbf{k}) c_{\mathbf{k}\sigma}^\dagger c_{\mathbf{k}\sigma}. \quad (3.5)$$

The creation (annihilation) operators for quasiparticles with wavevector $\mathbf{k}$, (pseudo) spins $\sigma = \pm 1$ and energy $E(\mathbf{k})$ are denoted by $c_{\mathbf{k}\sigma}^\dagger$ ($c_{\mathbf{k}\sigma}$). The energies which are measured relative to the Fermi level are calculated within the Renormalized Band Scheme. The residual interactions in the strongly renormalized Fermi liquid are assumed to be repulsive for short separations while being attractive for two quasiparticles of opposite momenta on neighboring sites. The former favors

---

[86] M. Neef, *Master Thesis*, Technische Universität Braunschweig (2004).
[87] M. Neef, M. Reese, and G. Zwicknagl, preprint (2004).

the formation of a SDW while the latter gives rise to a superconducting instability. Adopting a mean-field approximation yields

$$H_{\text{int}} \to H_{\text{SDW}} + H_{\text{SC}} \tag{3.6}$$

where

$$H_{\text{SDW}} = -\sum_{\mathbf{k}\sigma} \frac{\sigma}{2} \sum_{\mathbf{Q}_j} \left(h(\mathbf{Q}_j) c_{\mathbf{k}\sigma}^{\dagger} c_{\mathbf{k}+\mathbf{Q}_j\sigma} + h.c. \right) \tag{3.7}$$

and

$$H_{\text{SC}} = \frac{1}{2} \sum_{\mathbf{k}\sigma\sigma'} \left(\Delta_{\sigma\sigma'}(\mathbf{k}) c_{\mathbf{k}\sigma}^{\dagger} c_{-\mathbf{k}\sigma'}^{\dagger} + h.c. \right). \tag{3.8}$$

The periodically modulated magnetization associated with the SDW with propagation vectors $\mathbf{Q}_j$ as well as the superconducting pair potential $\Delta_{\sigma\sigma'}$ have to be determined selfconsistently

$$h(\mathbf{Q}_j) = \frac{U}{L} \sum_{\mathbf{k}\sigma} \frac{\sigma}{2} \langle c_{\mathbf{k}+\mathbf{Q}_j\sigma}^{\dagger} c_{\mathbf{k}\sigma} \rangle \tag{3.9}$$

and

$$\Delta_{\sigma\sigma'}(\mathbf{k}) = \frac{1}{L} \sum_{\mathbf{k}'\sigma''\sigma'''} g_{s\sigma\sigma';\sigma''\sigma'''}(\mathbf{k}, \mathbf{k}') \langle c_{-\mathbf{k}'\sigma''} c_{\mathbf{k}'\sigma'''} \rangle, \tag{3.10}$$

where the strength U of the local Hubbard-type repulsion is of the order of the quasiparticle band width $k_B T^*$ and $g_{\sigma\sigma';\sigma''\sigma'''}(\mathbf{k}, \mathbf{k}')$ is the effective pair attraction. The $\mathbf{k}$-summation runs over the entire paramagnetic Brillouin zone and L denotes the number of lattice sites. The expectation values denoted by $\langle \cdots \rangle$ have to evaluated with the eigenstates of the mean-field Hamiltonian $H_{\text{MF}} = H_0 + H_{\text{SDW}} + H_{\text{SC}}$, and consequently depend upon the order parameters. Therefore the self-consistency equations are coupled.

The mean-field Hamiltonian implicitly assumes that the amplitudes of both order parameters are small. In particular, we neglect here the pairing amplitudes of the form $\langle c_{-\mathbf{k}\sigma''} c_{\mathbf{k}+\mathbf{Q}_j\sigma'''} \rangle$. The latter are important when the gaps introduced by the antiferromagnetic order into the quasiparticle spectrum are large on the scale set by superconductivity. For a discussion of this point we refer to Refs. [88,89].

---

[88] G. Zwicknagl and P. Fulde, *Z. Phys. B* **43**, 23 (1981).
[89] P. Fulde and G. Zwicknagl, *J. Appl. Phys.* **53**, 8064 (1982).

The periodically modulated magnetization associated with the SDW acts on the conduction electrons like a periodic spin-dependent potential which we approximate by

$$h(\mathbf{r}) = \sum_{\mathbf{Q}_j} h(\mathbf{Q}_j) e^{i\mathbf{Q}_j \cdot \mathbf{r}} \quad (3.11)$$

with the same amplitudes $h(\mathbf{Q}_j) = h_0$ for the eight commensurate wave vectors $\mathbf{Q}_j \in \{(\pm\frac{\pi}{2a}, \pm\frac{\pi}{2a}, \pm\frac{\pi}{c})\}$. The magnetic superstructure breaks the translational invariance of the underlying lattice but preserves the point group symmetry. The mean-field Hamiltonian, however, is invariant under translations with

$$\mathbf{a}'_1 = (2a, 2a, 0), \quad \mathbf{a}'_2 = (2a, -2a, 0), \quad \mathbf{a}'_3 = (2a, 0, c). \quad (3.12)$$

The volume of the magnetic supercell is 16 times the volume of the paramagnetic unit cell. As a result the Brillouin zone is reduced and the quasiparticle states are modified by extra Bragg reflections. The opening of new gaps is important at sufficiently low temperatures $T \ll T_N$ where T_N is the ordering (Néel) temperature.

The order parameter $\Delta_{\sigma\sigma'}$ behaves as a two-fermion wave function in many respects. This is expressed by the fact that an off-diagonal long-range order (ODLRO) parameter is not the thermal expectation value of a physical observable but rather a complex pseudo- wave function describing quantum-phase correlations on the macroscopic scale of the superconducting coherence length. Its phase is a direct signature of the broken gauge invariance in the superconducting condensate.

Experiment (strongly) suggests that superconductivity in $CeCu_2Si_2$ occurs with anisotropic even-parity (pseudo-) spin singlet pairing. We therefore restrict ourselves to this case characterized by a scalar order parameter

$$\Delta_{\sigma\sigma'}(\mathbf{k}) = \phi(\mathbf{k})(i\sigma_2)_{\sigma\sigma'} \quad (3.13)$$

were $\phi(\mathbf{k})$ is a complex amplitude and σ_2 denotes the Pauli matrix in spin space. Since the two ordering temperatures, i.e., the antiferromagnetic Néel temperature T_A and the superconducting T_c are very close to each other we focus on pair states which are compatible with the translational symmetry of the paramagnetic lattice. The corresponding functions are listed in Refs. [68,90–92]. In the explicit calculations we restrict ourselves to one-dimensional representations for simplicity. The generalization to multi-dimensional representations is rather straightforward.[86]

Finally, the variation with momenta $\mathbf{k}$ and $\mathbf{k}'$ of the quasiparticle attraction $g(\mathbf{k}, \mathbf{k}')$ is expanded in terms of the basis functions belonging to the κ-th row, $\kappa =$

---

[90] M. Sigrist and K. Ueda, *Rev. Mod. Phys.* **63**, 239 (1991).
[91] R. Konno and K. Ueda, *Phys. Rev. B* **40**, 4329 (1989).
[92] M. Ozaki and K. Machida, *Phys. Rev. B* **39**, 4145 (1989).

$1, \ldots, d(\Gamma)$, of the $d(\Gamma)$-dimensional irreducible representation of the symmetry group

$$g(\mathbf{k}, \mathbf{k}') = \sum_{\Gamma} g_{\Gamma} \sum_{\kappa=1}^{d(\Gamma)} \varphi_{\Gamma_\kappa}(\mathbf{k}) \varphi^*_{\Gamma_\kappa}(\mathbf{k}'). \quad (3.14)$$

We further simplify the problem by focusing on the states $\phi_\Gamma(\mathbf{k})$ with the symmetry which yields the strongest quasiparticle attraction g_Γ. Assuming that this most stable order parameter is non-degenerate the self-consistency condition is

$$\phi(\mathbf{k}) = -g_{\Gamma_0} \varphi_{\Gamma_0}(\mathbf{k}) \frac{1}{L} \sum_{\mathbf{k}'} \varphi^*_{\Gamma_0}(\mathbf{k}') \langle c_{-\mathbf{k}'\downarrow} c_{\mathbf{k}'\uparrow} \rangle. \quad (3.15)$$

To solve the mean-field Hamiltonian Eqs. (3.7) and (3.8) we adopt the Nambu formalism which allows us to reduce the mean-field Hamiltonian to single-particle form in particle–hole space, i.e.,

$$H = \sum_{\mathbf{k}}^{\text{AFBZ}} \Psi^\dagger_{\mathbf{k}} \{ \hat{\mathbf{E}}(\mathbf{k}) \hat{\tau}_3 + \hat{\mathbf{h}} \hat{1} + \hat{\boldsymbol{\Delta}}(\mathbf{k}) \hat{\tau}_1 \} \Psi_{\mathbf{k}}. \quad (3.16)$$

The Nambu spinors $\Psi_{\mathbf{k}}$ have 32 components and are defined as

$$\Psi^\dagger_{\mathbf{k}} = \left(c_{\mathbf{k}\uparrow}, c_{\mathbf{k}+\mathbf{Q}_1\uparrow}, \ldots, c^\dagger_{-\mathbf{k}\downarrow}, c^\dagger_{-\mathbf{k}-\mathbf{Q}_1\downarrow}, \ldots \right). \quad (3.17)$$

They account for the coherent superposition of particles and holes which is the characteristic feature of the superconducting state. Here $\hat{1}$, $\hat{\tau}_1$ and $\hat{\tau}_3$ denote the unit matrix and the Pauli matrices in particle–hole space. The sixteen wave vectors $\mathbf{Q}_0 = 0, \mathbf{Q}_1, \ldots, \mathbf{Q}_{15}$ are the reciprocal lattice vectors appearing in the antiferromagnetic phase. The set includes the eight propagation vectors of the SDW and their higher harmonics. The $\mathbf{k}$-summation is restricted to the reduced Brillouin zone (AFBZ) of the antiferromagnetic state defined by the SDW. The structure of the Hamiltonian in particle–hole space is

$$\hat{H}(\mathbf{k}) = \begin{pmatrix} \hat{\mathbf{E}}(\mathbf{k}) + \hat{\mathbf{h}} & \hat{\boldsymbol{\Delta}} \\ \hat{\boldsymbol{\Delta}} & -\hat{\mathbf{E}}(\mathbf{k}) + \hat{\mathbf{h}} \end{pmatrix} \quad (3.18)$$

where the 16×16-diagonal matrix contains the quasiparticle energies of the paramagnetic normal phase

$$\left(\hat{\mathbf{E}}(\mathbf{k}) \right)_{\mathbf{Q}_i \mathbf{Q}_j} = \delta_{\mathbf{Q}_i \mathbf{Q}_j} E(\mathbf{k} + \mathbf{Q}_i). \quad (3.19)$$

The SDW acts like an effective magnetic field. The modulated spin density leads to Umklapp scattering which is accounted for by the matrix

$$\hat{\mathbf{h}} = -h_0(T) \hat{\mathbf{m}}. \quad (3.20)$$

The 16×16 matrix $\hat{\mathbf{m}}$ is a purely geometric quantity specifying the possible Umklapp processes while the temperature-dependent amplitude $h_0(T)$ has to be determined self-consistently. The diagonal matrix

$$\left(\hat{\mathbf{\Delta}}(\mathbf{k})\right)_{\mathbf{Q}_i\mathbf{Q}_j} = \delta_{\mathbf{Q}_i\mathbf{Q}_j}\Delta_0(T)\varphi_{\Gamma_0}(\mathbf{k}+\mathbf{Q}_i)$$
$$\equiv \Delta_0(T)\left(\hat{\mathbf{\Phi}}(\mathbf{k})\right)_{\mathbf{Q}_i\mathbf{Q}_j} \quad (3.21)$$

contains the superconducting order parameters with the given $\mathbf{k}$-dependent function $\phi_\Gamma(\mathbf{k})$ and a temperature-dependent amplitude $\Delta_0(T)$.

The self-consistency Eqs. (3.9) and (3.10) can be formulated in terms of the off-diagonal elements of the 32×32-matrix Green's function

$$\hat{G}(i\epsilon_n,\mathbf{k}) = \left(i\epsilon_n\hat{1} - \hat{H}(\mathbf{k})\right)^{-1} \quad (3.22)$$

according to

$$h_0(T) = \frac{U}{L}T\sum_{\epsilon_n}^{\epsilon_c}\sum_{\mathbf{k}}^{\text{AFBZ}}\frac{1}{16}\text{Tr}\left[\hat{\mathbf{m}}\hat{1}\hat{G}(i\epsilon_n,\mathbf{k})\right]$$

$$\Delta_0(T) = -\frac{g_\Gamma}{L}T\sum_{\epsilon_n}^{\epsilon_c}\sum_{\mathbf{k'}}^{\text{AFBZ}}\frac{1}{2}\text{Tr}\left[\hat{\mathbf{\Phi}}(\mathbf{k'})\hat{\tau}_1\hat{\tau}_3\hat{G}(i\epsilon_n,\mathbf{k'})\hat{\tau}_3\right]. \quad (3.23)$$

Here $\epsilon_n = \pi T(2n+1)$ denote the T-dependent Matsubara frequencies and ϵ_c is the energy cut-off required in weak-coupling theory. The coupling constants U and g_Γ as well as the cut-off ϵ_c are eliminated in the usual way in favor of the observable quantities $T_N^{(0)}$ and $T_c^{(0)}$. Solving the self-consistency equations for $\phi_{\Gamma_3}(\mathbf{k}) \sim \cos k_x a - \cos k_y a$ yields the results displayed in Figure III.7. It shows that both order parameters coexist in a finite temperature interval. This is due to the fact they have their maximum amplitudes on different parts of the Fermi surface (see Figure III.8). At sufficiently low temperatures the A-phase is finally expelled by superconductivity as shown in Figure III.7. This was confirmed by neutron diffraction which shows a suppression of magnetic Bragg peaks further below T_c.

IV. Quantum Phase Transitions

Most phase transitions in condensed matter are governed by the appearance of a spontaneously broken symmetry below a certain transition temperature T_c. The low temperature phase is then characterized by an order parameter belonging to a single nontrivial representation of the high temperature symmetry group. In the simplest case the order parameter is of the density type (diagonal long range order), for instance charge density n($\mathbf{r}$) or spin density $\mathbf{m}(\mathbf{r})$. Approaching T_c from

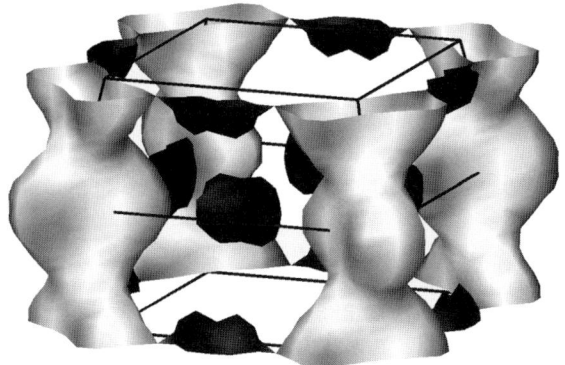

FIG. III.8. CeCu$_2$Si$_2$: Variation of the SC gap function amplitude $|\Delta(\mathbf{k})|/\Delta_0(T) = |\phi_{\Gamma_3}(\mathbf{k})|$ for the (pseudo-) singlet wave function with Γ_3-symmetry $\phi_{\Gamma_3}(\mathbf{k}) \sim \cos k_x a - \cos k_y a$ on the heavy quasiparticle sheet of the paramagnetic Fermi surface. The amplitude of this SC order parameter is maximal on the kidney-shaped surfaces centered along the Σ direction which are almost unaffected by the formation of the A phase. The dominant contributions to the latter come from the nesting parts on the heavy columns where the superconducting amplitude is small. Dark and light grey indicate large and small amplitudes, respectively.

above the correlation length of spatial order parameter fluctuations, the associated order parameter susceptibility and other thermodynamic quantities diverge. The divergence is characterized by critical exponents that depend only on spatial dimension d, on the range of interactions and number of order parameter components n which define the universality class of the model. In such finite temperature or 'classical' phase transitions the underlying microscopic quantum fluctuations of charge and spin densities etc. are not important ingredients for the long-range order because their coherence is destroyed by thermal fluctuations over time scales longer than $\hbar/kT_c$.

However, the broken symmetry state may not only be reached by lowering the temperature. Instead at $T = 0$ the tuning of a physical control parameter X, e.g., due to applied hydrostatic or chemical pressure via doping may drive the compound from the disordered to the ordered state and vice versa (see Figure IV.6). The corresponding value X_c defines the quantum critical point (QCP) where a quantum phase transition (QPT) takes place. In the latter the $T = 0$ quantum fluctuations can be coherent over arbitrary long time scales. For this reason the effective dimension for order parameter fluctuations close to a QPT is given by $d_{\text{eff}} = d + z$ where z is the dynamic exponent which characterizes the scaling of energies of quantum fluctuations with system size. The contribution of quantum fluctuations has therefore profound consequences for the critical exponents of thermodynamic and transport quantities. For example, in strongly correlated metals close to an antiferromagnetic QCP along the quantum critical line ($X = X_c$

or $|r| = 0$) in Figure IV.6 an anomalous non-Fermi liquid (NFL) temperature dependence of physical quantities like $\chi(T)$, $C(T)/T$ and $\rho(T)$ emerges. Its origin and theoretical description has been the subject of much recent investigations and controversy.[93,94] An additional important discovery is the observation that superconductivity often appears in a dome-like shape around the QCP. Viewed differently, it may be a successful strategy to look for unconventional superconducting states in searching around magnetic QCPs of suitable materials.

The presence of strong electronic correlations is by no means essential for the appearance of a quantum phase transition. Historically they have been first studied in weakly correlated metals without calling them explicitly with their modern name. As an introduction we will briefly discuss some early examples of QPTs. Subsequently well understood model theories for QPTs in local moment systems, notably the Ising model in a transverse field will be discussed. Then we come to QPTs in the strongly correlated systems where theoretical work has focused on the Kondo-lattice type models both with and without charge degrees of freedom. Finally the phenomenological scaling and Ginzburg–Landau theories applicable close to the QCP will be discussed. They are important for the interpretation of a large body of experimental work near the QCP of heavy-fermion compounds.

6. Quantum Phase Transition in Localized and Itinerant Magnets

The idea that tuning of a control parameter may drive an insulator or metal from the paramagnetic state to a magnetically ordered state at zero temperature is indeed a very old one. The classical Stoner–Wolfarth theory of itinerant ferromagnetism (FM)[95,96] identifies this control parameter as $X = IN(E_F)$ where I is the exchange integral of itinerant conduction electrons and $N(E_F)$ the conduction electron DOS per spin direction. In the paramagnetic regime with $X < X_c = 1$ the exchange interaction is too weak to cause an exchange splitting of conduction bands and stabilize a spin polarization. However the incipient FM order has its effect on the spin fluctuation spectrum. The typical life time τ_{sf} of a quantum fluctuation of magnetic moments diverges when $X \to X_c$. This may be seen from the dynamical susceptibility, calculated from the single-band Hubbard model for

---

[93] S. Sachdev, *Quantum Phase Transitions*, Cambridge University Press (1999).
[94] M. A. Continentino, *Quantum Scaling in Many-Body Systems*, World Scientific (2001).
[95] E. Stoner, *Proc. Roy. Soc.* **A165**, 372 (1938).
[96] E. Stoner, *Proc. Roy. Soc.* **A169**, 339 (1939).

a parabolic band within random-phase approximation (RPA). Close to the QCP ($X \lesssim X_c$) the spectrum of FM spin fluctuations with q → 0 is given by:[97,98]

$$\operatorname{Im}\chi_{-+}(\omega) = \frac{(\pi/4)N(E_F)\omega/qv_F}{[1-IN(E_F)]^2 + [(\pi/4)N(E_F)I\omega/qv_F]^2} \quad (4.1)$$

where v_F is the Fermi velocity. When $X \to X_c$ this spectrum is strongly peaked at the 'paramagnon' frequency ($S \gg 1$)

$$\omega_P = \frac{4}{\pi}\frac{v_F}{S}q. \quad (4.2)$$

Here $S = (1 - IN(E_F))^{-1}$ is the 'Stoner parameter' which governs the softening of the spin fluctuation spectrum in Eq. (4.1) on approaching the QCP. It also gives the enhancement of the static susceptibility $\chi = S\chi_0$ as compared to the free Pauli susceptibility χ_0. In addition the paramagnon excitations lead to a deviation from the linear specific heat behavior of the Fermi liquid described by[98,99]

$$C(T)/T = \gamma_0[m^*/m + S(T/T_{\text{SF}})^2 \ln(T/T_{\text{SF}})] \quad (4.3)$$

where m^*/m is the mass enhancement due to paramagnons[98] and T_{SF} is the spin fluctuation temperature given by $T_{SF} = E_F/S$.

Equivalently one may say that $\chi_{-+}(\omega)$ has a pole at the purely imaginary frequency $i\omega_P$. This represents a collective overdamped spin fluctuation mode. When X is tuned through the critical value $X_c = 1$ by pressure or alloying, the imaginary spin fluctuation pole moves to the real axis and becomes the FM spin-wave pole with frequency ($q \ll k_F$, $m_s \ll n$)

$$\omega(q) = Dq^2, \quad D \sim (I^2/E_F)k_F^{-2}nm_s. \quad (4.4)$$

Simultaneously a spontaneous FM moment caused by the spin polarization $m_s = n_\uparrow - n_\downarrow$ of conduction bands appears. Here D is the spin wave stiffness constant and $n = n_\uparrow + n_\downarrow$ is the number of conduction electrons per site. This collective spin excitation is undamped for small q because it is the Goldstone mode associated with the continuous SO(3) symmetry of the FM order parameter and therefore in the hydrodynamic limit it is protected by a conservation law against decay. For X only marginally above the FM QCP $X_c = 1$ one has only weak ferromagnetism (WFM) and anomalous thermodynamic and transport behavior due to paramagnon excitations was observed.[97,98] The first theory beyond the Hartree–Fock RPA level to address such quantum critical phenomena in itinerant

---

[97] N. F. Berk and J. R. Schrieffer, *Phys. Rev. Lett.* **17**, 1171 (1966).
[98] S. Doniach and S. Engelsberg, *Phys. Rev. Lett.* **17**, 750 (1966).
[99] C. P. Enz, *A Course on Many-Body Theory Applied to Solid-State Physics*, vol. 11 of *Lecture Notes in Physics*, World Scientific (1992).

ferromagnets was Moriya's self consistent renormalization (SCR) theory[48,100,101] for the WFM and the theory by Hertz[102] on which much of the later developments are based.

The best known example of an enhanced paramagnetic metal which is close to a FM quantum critical point is palladium metal. In this case one has a large susceptibility enhancement of $S \simeq 10$.[103] Indeed, alloying with only 0.5% Fe immediately leads to FM order.[104] According to band structure calculations a (linear) expansion of the lattice of pure Pd by $\sim 7\%$ should lead to a FM ground state. Experimentally Au–Pd–Au sandwiches have been prepared where an estimated volume expansion of 2.3% of Pd due to the larger lattice constant of Au leads to huge Pd Stoner factors of $S \sim 10^3$–10^4,[105] but FM order is still not achieved.

The classical, enhanced paramagnetism in Pd and associated QPT in Pd alloys has been complemented by Laves phase compounds AB_2 like $TiBe_2$[106] and $ZrZn_2$[107] which are slightly on the overcritical side, i.e., $IN(E_F) \geqslant 1$ and thus weak itinerant ferromagnets. By doping with Cu the alloy series $TiBe_{2-x}Cu_x$ exhibits a QCP at $x_c \simeq 0.155$ where ferromagnetism disappears. The compound $ZrZn_2$ has already been known for a long time and has recently been investigated with renewed interest because it exhibits a FM QCP as function of hydrostatic pressure at $p_c = 21$ GPa. As in UGe_2 it was found that surprisingly superconductivity coexists within the FM phase, albeit with a small T_c.[108]

Quite another quantum phase transition to a magnetically ordered state has been known since a long time in localized moment systems. It was studied under the name 'induced moment magnetism' without actually stressing that it is a generic type of quantum phase transition as we shall see. In localized ($4f$- or $5f$-) magnetic compounds with uniaxial symmetry the lowest CEF states of non-Kramers ions may consist of a nonmagnetic singlet ground state $|1\rangle$ and a nonmagnetic singlet (or doublet) excited state $|2\rangle$ at an energy Δ. They are both characterized by a vanishing moment, i.e., $\langle n|J_z|n\rangle = 0$ ($n = 1, 2$) with J_z denoting the total angular momentum component of the localized $4f$ or $5f$ states. Therefore there are no pre-existing localized moments that might order as in a conventional AF

---

[100] T. Moriya and A. Kawabata, *J. Phys. Soc. Jpn.* **35**, 669 (1973).
[101] T. Moriya, *Spin Fluctuations in Itinerant Electron Magnetism*, vol. 56 of *Springer Series in Solid-State Sciences*, Springer-Verlag, Berlin (1985).
[102] J. A. Hertz, *Phys. Rev. B* **14**, 1165 (1976).
[103] W. Gerhardt, F. Razavi, J. S. Schilling, D. Hüser, and J. A. Mydosh, *Phys. Rev. B* **24**, 6744 (1981).
[104] J. A. Mydosh, J. I. Budnick, M. P. Kawatra, and S. Skalski, *Phys. Rev. Lett.* **21**, 1346 (1968).
[105] M. B. Brodsky and A. J. Freeman, *Phys. Rev. Lett.* **45**, 133 (1980).
[106] F. Acker, Z. Fisk, J. L. Smith, and C. Y. Huang, *J. Magn. Magn. Mater.* **22**, 250 (1981).
[107] B. T. Matthias and R. M. Bozorth, *Phys. Rev.* **109**, 604 (1958).
[108] C. Pfleiderer, M. Uhlarz, S. M. Hayden, R. Vollmer, H. v. Löhneysen, N. R. Bernhoeft, and G. G. Lonzarich, *Nature* **412**, 58 (2001).

phase transition. The local moments themselves have to be induced at any given site at T_N. This is possible if a nondiagonal matrix element $\alpha = \langle 1|J_z|2\rangle$ exists. The effective RKKY inter-site exchange $J(\mathbf{q})$ then mixes the two states, thereby creating an *induced* ground state moment. If $J(\mathbf{q})$ is maximal at a wave vector $\mathbf{Q}$ this happens spontaneously at a temperature

$$T_N = \frac{\Delta}{2\tanh^{-1}(\frac{1}{\xi})}, \quad \text{here } \xi = \frac{\alpha^2 J(\mathbf{Q})}{2\Delta} \qquad (4.5)$$

is the control parameter of the quantum phase transition to an induced moment state. It takes place for $\xi > \xi_c = 1$. In general the ordered state has an incommensurate modulation with wave vector $\mathbf{Q}$. In the paramagnetic phase the magnetic singlet-singlet excitations disperse into a magnetic exciton band given by

$$\omega(\mathbf{q}) = \Delta \left[1 - \frac{\alpha^2 J(\mathbf{q})}{2\Delta} \right] \qquad (4.6)$$

The onset of the QFT is signified by a softening of the exciton mode as function of ξ at the incipient ordering vector $\mathbf{Q}$ which is given by $\omega(\mathbf{Q}) = \Delta[1 - \xi]$ and vanishes at $\xi_c = 1$. The control parameter contains the CEF splitting Δ which is susceptible to pressure. The latter may therefore be used to tune the singlet-singlet system through the QCP. Phase transitions of the induced moment type under ambient conditions as well as under pressure have been found in a number of rare earth systems like Pr_3Tl, TbSb (for a review see Ref. [109]) and dhcp Pr metal.[110,111] It also describes the AF order in the actinide compound UPd_2Al_3 with partly itinerant and partly localized 5f-electrons (see Section V). In fact dhcp Pr, which is approximately a singlet-doublet CEF system, is of special interest. There the critical mode softening and modulated moment appearance under uniaxial pressure shown in Figure IV.1 correspond to the modern concept of a quantum phase transition due to variation of a microscopic control parameter. However, this aspect was not stressed or realized at that time.

This is remarkable because it was found earlier that the simplest Hamiltonian which describes induced moments in the singlet-singlet system (though not precisely dhcp Pr) is of the type

$$H = \Delta \sum_i T_i^x + I \sum_{\langle ij \rangle} T_i^z T_i^z \qquad (4.7)$$

---

[109] P. Fulde, in *Handbook of the Physics and Chemistry of Rare Earth*, vol. 2, eds. K. A. Gschneider, Jr. and L. Eyring, North-Holland, Amsterdam (1979), chap. 17, p. 295.

[110] J. Jensen, K. A. McEwen, and W. G. Stirling, *Phys. Rev. B* **35**, 3327 (1987).

[111] J. Jensen and A. R. Mackintosh, *Rare Earth Magnetism: Structures and Excitations*, Clarendon Press, Oxford (1991).

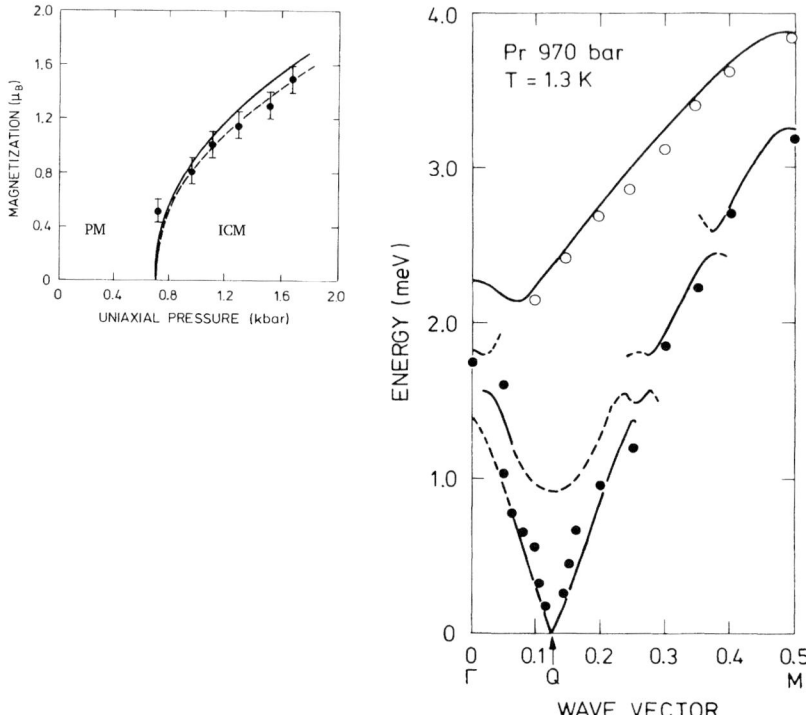

FIG. IV.1. Left panel: paramagnetic (PM) to incommensurate magnetic (ICM) quantum phase transition as function of uniaxial pressure in dhcp Pr. The quantum critical point is at $p_c \sim 0.7$ kbar. Circles are the experimental first harmonic of the ICM moment corresponding to modulation wave vector $\mathbf{Q} = 0.12\mathbf{a}^*$. Dashed and solid lines correspond to model calculations. Right panel: Optic (open circles) and acoustic (closed circles) exciton mode frequencies which give the energy scale of quantum fluctuations. Close to the quantum critical point at p_c and for the ordering wave vector $\mathbf{Q}$ this energy scale vanishes. (After Ref. [110].)

where the two orientations $|\uparrow\rangle, |\downarrow\rangle$ of the pseudo-spin T_z correspond to the CEF singlets $n = 1, 2$ respectively. Indeed, this is the n.n. Ising model in a transverse field (ITF), a genuine model for quantum phase transitions. The last term establishes AF order of Ising spins with a twofold degenerate ground state due to Z_2 symmetry. The first transverse field term introduces quantum fluctuations of the spins and destroys long-range order if the control parameter $\xi = I/(2\Delta)$ exceeds $\xi_c = 1$. This model is exactly solvable in 1D[112,113] and therefore is a reference point for the theory of QPTs. If one adds an infinitesimal staggered field term to

---

[112] E. Lieb, T. Schultz, and D. Mattis, *Ann. Phys. (N. Y.)* **16**, 407 (1961).
[113] P. Pfeuty, *Ann. Phys. (N.Y.)* **57**, 79 (1970).

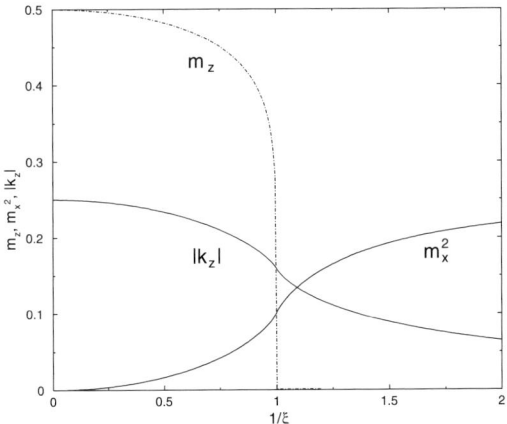

FIG. IV.2. Pseudo spin correlation functions for the ITF according to Eqs. (4.8), (4.9) as function of inverse control parameter $1/\xi = 2\Delta/I$. Order parameter $m_z = |\langle T_z\rangle|$ with critical (pressure) exponent 1/8 is also shown. (After Ref. [114].)

Eq. (4.7) one obtains a finite order parameter $\langle T_z\rangle$ given in Section VI.14 provided $\xi > \xi_c$. The critical exponent of the order parameter is 1/8. In addition the correlation functions may be calculated exactly. In good accuracy they are given by[112,113] ($j = i \pm 1$):

$$\langle T_i^z T_j^z\rangle = k_z, \qquad \langle T_i^x T_j^x\rangle = m_x^2, \qquad (4.8)$$

where

$$k_z = -\frac{1}{4\pi}\int_0^\pi dq\, \frac{\xi + \cos(q)}{\Lambda_q}, \quad m_x = \frac{1}{2\pi}\int_0^\pi dq\, \frac{1+\xi\cos(q)}{\Lambda_q} = \langle T_i^x\rangle \quad (4.9)$$

with $\Lambda_q = (1+\xi^2+2\xi\cos q)^{\frac{1}{2}}$. These correlation functions together with the order parameter are plotted in Figure IV.2. It shows nicely that long range order m_z is destroyed when the transverse quantum fluctuations characterized by m_x^2 overwhelm the longitudinal correlations given by k_z. The ITF is a generic model for QPTs that may be applied to quite different physical systems. For example, in the case of induced magnetic moment ordering discussed above the two pseudo-spin states correspond to the two CEF singlet states, the Ising interaction is due to the nondiagonal exchange between them and the transverse field is associated with the CEF splitting energy. On the other hand the same model may be applied to the problem of charge ordering in insulators with a 2D ladder type structure

---

[114] V. Yushankhai and P. Thalmeier, *Phys. Rev. B* **63**, 064402 (2001).

like α'-NaV$_2$O$_5$. In this case the pseudo-spin describes resonating, singly occupied $3d$ states within the rung of a ladder, the Ising interaction corresponds to the inter-site Coulomb interaction between d-electrons in different rungs and the transverse field is provided by the intra-rung kinetic (hopping) energy. The ITF then describes a quantum phase transition where charge ordering in the rungs is destroyed by the increase of the intra-rung kinetic energy. This will be discussed in detail in Section VI.14.

The induced moment magnetism can also appear in a different context in compounds with localized $3d$-electrons. Instead of singlet-singlet CEF states as before one may have here singlet-triplet level systems with a splitting Δ due to preformed dimers of $S = 1/2$ $3d$-spins. When the inter-dimer exchange coupling is slightly subcritical the magnetic exciton mode with a minimum at the wave vector $\mathbf{Q}$ has a finite but small energy and the compound is paramagnetic. Application of a magnetic field splits off one triplet component and the excitation energy at $\mathbf{Q}$ is driven to zero at a critical field H_c. There a quantum phase transition to an incommensurate magnetic phase takes place. An example is TlCuCl$_3$ where the Cu^{2+} spins form dimers.[115] Since for $3d$ spins the orbital degrees are quenched, the inter-dimer coupling is of the Heisenberg- rather than Ising type where the latter has a discrete Z_2 symmetry as discussed above. This makes an essential difference since the continuous SU(2) symmetry of the former allows the exchange Hamiltonian to be mapped to a hard-core boson Hamiltonian for the singlet-triplet boson excitations.[116] Then the field-driven QPT in TlCuCl$_3$ can be interpreted as Bose–Einstein condensation (BEC) of the singlet-triplet excitations. Another, even cleaner example of magnetic order through BEC, though with different microscopic details was recently identified in Cs$_2$CuCl$_4$.[117] Actually the mapping to the boson model is already possible for the xy-exchange model with U(1) symmetry in a transverse field which applies to dhcp Pr.[111] Thus the appearance of an incommensurate phase under pressure in Figure IV.1 may perhaps also be described within the BEC framework. This has not been investigated yet.

7. Quantum Criticality in the Kondo Lattice

As mentioned in the introduction a great part of the interest on QPTs is focused on strongly correlated metallic systems which exhibit pronounced non-Fermi liquid

---

[115] C. Ruegg, N. Cavadini, A. Furrer, H. U. Güdel, K. Krämer, H. Mutka, A. Wildes, K. Habicht, and P. Vorderwisch, *Nature* **423**, 62 (2003).

[116] M. Matsumoto, B. Normand, T. M. Rice, and M. Sigrist, *Phys. Rev. B* **69**, 054423 (2004).

[117] T. Radu, H. Wilhelm, V. Yushankhai, D. Kovrizhin, R. Coldea, T. Lühmann, and F. Steglich, *Phys. Rev. Lett.* **95**, 127202 (2005).

(NFL) behavior. This is found frequently close to pressure- (hydrostatic or chemical) and field-induced QPTs from the paramagnetic to the antiferromagnetic phase of Ce- or Yb-based heavy fermion metals. The schematic phase diagram for such compounds is shown in Figure IV.6. Prominent examples are found among the class of Ce122, Ce115 and Ce218 intermetallic compounds and alloys.[54] Classical cases are $CePd_2Si_2$,[118] $CeNi_2Ge_2$[119] and more recently $YbRh_2Si_2$.[120] Most Ce compounds also exhibit dome-shaped superconductivity, sometimes with very small T_c. It appears in the NFL regime around the quantum critical point of AF order. For an example see Figure IV.6 (right panel).

The generic model to describe the quantum critical Ce HF compounds is the Kondo lattice model given by

$$H_{KL} = \sum_{\mathbf{k}\sigma} \epsilon_{\mathbf{k}} c^\dagger_{\mathbf{k}\sigma} c_{\mathbf{k}\sigma} + J_K \sum_i \mathbf{s}_i \mathbf{S}_i. \qquad (4.10)$$

The first term describes conduction electrons with a dispersion $\epsilon_{\mathbf{k}}$ and bandwidth W. The second term is a local AF ($J_K > 0$) coupling of conduction electron spins $\mathbf{s}_i$ to localized spins $\mathbf{S}_i$. There are two competing effects. Firstly, the solutions of the Kondo impurity model (a single spin $\mathbf{S}_{i=0}$ coupled to the Fermi sea) shows that below the Kondo temperature $T_K = W\exp(-1/J_K N(0))$ the local moment is screened. A singlet is formed which extends to a distance $\xi_K \sim \hbar v_F/T_K$ from the impurity site.[50] Secondly, in the lattice the polarization of conduction electrons due to an on-site exchange induces an effective RKKY-type interaction between localized spins. This leads to a tendency to magnetic order at a temperature of the order $T_{RKKY} \sim J_K^2 \chi(2k_F)$ where $\chi(\mathbf{q})$ is the conduction electron susceptibility. When $T_{RKKY} \ll T_K$ local but overlapping singlets form a nonmagnetic state and below a coherence temperature $T_{coh} < T_K$ they disperse into quasiparticle bands as indicated in Figure IV.6 (see also Section III). For $T_{RKKY} \gg T_K$ the singlet formation is inhibited and magnetic order sets in at T_N. When the two temperature scales are about the same size one expects a quantum phase transition between the magnetically ordered and nonmagnetic heavy Fermi liquid state. This criterion is only a heuristic guide because T_K is the nonperturbative energy scale (the singlet binding energy) of the impurity problem, whereas T_{RKKY} is the perturbative energy scale on the lattice. The control parameter of the model is $X = J_K/W$ which may be assumed to vary linearly with pressure. This behavior is illustrated in the Doniach-type phase diagram around

---
[118] F. M. Grosche, I. R. Walker, S. R. Julian, N. D. Mathur, D. M. Freye, M. J. Steiner, and G. G. Lonzarich, *J. Phys. Condens. Matter* **13**, 2845 (2001).
[119] R. Küchler, N. Oeschler, P. Gegenwart, T. Cichorek, K. Neumaier, O. Tegus, C. Geibel, J. A. Mydosh, F. Steglich, L. Zhu, and Q. Si, *Phys. Rev. Lett.* **91**, 066405 (2003).
[120] J. Custers, P. Gegenwart, H. Wilhelm, K. Neumaier, Y. Tokiwa, O. Trovarelli, C. Geibel, F. Steglich, C. Pepin, and P. Coleman, *Nature* **424**, 524 (2003).

the QCP (Figure IV.6). The wedge above the QCP is the region where NFL behavior of thermodynamic coefficients and transport properties is observed.

This qualitative picture is hard to quantify. In fact the Kondo lattice model is an unsolved problem and only various approximative and numerical methods have been applied to it. The problem may be somewhat simplified by eliminating the charge degrees of freedom. This was proposed in Ref. [121] for the 1D Kondo chain. Using a Jordan–Wigner transformation the 1D conduction electrons may be replaced by a second spin system with xy-type inter-site coupling in addition to the local Kondo spins. The model then reads

$$H_{KN} = 2J \sum_{\langle ij \rangle} \left(\tau_i^x \tau_j^x + \tau_i^y \tau_j^y + \delta \tau_i^z \tau_j^z \right) + J_K \sum_i \boldsymbol{\tau}_i \mathbf{S}_i . \quad (4.11)$$

The 1D 'Kondo necklace' model based on the Jordan–Wigner transformation of H_{KL} has $\delta = 0$. But later this was generalized to $\delta > 0$ in arbitrary dimension and treated as a model for the competition of AF order (J) and local singlet formation (J_K) in its own right. The control parameter is now $X = J_K/2J$. The constant $2J$ associated with interacting spins corresponds to the bandwidth W of conduction electrons in the original H_{KL}. The approximate quantum critical phase diagram of this model may be obtained with the help of the bond operator method.[122] It starts from the observation that the four singlet-triplet basis states of a local pair of spins $\boldsymbol{\tau}_i, \mathbf{S}_i$ may be represented by singlet-triplet boson creation operators according to $|s\rangle = s^\dagger |0\rangle$ and $|t_\alpha\rangle = t_\alpha^\dagger |0\rangle$ ($\alpha = x, y, z$). The spin operators may then be expressed in terms of singlet and triplet boson operators:

$$S_{n,\alpha} = \frac{1}{2} \left(s_n^\dagger t_{n,\alpha} + t_{n,\alpha}^\dagger s_n - i\epsilon_{\alpha\beta\gamma} t_{n\beta}^\dagger t_{n\gamma} \right)$$

$$\tau_{n,\alpha} = \frac{1}{2} \left(-s_n^\dagger t_{n,\alpha} - t_{n,\alpha}^\dagger s_n - i\epsilon_{\alpha\beta\gamma} t_{n\beta}^\dagger t_{n\gamma} \right). \quad (4.12)$$

These operators have to fulfill the local constraint $s_n^\dagger s_n + \sum_\alpha t_{n\alpha}^\dagger t_{n\alpha} = 1$ at every site n. Using the above transformation in H_{KN} one obtains various bosonic interaction terms which may be decoupled by a mean field approximation both in the Hamiltonian and in the constraint. In the strong coupling region where $X = J_K/2J$ is large the ground state is characterized by the molecular field $\bar{s} = \langle s \rangle$ corresponding to a condensation of the local singlet bosons. The decoupling then leads to a bilinear Hamiltonian in the triplet bosons which may be diagonalized and yields the triplet excitation energies $\omega_\mathbf{k}$. From the excitation spectrum the singlet amplitude $\bar{s}(X)$ and the chemical potential $\mu(X)$ (to satisfy the mean-field constraint) are determined self-consistently. Then the minimum

---

[121] S. Doniach, *Physica B* **91**, 231 (1977).
[122] G. M. Zhang, Q. Gu, and L. Yu, *Phys. Rev. B* **62**, 69 (2000).

triplet excitation energy is at the AF zone boundary vector $\mathbf{q} = \mathbf{Q}$, and it is equivalent to the spin gap in the Kondo singlet phase. It is given by ($z =$ coordination number)

$$\Delta_{SP} = J_K \left(\frac{1}{4} + \mu/J_K\right)\sqrt{1 - zd/2}$$

$$d = \frac{2J}{J_K} \frac{\bar{s}^2}{(\frac{1}{4} + \mu/J_K)} \quad (4.13)$$

where the dimensionless parameter $d(X)$ is determined by the self-consistent equation

$$d = \frac{4J}{J_K}\left[1 - \frac{1}{2N}\sum_{\mathbf{k}} \frac{\omega_0}{\omega_{\mathbf{k}}}\right]$$

$$\frac{\omega_{\mathbf{k}}}{\omega_0} = \sqrt{1 + d\gamma_{\mathbf{k}}} \quad \text{with} \quad \gamma_{\mathbf{k}} = \sum_{\alpha} \cos k_\alpha. \quad (4.14)$$

When the intermediate coupling regime is approached by decreasing $X^{-1} = 2J/J_K$ the spin gap eventually collapses and a magnetically ordered phase is established. The solution of the above equations is shown in Figure IV.3 for the

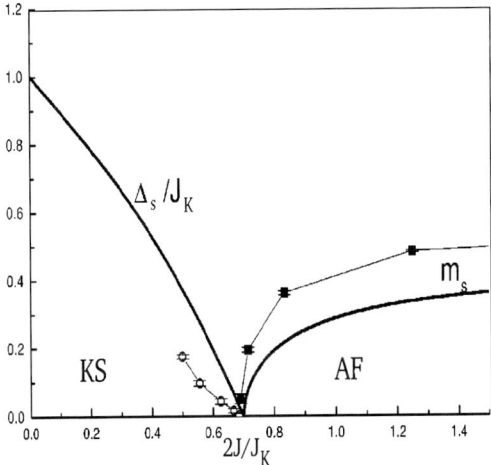

FIG. IV.3. Spin gap Δ_s in the Kondo singlet (KS) phase and staggered magnetization m_s in the AF phase as function of the inverse control parameter $1/X = 2J/J_K$. Full line is the mean field result for H_{KN} in 2D for the xy case ($\delta = 0$). The dotted line is obtained from Monte Carlo simulations. The QCP is at $1/X_c \simeq 0.7$. (After Ref. [122].)

2D case. Indeed it exhibits a QPT at a critical value $X_c \sim 1.43$. For larger X, i.e., in the AF regime an analogous calculation may be performed leading to the staggered magnetization m_s shown in the previous figure. In 3D results are similar but the scaling exponents for $\Delta_{SP}(X)$ and $m_s(X)$ close to the QCP are different. We note that the boson representation employed here for the Kondo-necklace type model is identical to the one used in the spin-dimer problem in TlCuCl$_3$.[116] Therefore the transition from spin gap to AF phase may also be interpreted as a BEC of triplet bosons. The mean-field boson treatment confirms the qualitative conjectures made above on the quantum critical phase diagram by simply comparing the energy scales of singlet formation and magnetic order. Because it respects the local constraint for bosons only on the average this method is however completely inadequate to investigate how the on-site singlet and inter-site magnetic correlations compete as function of control parameter X and also as function of temperature. For this purpose one has to use advanced numerical approaches like the finite temperature Lanczos method[123] for finite size clusters. This method has recently been employed to Kondo lattice like models given by H_{KL} and also H_{KN} with $\delta = 1$, i.e., in the Heisenberg limit for the interacting τ-spins.[124] Because each site has 4 states, the possible cluster sizes for exact diagonalization are limited. The 8-site cluster of the square lattice with periodic boundary conditions was investigated and correlation functions and specific heat were calculated.[124] The Lanczos procedure is repeated 400 times for random starting vectors in the 4^N ($N = 8$)-dimensional Hilbert space until convergence for thermodynamic quantities is achieved.

The specific heat results are shown in Figure IV.4 (upper panel) for various values of the control parameter $X = J_K/2J$. For $J_K = 0$ the broad upper maximum corresponds to the AF correlations of the 'itinerant' τ-spins. When J_K is turned on a lower much sharper maximum rapidly evolves which shifts to higher temperatures and for $X \sim 1.5$ eventually merges with the upper maximum. The origin of the lower sharp maximum becomes clear if one monitors the on-site singlet correlation $\langle \tau_1 \cdot S_1 \rangle$ and the induced 'RKKY' correlations $\langle S_1 \cdot S_2 \rangle$ between the localized (but non-interacting) spins at n.n. sites 1 and 2. These correlations are shown in Figure IV.5 as function of temperature. For $J_K = 0$ both correlations are absent. When J_K is turned on $\langle \tau_1 \cdot S_1 \rangle$ develops AF on-site singlet correlations with a limiting value of -0.75 in the strong coupling limit (upper panel). Due to the inter-site coupling of τ-spins the local singlet formation also induces AF inter-site correlations of the previously uncoupled S_i spins (lower panel). In the strong coupling limit when on-site singlets are formed, however, the inter-site correlations of S_i spins are diminished again. The sharp maximum in C_V is well correlated with the inflection point of the induced inter-site correlations. Therefore

---

[123] G. Jaclic and N. B. Prelovsek, *Adv. Phys.* **49**, 1 (2000).
[124] I. Zerec, B. Schmidt, and P. Thalmeier, *Physica B*, in press.

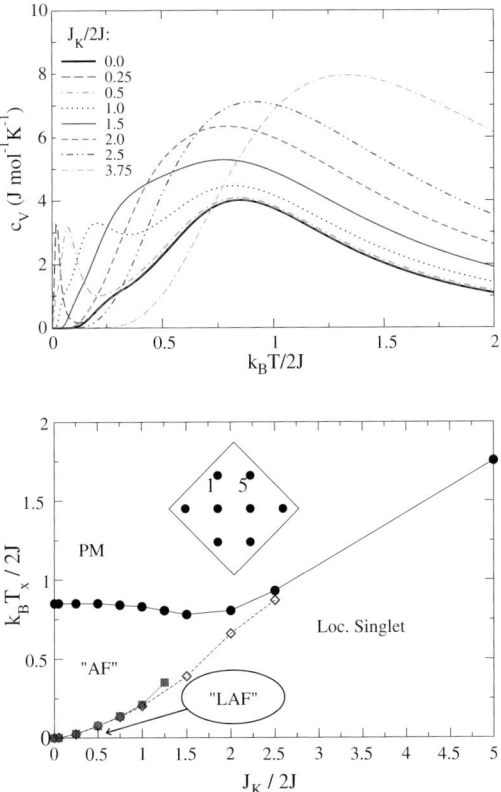

FIG. IV.4. Upper panel: The specific heat C_V of the 8-site cluster. Lower panel: 'Phase diagram' of the 8-site cluster (inset) obtained from tracing the characteristic maxima of C_V (full lines) and inflection points of correlation functions (from Figure IV.5) as function of control parameter $X = J_K/2J$. PM = paramagnetic phase, 'AF' = phase with AF correlations of τ-spins. 'LAF' = phase with AF correlations of local spins $\mathbf{S}$. 'Local Singlet'—phase with Kondo spin gap. (After Ref. [124].)

in the thermodynamic limit it may be interpreted as the C_V-anomaly due to AF order of partially Kondo-screened local moments. For larger J_K the sharp lower C_V maximum merges with the broad upper maximum but the inflection point in $\langle \mathbf{S}_1 \cdot \mathbf{S}_2 \rangle$ may still be identified. Following these characteristic temperatures as function of X a 'phase diagram' of the Kondo-lattice type model H_{KN} ($\delta = 1$) may be constructed as shown in Figure IV.4 (lower panel). For small values of J_K one finds a correlated state of 'itinerant' τ spins ('AF') below the temperature of the broad maximum in C_V. Below the sharp maximum temperature one has induced antiferromagnetic correlations of the local $\mathbf{S}$ spins ('LAF'). For larger

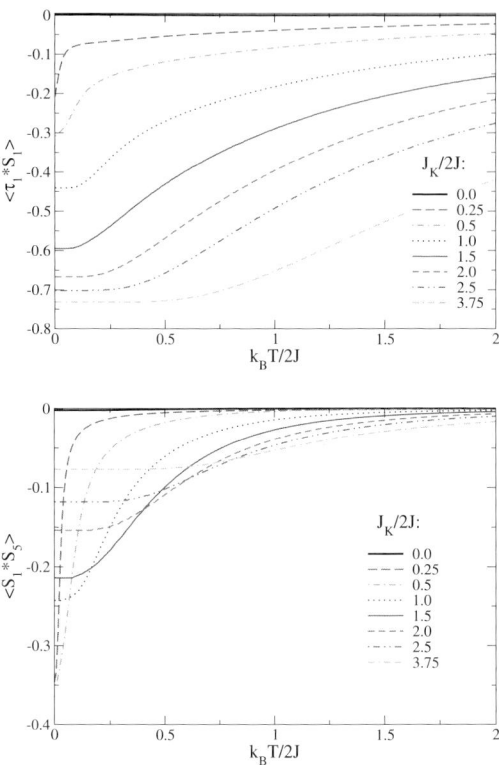

FIG. IV.5. Upper panel: On-site correlations show singlet formation at low temperatures. Lower panel: Simultaneously inter-site 'RKKY' correlations are induced for moderate J_K and suppressed again for large J_K. (After Ref. [124].)

values of J_K both maxima merge but correlations of the LAF state are still visible (dotted line). For even larger J_K the 'Local Singlet' formation dominates correlation and specific heat behavior. Of course, due to the finite cluster size one may not strictly speak about thermodynamic phases and phase boundaries. However, the qualitative evolution of correlations and thermodynamic anomalies as function of control parameter X may be expected to survive in the thermodynamic limit. It is interesting to compare these findings with the previous mean-field calculation for the quantum phase transition as function of X. In the cluster calculation one would identify the 'QCP' with the value of $X = J_K/2J$ where the correlations show a cross-over from inter-site ('LAF') to 'Local Singlet' type behavior along the $T = 0$ line in Figure IV.4 (lower panel). This is found to be at $X_c \sim 1$–1.5 and compares reasonably well with the value $X_c = 1.43$ of the mean-field calculation (where however $\delta = 0$ was taken).

8. SCALING THEORY CLOSE TO THE QUANTUM CRITICAL POINT

The previous calculations give an insight into the microscopic mechanism of singlet formation vs magnetic order in the Kondo lattice. These results are, however, still far removed from explaining the most common experiments around the QCP, notably the temperature and field scaling for specific heat, susceptibility, resistivity, etc. Assuming the existence of a QCP some insight into the dependence of physical quantities on temperature and control parameters close to it (see Figure IV.6) may be obtained within a simple phenomenological scaling theory. A Kondo impurity in a metallic host shows all the signatures of a local Landau Fermi liquid state[50] at temperatures $T \ll T^*$, notably a scaling of the free energy density with T/T^*. This leads to universal relations among low temperature thermodynamic quantities irrespective of the microscopic details. This idea has been successfully extended to the Fermi liquid phase of heavy fermion and

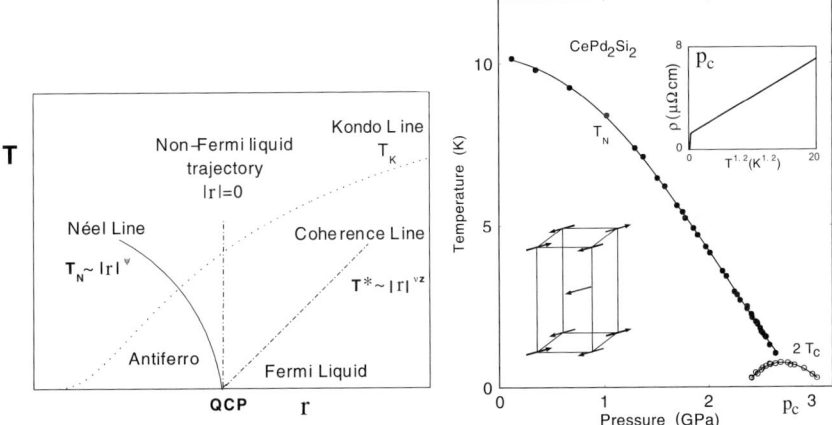

FIG. IV.6. Left panel: Schematic phase diagram for Kondo compounds with a QCP ($r = 0$) separating AF (left) and LFL (right) phases (full line). Scaling of characteristic temperatures is indicated (broken lines). (After Ref. [125].) Right panel: Hydrostatic pressure induced AF to HF liquid QPT in $CePd_2Si_2$. The AF structure is indicated. At the critical pressure $p_c = 2.86$ GPa the resistivity shows pronounced NFL behavior $\Delta\rho(T) \sim T^n$ with $n = 1.2$ down to the superconducting transition (inset). (After Ref. [118].)

---

[125] M. A. Continentino, *Brazilian Journal of Physics* **35**, 197 (2004).

mixed-valent compounds in a phenomenological scaling ansatz.[126–128] The aim was to explain observed relations between quantities like specific heat, thermal expansion, magnetostriction and others. It is natural to apply these ideas also to the vicinity of the QCP, where the characteristic energies $T^*(p)$ and $T_c(p)$ themselves depend on the distance r to the QCP which then appears as a further scaling variable.[129] The associated correlation length (ξ) and time scales of quantum fluctuations (τ) diverge on approaching the phase transition. Their critical exponents are universal, depending only on dimension and the degrees of freedom of the order parameter. We define the quantities

$$r = \frac{X - X_c}{X_c}, \quad t = \frac{T - T_c}{T_c} \quad (X = p \text{ or } H) \tag{4.15}$$

which measure the distance to the critical control parameter X_c and the transition temperature T_c respectively. On approaching the QCP at $T = 0$, $r = 0$ the correlation length, fluctuation time and free energy scale like[94,125]

$$\xi \sim |r|^{-\nu}, \quad \tau \sim |r|^{-\nu z}, \quad f \sim |r|^{2-\alpha} \tilde{f}\left(\frac{T}{T^*}, \frac{H}{H^*}\right). \tag{4.16}$$

Here H^* has the meaning of 'metamagnetic' field scale. For fields $H \gg H^*$ the heavy quasiparticle state is destroyed by breaking the Kondo singlet state. For QPTs the hyperscaling relation which relates critical exponents to the effective dimension is given by[94]

$$2 - \alpha = \nu d_{\text{eff}}, \quad d_{\text{eff}} = d + z. \tag{4.17}$$

In the case of a Gaussian fix point appropriate for $d_{\text{eff}} > 4$ one has $\nu = \frac{1}{2}$. In the free energy of Eq. (4.16) which is a generalization of the one used in[126–128] the characteristic temperature (T^*) and metamagnetic field (H^*) have scaling relations

$$T^* \sim |r|^{\nu z}, \quad H^* \sim |r|^{\phi_h}. \tag{4.18}$$

In the magnetically ordered regime T^* has to be replaced by the magnetic transition temperature which scales as $T_c \sim |r|^\psi$ where ψ is the shift exponent. Below the upper critical dimension, i.e., for $d_{\text{eff}} < d_c = 4$ the hyperscaling relation Eq. (4.17) is equivalent to the assumption $\psi = \nu z$.[125] In this case $T_c(r)$ and $T^*(r)$

---

[126] R. Takke, M. Niksch, W. Assmus, B. Lüthi, R. Pott, R. Schefzyk, and D. K. Wohlleben, *Z. Phys. B* **44**, 33 (1981).
[127] P. Thalmeier and P. Fulde, *Europhys. Lett.* **1**, 367 (1986).
[128] A. B. Kaiser and P. Fulde, *Phys. Rev. B* **37**, 5357 (1988).
[129] L. Zhu, M. Garst, A. Rosch, and Q. Si, *Phys. Rev. Lett.* **91**, 066404 (2003).

scale symmetrically around the QCP (Figure IV.6), however for $d_{\text{eff}} > 4$ in general one has $\psi \neq \nu z$. This is known as 'breakdown of hyperscaling'. Within a generalized Landau–Ginzburg–Wilson approach this may be understood as the effect of a dangerously irrelevant quartic interaction. Although it scales to zero for $d_{\text{eff}} > 4$, it changes nevertheless the scaling behavior at finite T leading to a modified shift exponent $\psi = z/(d_{\text{eff}} - 2)$ for $d_{\text{eff}} > d_c$.

From an experimental viewpoint the most interesting aspect is the temperature dependence of physical properties at the QCP ($r = 0$) in the non-Fermi liquid regime. Very useful quantities are specific heat $C = (T/V)(\partial S/\partial T)_p$ and thermal expansion $\alpha = (1/V)(\partial V/\partial T)_p$.[119,129] At the QCP ($r = 0$) they scale with temperature like

$$C(T) \sim T^{d/z}, \quad \alpha(T) \sim T^{(d-\frac{1}{\nu})/z} \quad \text{and} \quad \Gamma = \frac{\alpha}{C} \sim T^{-\frac{1}{\nu z}}. \tag{4.19}$$

This means that the temperature dependence of the critical 'Grüneisen ratio' Γ ($r = 0$) is controlled by the exponent which directly determines the time scale of quantum fluctuations in Eq. (4.16). Using this important relation a consistent explanation of experiments in the NFL compound $CeNi_2Ge_2$ can indeed be given (Figure IV.7). Tables of the scaling behavior of the quantities in Eq. (4.19) for various d, z have been given in Ref. [129].

The exponent νz determines at the same time pressure scaling (Eq. (4.18)) of the characteristic temperature T^* on the nonmagnetic side of the QCP. On the other hand the pressure scaling exponent ϕ_h of the characteristic field H^* is an independent quantity within the scaling ansatz. Experimentally it has been investigated in detail for $CeRu_2Si_2$ which has a metamagnetic field scale $H^*(p=0) = 7.8T$ ($\mathbf{H} \parallel c$). It was found empirically that $\phi_h = 2 - \alpha = \nu z$ is fulfilled. According to the free energy in Eq. (4.16) this implies with $m = (\partial f/\partial H)$ that $m(H^*) = $ const independent of pressure. This was indeed found experimentally.[130] The empirical relation $2 - \alpha = \nu z$ may be interpreted as quantum hyperscaling relation with dimension $d = 0$ according to Eq. (4.17). The empirical validity of such a relation points to a dimensional crossover as function of pressure close to the QCP which is caused by the different divergence of spatial and temporal correlations.[125]

An explicit calculation of scaling exponents close to the QCP demands the use of effective field theories based on Ginzburg–Landau type action functionals for the spatial and temporal order parameter fluctuations. Such theories are not specific for strongly correlated electron systems and therefore are beyond the scope of this review. As mentioned before they have indeed first been constructed for

---

[130] A. Lacerda, A. de Visser, P. Haen, P. Lejay, and J. Flouquet, *Phys. Rev. B* **40**, 8759 (1989).

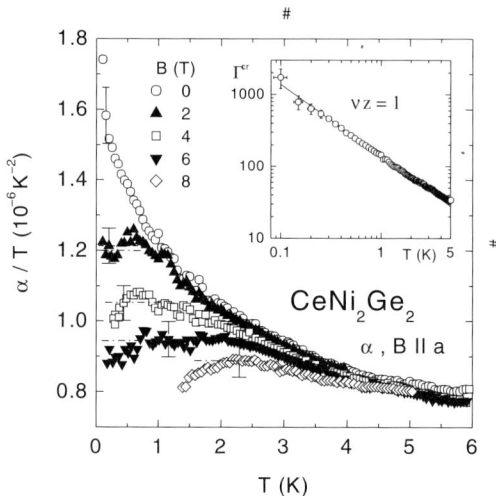

FIG. IV.7. Thermal expansion showing the suppression of NFL behavior as function of field. The inset shows that for $B = 0$ the critical contribution to the Grüneisen ratio Γ of $CeNi_2Ge_2$ scales like $\Gamma \sim 1/T^{\nu z}$ with $\nu z = 1$. According to Eq. (4.19) this means (assuming $z = 2$ for AF SDW) a mean field correlation length exponent $\nu = 1/2$ which is in agreement with $d_{\text{eff}} = d + z = 5$ for the effective dimension. (After Ref. [119].)

weakly correlated metals in Refs. [48,100,101] and Ref. [102] and have been reinvestigated later in hindsight of QPTs in strongly correlated metals.[131] Until now the results and even starting assumptions of these field theoretical approaches are controversial and will not be discussed here.

V. Partial Localization

The concept of orbital-selective localization applies to correlated systems with orbital degeneracies. Important examples are transition metal oxides[132–135] and $5f$ compounds. In these materials, the intra-atomic correlations as described by Hund's rules play an important role. Nevertheless the physics of partial localization in transition metal oxides and $5f$ systems is quite different. In compounds

---

[131] A. J. Millis, *Phys. Rev. B* **48**, 7183 (1993).
[132] V. I. Anisimov, I. A. Nekrasov, D. E. Kondakov, T. M. Rice, and M. Sigrist, *Eur. Phys. J. B* **25**, 191 (2002).
[133] A. Koga, N. Kawakami, T. M. Rice, and M. Sigrist, *Phys. Rev. Lett.* **92**, 216402 (2004).
[134] A. Liebsch, *Phys Rev. Lett.* **95**, 116402 (2005).
[135] M. S. Laad, L. Craco, and E. Müller-Hartmann, *Europhys. Lett.* **69**, 984 (2005).

with d electrons the large crystalline electric field (CEF) set up by the surroundings of a transition metal ion plays a major role. It is often larger than the bandwidth as, for example in the manganites (Section VI.15). In case of a cubic lattice it splits the five d orbitals into a t_{2g} triplet and an e_g doublet and the corresponding subbands are well separated. When the Hund's rule energy is larger than this splitting and when the orbital energy of the t_{2g} is lower than that of the e_g states the t_{2g} states will be occupied by the first three d electrons. Those three d electrons remain localized in a high-spin state with $S = 3/2$. Additional d electrons occupy e_g states and remain delocalized. The situation differs when the CEF splitting is larger than Hund's rule coupling. In that case the t_{2g} subband can accommodate six electrons. When the d electron count per ion n_d is larger than six, i.e., $n_d > 6$ only $(n_d - 6)$ d electrons will be itinerant and contribute to metallic behavior.

In $5f$ compounds we are facing a different situation. Since the $5f$ atomic wavefunctions are closer to the nuclei than d electron wavefunctions are, CEF splittings are smaller and less important. But Hund's rule energies are larger. Therefore when we deal with a situation where the $5f$ count per actinide ion n_f exceeds two, i.e., $n_f > 2$ only those $5f$ electrons will delocalize which enable the remaining ones to form a Hund's rule state. Otherwise the Coulomb interaction is increased so much that delocalization is disadvantageous as far as energy is concerned. Therefore Hund's rule correlations may strongly enhance anisotropies in the kinetic energy and eventually lead to the co-existence of band-like itinerant $5f$ states with localized atomic-like ones.

The central focus of the present section is the dual model for actinide-based heavy fermion compounds which assumes the co-existence of delocalized and localized $5f$ electrons.

Initially, the dual character has been conjectured for UPd_2Al_3 where the variation with temperature of the magnetic susceptibility points to the coexistence of CEF-split localized $5f$ states in a heavy fermion system with $5f$-derived itinerant quasiparticles. Direct experimental evidence for the co-existence of $5f$-derived quasiparticles and local magnetic excitations is provided by recent neutron scattering experiments.[136] There is clear evidence that the presence of localized $5f$ states is responsible for the attractive interaction leading to superconductivity.[137] In addition the dual model could allow for a rather natural description of heavy fermion superconductivity co-existing with $5f$-derived magnetism. For a recent review of experimental facts see Refs. [53,54].

Heavy quasiparticles have been observed by de Haas–van Alphen (dHvA) experiments in a number of U compounds. The experiments unambiguously confirm

---

[136] A. Hiess, N. Bernhoeft, N. Metoki, G. H. Lander, B. Roessli, N. K. Sato, N. Aso, Y. Haga, Y. Koike, T. Komatsubara, et al., cond-mat/0411041 (2004).

[137] N. Sato, N. Aso, K.. Miyake, R. Shiina, P. Thalmeier, G. Varelogiannis, C. Geibel, F. Steglich, P. Fulde, and T. Komatsubara, *Nature* **410**, 340 (2001).

that some of the U $5f$ electrons must have itinerant character. It has been known for quite some time that the $5f$-states in actinide intermetallic compounds cannot be considered as ordinary band states. Standard bandstructure calculations based on the Local Density Approximation (LDA) to Density Functional Theory fail to reproduce the narrow quasiparticle bands. On the other hand the LDA bandwidths are too small to explain photoemission data.[138,139] These shortcomings reflect the inadequate treatment of local correlations within ordinary electron structure calculation. Theoretical studies aiming at an explanation of the complex low-temperature structures lay emphasis on the partitioning of the electronic density into localized and delocalized parts.[140,141] Concerning the low-energy excitations it has been shown that the dual model allows for a quantitative description of the renormalized quasiparticles—the heavy fermions—in UPd_2Al_3. The measured dHvA frequencies for the heavy quasiparticle portions as well as the large anisotropic effective masses can be explained very well by treating two of the $5f$ electrons as localized.

The central goal of the present section is (1) to demonstrate that the dual model allows to determine the heavy quasiparticles in U compounds without adjustable parameters and (2) to give a microscopic justification for the underlying assumptions.

Before turning to a discussion of the dual model, its results and their implications we should like to add a few comments. In referring to the dual model one has to keep in mind that the latter provides an effective Hamiltonian designed exclusively for the low-energy dynamics. As such it seems appropriate for typical excitation energies $\hbar\omega$ below ~ 10 meV. In general, effective low-energy models are derived from the underlying microscopic Hamiltonians—to borrow the language of Wilson's renormalization group—by integrating out processes of higher energies. In the case of $5f$ systems the conjecture is that the hybridization between the conduction electrons and the $5f$ states effectively renormalizes to zero for some channels while staying finite for others. We shall show that the physical mechanism leading to the orbital-dependent renormalization of the hybridization matrix elements are the intra-atomic correlations which are often described by Hund's rules. To focus on the role of the intra-atomic correlations we consider model Hamiltonians for the $5f$ subsystem where the hybridization with the conduction

---

[138] J. W. Allen, in *Synchrotron Radiation Research: Advances in Surface and Interface Science*, Plenum Press, New York (1992), vol. 1, chap. 6, p. 253.

[139] S. Fujimori, Y. Saito, M. Seki, K. Tamura, M. Mizuta, K. Yamaki, K. Sato, T. Okane, A. Tanaka, N. Sato, et al., *J. Electron Spectrosc. Relat. Phenom.* **101–103**, 439 (1999).

[140] L. Petit, A. Svane, W. M. Temmerman, Z. Szotek, and R. Tyer, *Europhys. Lett.* **62**, 391 (2003).

[141] J. M. Wills, O. Eriksson, A. Delin, P. H. Andersson, J. J. Joyce, T. Durakiewicz, M. T. Butterfield, A. J. Arko, D. P. Moore, and L. A. Morales, *J. Electron Spectrosc. Relat. Phenom.* **135**, 163 (2004).

electrons is accounted for by introducing effective $5f$ hopping. The orbital-dependent suppression of hybridization then translates into orbital-selective localization.

The concept of correlation-driven partial localization in U compounds has been challenged by various authors (see, e.g., Ref. [142]). The conclusions are drawn from the fact that conventional band structure calculations within the Local Density Approximation (LDA) which treat all $5f$-states as itinerant can reproduce ground state properties like Fermi surface topologies and densities. The calculation of ground state properties, however, cannot provide conclusive evidence for the delocalized or localized character of the $5f$-states in actinides. First, the presence of localized states can be simulated in standard band calculations by filled bands lying (sufficiently far) below the Fermi level. Second, the Fermi surface is mainly determined by the number of particles in partially filled bands and the dispersion of the conduction bands which, in turn, depends mainly on the geometry of the lattice. A change in the number of band electrons by an even amount does not necessarily affect the Fermi surface since a change by an even number may correspond to adding or removing a filled band. As such, the Fermi surface is not a sensitive test of the microscopic character of the states involved. Unambiguous proof of the dual character can be provided by an analysis of the spectral function. Of particular importance are characteristic high-energy features associated with transitions into excited local multiplets. A detailed discussion of these features will be given in Section VIII.

9. Heavy Quasiparticles in UPd_2Al_3

Within the dual model the strongly renormalized quasiparticles in U-based heavy fermion compounds are described as itinerant $5f$ electrons whose effective masses are dressed by low-energy excitations of localized $5f$ states. We refer to Refs. [33,54] for a detailed description which proceeds in three steps. The latter include (a) a band structure calculation to determine the dispersion of the bare itinerant $5f$ states (b) a quantum chemical calculation which yields the localized $5f$ multiplet states and, in particular, their coupling to their itinerant counterparts and, finally, (c) a standard (self-consistent) many-body perturbation calculation to determine the renormalized effective mass. We should like to emphasize, however, that we treat all $5f$ electrons as quantum mechanical particles obeying Fermi anticommutation relations.

The scheme has been successfully applied to UPd_2Al_3 and UPt_3 as can be seen from the comparison between the calculated and measured dHvA frequencies and

---

[142] I. Opahle, S. Elgazzar, K. Koepernik, and P. M. Oppeneer, *Phys. Rev. B* **70**, 104504 (2004).

the effective masses in Figure V.1 and Table V.1, respectively. It is important to note that the data are derived from a parameter-free calculation. To show this we examine the individual steps as described above.

First, the bare $5f$ band dispersion is determined from a parameter-free ab initio calculation by solving the Dirac equation for the self-consistent LDA potentials

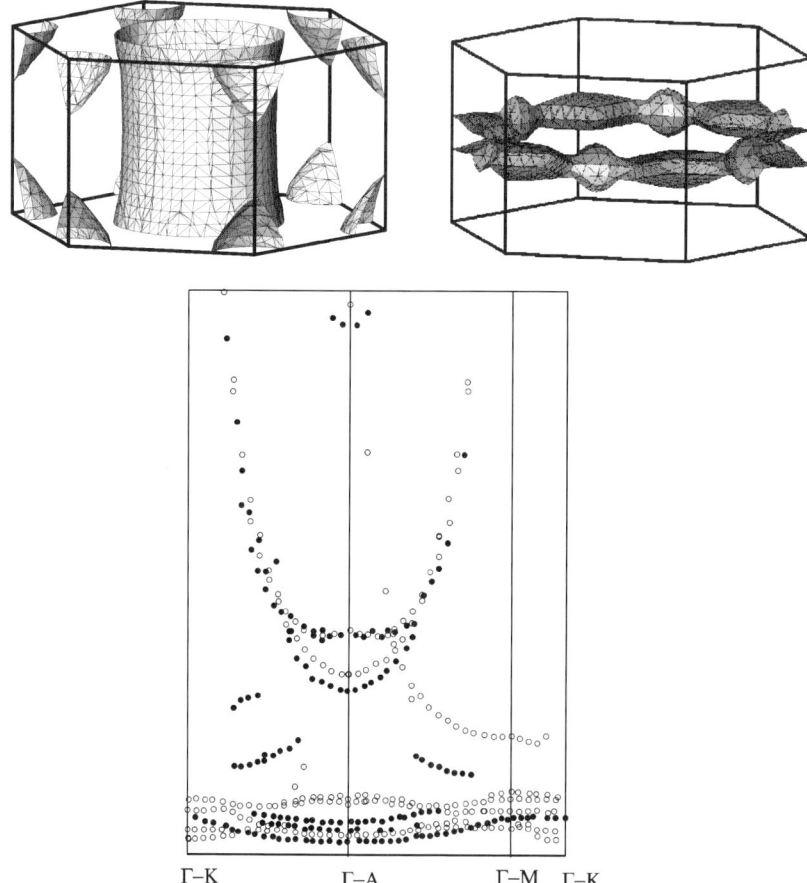

FIG. V.1. Upper panel: Fermi surface of UPd_2Al_3 calculated within the dual model.[33] The main cylinder part has effective masses with $m^* = 19$–$33m$, the highest masses are found on the torus. Lower panel: Comparison of experimental dHvA frequencies (black symbols) from Ref. [143] and calculated frequencies (open symbols) (After Ref. [33]). The large parabola corresponds to the main FS cylinder.

---

[143] Y. Inada, H. Yamagani, Y. Haga, K. Sakurai, Y. Tokiwa, T. Hinma, E. Yamamoto, Y. Onuki, and T. Yanagisawa, *J. Phys. Soc. Jpn.* **68**, 3643 (1999).

TABLE V.1. EFFECTIVE MASSES IN UPd$_2$Al$_3$ FOR **H** ∥ c. NOTATION FOR FS SHEETS AND EXPERIMENTAL VALUES FROM REF. [143], m_0 IS THE FREE ELECTRON MASS. THEORETICAL VALUES FROM REF. [33]

FS sheet	m^*/m (exp.)	m^*/m (theory)
ζ	65	59.6
γ	33	31.9
β	19	25.1
ϵ_2	18	17.4
ϵ_3	12	13.4
β	5.7	9.6

but excluding two U $5f$ ($j = \frac{5}{2}$) states from forming bands. The apparent absence of Kramers' degeneracy in this compound suggests to treat an *even* number of $5f$ electrons as localized. The calculations yield the dHvA frequencies which can be directly compared with experimental data. At this point we should like to briefly comment on the strategy to account for long-range antiferromagnetic order. The two examples, UPd$_2$Al$_3$ and UPt$_3$, represent two different categories. In UPd$_2$Al$_3$ localized $5f$ moments order antiferromagnetically at $T_N \simeq 14.5$ K with the induced moment being $\simeq 0.83\mu_B$ per U. The heavy quasiparticles form in the magnetically ordered state. As a consequence, the calculation of the bare $5f$ bands employs the experimentally observed antiferromagnetic structure. The superstructure strongly affects the Fermi surface topology for the heavy quasiparticles. The corresponding paramagnetic model cannot reproduce the heaviest orbit.

In the second step, the localized U $5f$ states are calculated by diagonalizing the Coulomb matrix in the restricted subspace of the localized $5f$ states. Assuming the j-j coupling scheme, the Coulomb matrix elements are evaluated using the radial functions of the ab initio band structure potentials. The coupling between the localized and delocalized $5f$ electrons is directly obtained from the expectation values of the Coulomb interaction in the $5f^3$ states.

Finally, the renormalization of the effective masses which results from the coupling between the two $5f$ subsystems is estimated. The itinerant $5f$ states scatter off the low-energy excitations of the localized $5f^2$ configurations. The situation resembles that in Pr metal where a mass enhancement of the conduction electrons by a factor of 5 results from virtual crystal field (CEF) excitations of localized $4f^2$ electrons.[144] The effective masses in Table V.1 are obtained from an isotropical renormalization of the band mass m_b is given by

$$\frac{m^*}{m_b} = 1 - \frac{\partial \Sigma}{\partial \omega}|_{\omega=0}. \tag{5.1}$$

---

[144] R. White and P. Fulde, *Phys. Rev. Lett.* **47**, 1540 (1981).

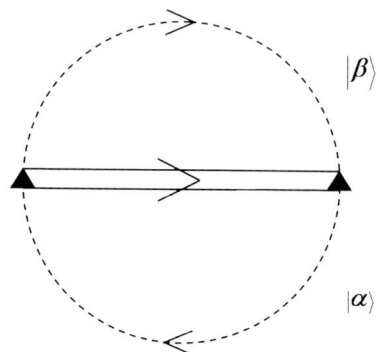

FIG. V.2. Self-energy diagram due to local intra-atomic excitations leading to a mass enhancement. Solid double line: full Green's function of the conduction electrons. Dashed lines: effective intra-atomic two-level states $|\alpha\rangle$, $|\beta\rangle$ separated by an excitation energy $\bar{\delta} = 7$ meV. Triangles: matrix elements $a|M|$.

The local self-energy of the delocalized $5f$ states $\Sigma(\omega)$ is displayed in Figure V.2. The explicit expressions are given in.[33] The mass enhancement is calculated self-consistently inserting values for the density of states at the Fermi level $N(0) = 2.76$ states/(eV cell spin) obtained from the bandstructure, when two $5f$ electrons are kept localized. The vertex is given by $a|M| = 0.084$ eV where the prefactor a denotes the $5f$ weight per spin and U atom of the conduction electron states near E_F. The matrix element M describes the transition between the localized states $|\Gamma_4\rangle$ and $|\Gamma_3\rangle$ due to the Coulomb interaction U_{Coul} with the delocalized $5f$ electrons. These are the two lowest eigenstates of the localized $5f^2$ system in the presence of the CEF. They have $J = 4$ in accordance with Hund's rule and are combinations of $|J_z\rangle = |\pm 3\rangle$. Finally, the dynamical susceptibility is approximated by that of an effective two-level system with an excitation energy $\bar{\delta} \simeq 7$ meV.

10. MICROSCOPIC MODEL CALCULATION

To illustrate the orbital selection by intra-atomic correlations we consider a simple molecular model consisting of two actinide atoms at sites a and b. The strong Coulomb interaction among the $5f$ electrons at the same site leads to well-defined ionic configurations f^n with energies $E(f^n)$. To model U compounds we assume that the total number of $5f$ electrons in the cluster be five corresponding to an averaged f-occupation of 2.5 per U site. The ground state will be a linear combination of states $|a; f^3\rangle|b; f^2\rangle$ and $|a; f^2\rangle|b; f^3\rangle$, respectively. These two sets of states are coupled by the hopping term. Since both atoms have more than one electron in their $5f$ shells intra-atomic correlations come into play. The two sets of basis functions split into groups of states characterized by the total angular momenta $J(a)$ and $J(b)$, respectively, the energy differences being of the order of the $5f$ exchange constant, i.e., approximately 1 eV. Since the spin-orbit interaction is large we use j-j coupling and restrict ourselves to $5f$ states with $j = 5/2$. We are aiming at the low-energy subspace which is spanned by the states $|a; f^3, J(a) = 9/2\rangle|b; f^2, J(b) = 4\rangle$ and $|a; f^2, J(a) = 4\rangle|b; f^3, J(b) = 9/2\rangle$, in close analogy to Hund's rules. Transferring an electron from site a to site b changes the local f occupation and the total angular momenta

$$|a; f^3, J(a) = 9/2\rangle|b; f^2, J(b) = 4\rangle \to |a; f^2, J'(a)\rangle|b; f^3, J'(b)\rangle \quad (5.2)$$

and the resulting final state will usually contain admixtures from excited multiplets. The transfer of a $5f$ electron from site a to site b causes intra-atomic excitations against which the gain in kinetic energy has to be balanced. The crucial point is that the overall weight of the low-energy contributions to the final state depends upon (a) the orbital symmetry of the transferred electron, i.e., on j_z and (b) on the relative orientation of $\mathbf{J}(a)$ and $\mathbf{J}(b)$. The latter effect closely parallels the "kinetic exchange" well-known from transition metal compounds. The requirement that the gain in energy associated with the hopping be maximal leads to orbital selection. The dynamics in the low-energy subspace is described by an effective single-particle Hamiltonian where some of the transfer integrals are renormalized to zero while others are reduced yet remain finite.

These qualitative considerations are the basis for microscopic model calculations which proceed from the simple model Hamiltonian

$$H = H_{\text{band}} + H_{\text{Coul}}. \quad (5.3)$$

The local Coulomb repulsion part

$$H_{\text{Coul}} = \frac{1}{2} \sum_a \sum_{j_{z_1},\ldots,j_{z_4}} U_{j_{z_1},j_{z_2} j_{z_3} j_{z_4}} c^\dagger_{j_{z_1}}(a) c^\dagger_{j_{z_2}}(a) c_{j_{z_3}}(a) c_{j_{z_4}}(a) \quad (5.4)$$

is written in terms of the usual fermionic operators $c^\dagger_{j_z}(a)$ ($c_{j_z}(a)$) which create (annihilate) an electron at site a in the $5f$-state with total angular momentum $j = 5/2$ and z-projection j_z. Considering the fact that the spin-orbit splitting is large we neglect contributions from the excited spin-orbit multiplet $j = 7/2$ and adopt the j-j coupling scheme. The Coulomb matrix element $U_{j_{z_1} j_{z_2} j_{z_3} j_{z_4}}$ for $j_{zi} = -5/2, \ldots, 5/2$

$$U_{j_{z_1} j_{z_2} j_{z_3} j_{z_4}} = \sum_J U_J C^{JJ_z}_{5/2, j_{z1}; 5/2, j_{z2}} C^{JJ_z}_{5/2, j_{z3}; 5/2, j_{z4}} \quad (5.5)$$

are given in terms of the usual Clebsch–Gordan coefficients $C^{\cdots}_{\cdots}$ and the Coulomb parameters U_J. Here J denotes the total angular momentum of two electrons and $J_z = j_{z1} + j_{z2} = j_{z3} + j_{z4}$. The sum is restricted by the antisymmetry of the Clebsch–Gordan coefficients to even values $J = 0, 2, 4$.

The kinetic energy operator describes the hopping between all pairs at neighboring sites $\langle ab \rangle$

$$H_{\text{band}} = -\sum_{\langle ab\rangle, j_z} t_{j_z}\left(c^\dagger_{j_z}(a)c_{j_z}(b) + h.c.\right) + \sum_{a, j_z} \epsilon_f c^\dagger_{j_z}(a)c_{j_z}(a). \tag{5.6}$$

We assume the transfer integrals t_{j_z} to be diagonal in the orbital index j_z. While this is certainly an idealization it allows us to concentrate on our main interest, i.e., the interplay of intra-atomic correlations and kinetic energy. Finally, we account for the orbital energy ϵ_f which determines the f-valence of the ground state.

Due to the local degeneracy, the Hilbert space increases rapidly with the number of lattice sites and exact (numerical) solutions are possible only for relatively small clusters. For extended systems, i.e., for periodic solids the ground state and the low-lying excitations can be determined within a mean-field approximation.[145] The slave-boson functional integral method allows for a discussion of various ground states and co-operative phenomena starting from realistic bare electronic band structures. The orbital-dependent separation of the low-energy excitations into dispersive quasiparticle bands and incoherent background is observed in the spectral functions of a linear chain calculated by means of Cluster Perturbation Theory.[146] Itineracy is reflected in a discontinuity of the orbital-projected momentum distribution function

$$n_{j_z}(\mathbf{k}) = \int d\omega\, f(\omega) A_{j_z}(\mathbf{k}, \omega) \tag{5.7}$$

where $A_{j_z}(\mathbf{k}, \omega)$ is the single-particle spectral function while $f(\omega)$ denotes the Fermi distribution. Here we discuss the qualitative features derived for two-site clusters where we can find simple approximate forms for the ground-state wavefunction in limiting cases.

In order to quantify the degree of localization or, alternatively, of the reduction of hopping of a given j_z orbital by local correlations, the ratio of the j_z-projected kinetic energy T_{j_z} and the bare matrix element t_{j_z}

$$\frac{T_{j_z}}{t_{j_z}} = \sum_{\langle ab\rangle,} \langle \Psi_{gs}|\left(c^\dagger_{j_z}(a)c_{j_z}(b) + h.c.\right)|\Psi_{gs}\rangle \tag{5.8}$$

is calculated.[33] The ground-state wavefunction $|\Psi_{gs}\rangle$ contains the strong on-site correlations. A small ratio of T_{j_z}/t_{j_z} indicates partial suppression of hopping for electrons in the $\pm j_z$ orbitals. Two kinds of correlations may contribute to that

---

[145] J. Jedrak and G. Zwicknagl, preprint (2005).
[146] F. Pollmann, and G. Zwicknagl, *Phys Rev. B* **73**, 035121 (2006).

process. The first one is based on the reduction of charge fluctuations due to the large values of the isotropically averaged Coulomb repulsion which results in an isotropic renormalization of the kinetic energy. As this is a typical high-energy effect we defer the discussion to Section VIII. In the strong-coupling limit the reduction of charge fluctuation is accounted for by restricting the ground state to the well-defined atomic configurations. The quantity of interest here is the orbital-dependent reduction T_{j_z}/t_{j_z} which is due to intra-atomic correlations. As the latter are local in nature, even small clusters should adequately describe the important qualitative features. The results for T_{j_z}/t_{j_z}[33]—initially obtained perturbatively for a two-site cluster—as well as their interpretation are confirmed by detailed calculations based on exact diagonalization for small clusters.[147] Figure V.3 displays the reduction factors for a two-site cluster. The model parameters

$$\Delta U_4 = U_{J=4} - U_{J=0} = -3.79 \text{ eV}$$
$$\Delta U_2 = U_{J=2} - U_{J=0} = -2.72 \text{ eV}. \quad (5.9)$$

are chosen appropriate for UPt$_3$. These findings demonstrate that in particular Hund's rule correlations strongly enhance anisotropies in the hopping. For a certain range of parameters this may result in a complete suppression of the effective hopping except for the largest one, which remains almost unaffected. This provides a microscopic justification of partial localization of $5f$ electrons which is observed in a number of experiments on U intermetallic compounds.

As the relevant correlations are local, the general results qualitatively agree with those found for a three-site cluster and four-site clusters.[148] The magnetic character, however, is affected by finite size effects. This can be seen by varying the cluster sizes and the boundary conditions. Although the total angular momentum component $\mathcal{J}_z$ may be different for periodic and open boundary conditions we can identify the following different regimes in the strong-coupling limit. In the strongly anisotropic limit with dominating transfer integral $|t_{3/2}| \gg |t_{1/2}| = |t_{5/2}|$ the high-spin states with ferromagnetic inter-site correlations are energetically most favorable. In the two-site cluster, the ground state has a very simple form

$$|\Psi\rangle = \frac{1}{\sqrt{2}}\left(c^\dagger_{3/2}(a) + \frac{t}{|t|}c^\dagger_{3/2}(b)\right)c^\dagger_{5/2}(a)c^\dagger_{1/2}(a)c^\dagger_{5/2}(b)c^\dagger_{1/2}(b)|0\rangle \quad (5.10)$$

being simultaneously an eigenstate of the Coulomb energy and of the kinetic energy. It can be considered as a bonding $j_z = 3/2$ state in a ferromagnetically aligned background. The high-spin phases are followed by complicated

---
[147] D. V. Efremov, N. Hasselmann, E. Runge, P. Fulde, and G. Zwicknagl, *Phys. Rev. B* **69**, 115114 (2004).
[148] M. Reese, F. Pollmann, and G. Zwicknagl (2006), unpublished.

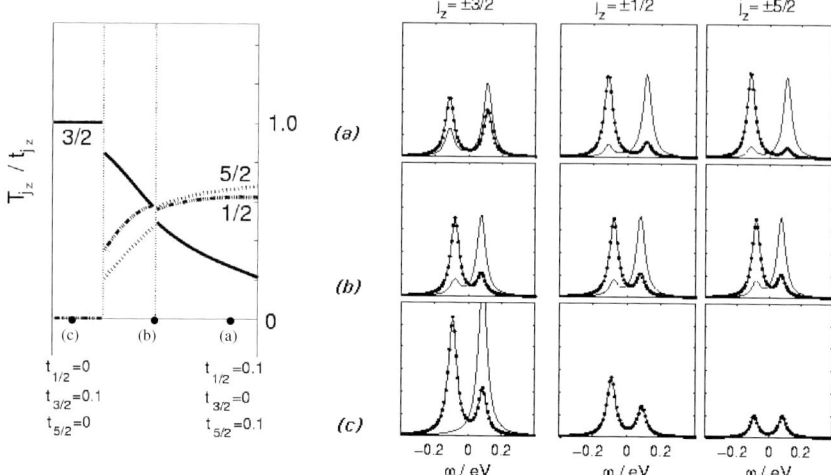

FIG. V.3. Left panel: Values T_{j_z}/t_{j_z} for a two-site cluster along a line connecting linearly the points written below the figure. (After Ref. [147].) Right panel: Variation with wave number of $(A_{j_z}(k_\ell, \omega) + A_{-j_z}(k_\ell, \omega))$ in the low-energy regime calculated for a two-site cluster with five electrons. The corresponding values of the transfer integrals are those of the left panel. The full line and the dotted line refer to $k = 0$ and $k = \pi$, respectively. Spectral weight is transferred to local excitations (valence transitions and transitions into excited atomic multiplets) which are not displayed here. The Lorentzian broadening is $\eta = 0.03$ (After Ref. [146].)

intermediate-spin phases as the isotropic limit $t_{1/2} = t_{3/2} = t_{5/2}$ is approached. In the case with subdominant $|t_{3/2}| \ll |t_{1/2}| = |t_{5/2}|$ low-spin phases with antiferromagnetic intersite correlations are formed. In a two-site cluster, they involve linear combinations of

$$c^\dagger_{1/2}(a)c^\dagger_{\pm 5/2}(a)c^\dagger_{\pm 3/2}(a)c^\dagger_{\mp 5/2}(b)c^\dagger_{\mp 1/2}(b)|0\rangle,$$
$$c^\dagger_{\pm 5/2}(a)c^\dagger_{\pm 3/2}(a)c^\dagger_{\mp 5/2}(b)c^\dagger_{1/2}(b)|0\rangle.$$

The splitting of the low-energy excitations into dispersive quasiparticle states and incoherent background is reflected in the single-particle spectral functions $A_{j_z}(k_\ell, \omega)$ where the discrete set of quantum numbers $k_\ell = 0, \pi$ labels the single-particle eigenstates of the two-site cluster. The variation with the transfer integrals is displayed in Figure V.3. The j_z-channels with dominant hopping exhibit dispersive narrow peaks while those with subdominant hopping yield an incoherent background. Considerable spectral weight is transferred to the high-energy excitations not shown here.

11. Superconductivity Mediated by Intra-Atomic Excitations

Since the discovery of the isotope effect[149,150] and the work of Fröhlich[151] the electron–phonon interaction has been considered the main cause of Cooper-pair formation. By exchanging virtual phonons, electrons may attract each other and form Cooper pairs. Later it was pointed out that phonons need not be the only bosons the exchange of which results in electron attraction. Also magnetic excitations such as paramagnons were considered as candidates for generating superconductivity, although not necessarily in a conventional s-wave pairing state.[97,152] Also it had been pointed out that crystalline electric field (CEF) excitations in rare-earth ions like Pr have a pronounced effect on superconductivity[153–155] when such ions are added to a conventional superconductor and furthermore, that those excitations can be either pair-breaking or pair forming depending on matrix elements between different CEF levels. The experimental observation of those effects, e.g., in doped $LaPb_3$ and $LaSn_3$[156,157] demonstrated not only the reality but also the magnitude of that kind of boson exchanges between conduction electrons. After the discovery of high-temperature superconductivity in some of the cuprate perovskites and even before for some of the heavy-fermion superconductors numerous suggestions of non-phononic pairing interactions were made.[158–166] But they were mainly qualitative rather than quantitative and therefore remained inconclusive. For reviews see Refs. [54,167]. Therefore it is of interest that for UPd_2Al_3 experimental evidence exists for a non-phononic mechanism causing superconductivity. It has a transition temperature T_c of $T_c = 1.8$ K; which is

---

[149] E. Maxwell, *Phys. Rev.* **78**, 477 (1950).
[150] C. A. Reynolds, B. Serin, W. H. Wright, and L. B. Nesbitt, *Phys. Rev.* **78**, 487 (1950).
[151] H. Fröhlich, *Phys. Rev.* **79**, 1950 (845).
[152] D. Fay and J. Appel, *Phys. Rev. B* **16**, 2325 (1977).
[153] P. Fulde, L. L. Hirst, and A. Luther, *Z. Phys.* **238**, 99 (1970).
[154] J. Keller and P. Fulde, *J. Low Temp. Phys.* **4**, 289 (1971).
[155] J. Keller and P. Fulde, *J. Low Temp. Phys.* **12**, 63 (1973).
[156] F. Heiniger, E. Bucher, J. P. Maitra, and L. D. Longinotti, *Phys. Rev. B* **12**, 1778 (1975).
[157] R. W. McCallum, W. A. Fertig, C. A. Luengo, M. B. Maple, E. Bucher, J. P. Maitra, A. R. Schwedler, L. Mattix, P. Fulde, and J. Keller, *Phys. Rev. Lett.* **34**, 1620 (1975).
[158] D. J. Scalapino, E. Loh, and J. E. Hirsch, *Phys. Rev. B* **34**, 8190 (1986).
[159] K. Miyake, S. Schmidt-Rink, and C. M. Varma, *Phys. Rev. B* **34**, 6554 (1986).
[160] P. W. Anderson, *Science* **235**, 1196 (1987).
[161] J. Spalek, *Phys. Rev. B* **38**, 208 (1988).
[162] T. Moriya, Y. Takahashi, and K. Ueda, *J. Phys. Soc. Jpn.* **59**, 2905 (1990).
[163] P. Monthoux, A. V. Balatsky, and D. Pines, *Phys. Rev. Lett.* **67**, 3448 (1991).
[164] P. Monthoux and G. G. Lonzarich, *Phys. Rev. B* **59**, 14598 (1999).
[165] P. Monthoux and G. G. Lonzarich, *Phys. Rev. B* **63**, 054529 (2001).
[166] P. Monthoux and G. G. Lonzarich, *Phys. Rev. B* **66**, 224504 (2002).
[167] N. M. Plakida, *High-Temperature Superconductivity*, Springer, Berlin, Heidelberg (1995).

below the onset of antiferromagnetic (AF) order with a Néel temperature of $T_N = 14.3$ K. Some evidence for a non-phononic pairing mechanism is provided by UPd$_2$Al$_3$–Al$_2$O$_3$–Pb tunneling measurements.[168] The differential conductivity dI/dV shows structure in the regime of 1 meV demonstrating that there are low-energy bosons which result in a frequency dependence $\Delta(\omega)$ of the order parameter. For phonons this structure would be an order of magnitude higher in energy. The Debye energy of UPd$_2$Al$_3$ is 13 meV. In addition inelastic neutron scattering (INS) experiments show that the CEF based magnetic excitation energy $\omega_E(\mathbf{q})$ at $\mathbf{q} = \mathbf{Q}$ is between 1–1.5 meV depending on temperature.[136,137,169,170] It has been argued that these excitations show up in $\Delta(\omega)$ and cause superconductivity. This is seen in Figure V.4 which shows the INS data as well as the tunneling density of states. The strong coupling of those AF excitons to conduction electrons is also demonstrated in these experiments.

The AF structure of UPd$_2$Al$_3$ consists of ferromagnetic hexagonal planes with a moment of $\mu = 0.83\mu_B$ per U ion pointing in [100] direction and stacked antiferromagnetically along the c-axis.[171,172] This corresponds to an AF wave vector $\mathbf{Q} = (0, 0, 1/2)$. The large moment supports the dual model with two localized $5f$ electrons. As discussed before the Hund's rule ground-state multiplet of the $5f^2$ localized electrons is $J = 4$. In a CEF only the two lowest singlets $|\Gamma_3\rangle$ and $|\Gamma_4\rangle$ have to be taken into account. The Hamiltonian is then of the form

$$H = \sum_{\mathbf{k}\sigma} \epsilon_{\mathbf{k}\sigma} c^\dagger_{\mathbf{k}\sigma} c_{\mathbf{k}\sigma} + \delta \sum_i |\Gamma_4\rangle\langle\Gamma_4|_i$$
$$- J_{\text{ff}} \sum_{\langle ij \rangle} \mathbf{J}_i \mathbf{J}_j - 2I_0(g_{\text{eff}} - 1) \sum_i \mathbf{s}_i \mathbf{J}_i \quad (5.11)$$

where

$$\epsilon_{\mathbf{k}\sigma} = \epsilon_\perp(k_\perp \sigma) - 2t_\parallel \cos k_z \quad (5.12)$$

serves as a model for the Fermi surface of UPd$_2$Al$_3$ in the paramagnetic state.

Figure V.3 shows that the Fermi surface in the AF state consists of a cylindrical part and a torus. The torus has the highest effective mass. For simplicity we will neglect it here and model the paramagnetic Fermi surface by a cylinder. The

---

[168] M. Jourdan, M. Huth, and H. Adrian, *Nature* **398**, 47 (1999).

[169] N. Sato, N. Aso, G. H. Lander, B. Roessli, T. Komatsubara, and Y. Endoh, *J. Phys. Soc. Jpn.* **66**, 1884 (1997).

[170] N. Bernhoeft, N. Sato, B. Roessli, N. Aso, A. Hiess, G. H. Lander, Y. Endoh, and T. Komatsubara, *Phys. Rev. Lett.* **81**, 4244 (1998).

[171] A. Krimmel, P. Fischer, B. Roessli, H. Maletta, C. Geibel, C. Schank, A. Grauel, A. Loidl, and F. Steglich, *Z. Phys. B* **86**, 161 (1992).

[172] H. Kita, A. Dönni, Y. Endoh, K. Kakurai, N. Sato, and T. Komatsubara, *J. Phys. Soc. Jpn.* **63**, 726 (1994).

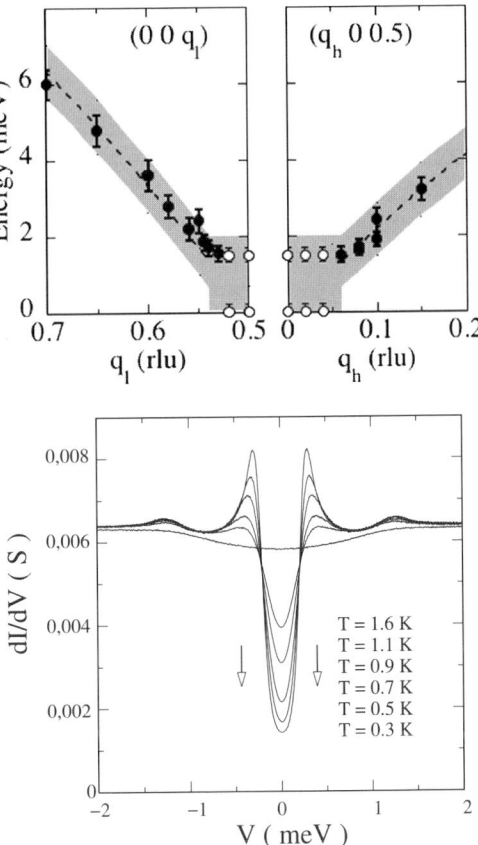

FIG. V.4. Upper panel: Magnetic exciton dispersion from INS along c^* $(0, 0, q_l)$ and a^* $(q_h, 0, 0.5)$ around the AF zone center $\mathbf{Q} = (0, 0, 0.5)$ for $T = 2$ K just above $T_c = 1.8$ K. A flat dispersion with a corresponding high DOS is observed for $\mathbf{q} \simeq \mathbf{Q}$. For $T < T_c$ the additional quasielastic response (open circles at zero energy) evolves into sharp resonance peaks within the gap. (After Ref. [136].) Lower panel: Differential conductivity dI/dV for tunneling current along c. The additional hump at $V \simeq 1.5$ meV has been associated with the magnetic exciton mode at $\mathbf{Q}$ in the upper panel. This feature is due to a frequency dependent gap $\Delta(\omega)$ caused by a strong coupling of quasiparticles to magnetic excitons. However the evaluation of the tunneling data needs additional justification (J. Geerk, private communication). (After Ref. [168].)

antiferromagnetic coupling to the conducting electrons is considerably larger than in systems like $TbMo_6S_8$[173] and $HoNi_2B_2$.[174] One should also keep in mind that

---

[173] M. Ishikawa and Ø. Fischer, *Solid State Comm.* **24**, 747 (1977).
[174] A. Amici, P. Thalmeier, and P. Fulde, *Phys. Rev. Lett.* **84**, 1800 (2000).

in accurate determination of T_c is anyway out of reach and that the aim is here to demonstrate that the non-phononic pairing mechanism yields the right order of magnitude for T_c.

Returning to Eq. (5.12) we note that $t_\parallel$ determines the amount of corrugation of the cylinder. The second term in Eq. (5.11) denotes the CEF splitting with $\delta = 6$–7 meV. The coupling between localized $5f$ electrons on nearest-neighbor sites is given by J_{ff} and the on-site exchange between localized and itinerant electrons is described by the last term of Eq. (5.11). The effective Landé factor g_{eff} refers to the localized $5f$ electrons. The total intersite exchange is therefore

$$J(\mathbf{q}) = J_{\text{eff}}(\mathbf{q}) + I_0^2 (g_{\text{eff}} - 1)^2 \chi_e(\mathbf{q}) \tag{5.13}$$

where $\chi_e(\mathbf{q})$ is the spin susceptibility of the itinerant electrons. Due to that intersite interaction the susceptibility of the system becomes

$$\chi(\mathbf{q}, \omega) = \frac{u(\omega)}{1 - J(\mathbf{q}) u(\omega)} \tag{5.14}$$

where $u(\omega)$ is the single-ion susceptibility[175] and the CEF excitation energy goes over into an excitation band (magnetic exciton). When $J(\mathbf{q})$ has its maximum value at $\mathbf{q} = \mathbf{Q}$ and $J(\mathbf{Q}) \equiv J_e$ exceeds a critical value, i.e., $J_e > J_{\text{crit}}$ the system becomes an induced AF. In that case the Néel temperature T_N is given by

$$T_N = \frac{\delta}{2 \tanh^{-1}(J_c/J_e)}. \tag{5.15}$$

UPd$_2$Al$_3$ is an induced AF and one finds that $J_e/J_c = 1.015$. The critical value is given by $J_c = \delta/2M^2$ where $M = 2\langle \Gamma_4 | J_x | \Gamma_3 \rangle_i$. For temperatures $T < T_N$ the susceptibility is again of a form similar to Eq. (5.14) but now the single-ion susceptibility contains the effect of the AF molecular field acting on the $|\Gamma_4\rangle$, $|\Gamma_3\rangle$ states. In that case the magnetic excitations form a band of AF magnons. For a review see, e.g., Refs. [109,176]. They have originally been measured[177] with relatively low resolution and later with much better one, see Refs. [136,137] and [178]. Their dispersion has also been derived theoretically in Ref. [179] by in-

---

[175] B. R. Cooper, *Phys. Rev.* **163**, 444 (1967).
[176] J. Jensen and A. R. Mackintosh, *Rare Earth Magnetism, Structures and Excitations*, Clarendon Press, Oxford (1991).
[177] T. E. Mason and G. Aeppli, *Matematisk-fysiske Meddelelser.* **45**, 231 (1997).
[178] N. Bernhoeft, *Eur. Phys. J. B* **13**, 685 (2000).
[179] P. Thalmeier, *Eur. Phys. J. B* **27**, 29 (2002).

cluding the molecular field as well as the anisotropic exchange and agreed nicely with the measured ones. An approximate form is

$$\omega_E(q_z) = \omega_{\text{ex}}\big[1 + \beta \cos(cq_z)\big] \tag{5.16}$$

with $\omega_{\text{ex}} = 5$ meV, $\beta = 0.8$ and c denoting the lattice constant perpendicular to the hexagonal planes. The corresponding boson propagator $K(q_z, \omega_\nu)$ in Matsubara frequency notation ($\omega_\nu = 2\pi T$, $\nu =$ integer) replaces the phonon propagator when the superconducting properties are calculated.[180] It is of the form

$$K(q_z, \omega_\nu) = g \frac{\omega_{\text{ex}}^2}{(\omega_E(q_z))^2 + \omega_\nu^2} \tag{5.17}$$

where g denotes the coupling constant between conduction electrons and magnetic excitons. Their interaction Hamiltonian H_{c-f} can be written in a pseudospin notation by introducing for the two levels $|\Gamma_3\rangle_i$ and $|\Gamma_4\rangle_i$ of a U site i the pseudospin τ_i, so that $\tau_{iz}|\Gamma_{3(4)}\rangle_i = {}^+_{(-)}\frac{1}{2}|\Gamma_{3(4)}\rangle_i$. Then we may write

$$H_{c-f} = I \sum_i \sigma_{iz} \tau_{ix} \tag{5.18}$$

where σ_i refers to the itinerant $5f$ electron. The two coupling constants g and I are related through

$$g = \frac{I^2}{4}\left(\frac{1}{c}\frac{p_0^2}{2\pi}\right)\frac{1}{\omega_{\text{ex}}} \tag{5.19}$$

where p_0 is the radius of the circle in the p_x, p_y plane which contains the same area as the hexagon defining the Brillouin zone.

We are now in the position to write down and solve Eliashberg's equations for the conduction electron self-energy $\Sigma(p_z, \omega_n)$ and order parameter $\Delta(p_z, \omega_n)$.

$$\Sigma(p_z, \omega_n) = \frac{T}{N_z} \sum_{p'_z, m} K(p_z - p'_z; \omega_n - \omega_m) \int \frac{dp'_\perp}{(2\pi)^2} G(p'_\perp, p'_z, \omega_m)$$

$$\Delta(p_z, \omega_n) = -\frac{T}{N_z} \sum_{p'_z, m} K(p_z - p'_z; \omega_n - \omega_m) \Delta(p'_z, \omega_m)$$

$$\times \int \frac{dp'_\perp}{(2\pi)^2} |G(p'_\perp, p'_z, \omega_m)|^2 \tag{5.20}$$

---

[180] P. McHale, P. Fulde, and P. Thalmeier, *Phys. Rev. B* **70**, 014513 (2004).

where N_z is the number of lattice sites along the z-axis and $\omega_n = 2\pi T(n + 1/2)$. It has been assumed that the order parameter has even parity (singlet channel). The electron Green function has the usual form

$$G^{-1}(\mathbf{p}, \omega_n) = i\omega_n - \epsilon_\mathbf{p} - \Sigma(p_z, \omega_n) \quad (5.21)$$

with $\epsilon_\mathbf{p}$ given by Eq. (5.12).

After the $dp'_\perp$ integration in Eq. (5.20) has been done the equations reduce to a one-dimensional problem. Thereby it is essential that the kernel $K(q_z, \omega_v)$ is strongly peaked at $q_z = \pi/c$ and $\omega_v = 0$. Therefore, loosely speaking the gap equation is of the form

$$\Delta(p_z, \pi T) = -C(p_z)\Delta\left(p_z - \frac{\pi}{c}, \pi T\right) \quad (5.22)$$

where $C(p_z)$ is a smooth positive function. This suggests the form

$$\Delta(\mathbf{p}) = \Delta \cos(cp_z) \quad (5.23)$$

with A_{1g} symmetry. The symmetry allows also for a multiplication of the right-hand side by a fully symmetric function $f(p_x, p_y)$. The order parameter has lines of nodes at the AF zone boundary $p_z = \pm Q_z/2 = \pm\pi/(2c)$.

One finds that Eliashberg's equations yield also an odd-parity solution of the form

$$\Delta(p_z) = \Delta \sin(cp_z) \quad (5.24)$$

with a spin part $|\chi\rangle = (2)^{-1/2}(|\uparrow\downarrow\rangle + |\downarrow\uparrow\rangle)$ and the same T_c. Note that because of the Ising-like interaction (5.18) rotational symmetry in spin space is broken. A more general study of possible order parameters due to pair potentials based on magnetic excitons was undertaken in Ref. [179]. It turned out that in the weak coupling limit, i.e., without taking retardations into account, one of the odd-parity triplet states has lower energy than the singlet one. However this could be a consequence of the weak coupling assumption which does not apply to UPd_2Al_3. There is strong experimental evidence discussed below that the order parameter has indeed A_{1g} symmetry (see Eq. (5.23)) and therefore we discard the A_{1u} solution (5.24).

In the following we discuss the parameters which are required within this simplified model to explain the anisotropic effective mass and the ones which are needed to explain the observed T_c. For the DOS at the Fermi energy a value of $N(E_F) \simeq 2$ states/(eV-cell-spin) seems appropriate. This includes also the torus. The local intra-atomic excitations responsible for the mass enhancement are characterized by $\delta \simeq 7$ meV[33] and $\omega_{ex} \simeq 5$ meV.[180] With these values it turns out that a value of

$$I^2 N(E_F) = 0.026 \text{ eV} \quad (5.25)$$

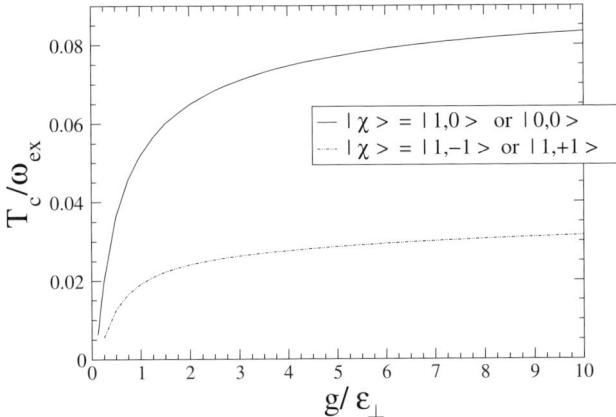

FIG. V.5. Left panel: The dependence of the superconducting T_c on the electron-magnetic exciton coupling constant g (Eq. 5.19). Parameter values are $\omega_{ex} = 0.01\epsilon_\perp$, $\alpha = -0.1$, $\beta = 0.8$. The full curve corresponds to opposite spin pairing states $|\chi\rangle$ ($S_z = 0$) A_{1g} and A_{1u} with gap functions given in Eqs. (5.23), (5.24) and node lines at $k_z = \pm\frac{\pi}{c}$ and $k_z = 0$ respectively. The dashed curve is a less favorable state with $|S_z| = 1$ and more node lines. For a value $g/\epsilon_\perp \sim 2$ which leads to the observed mass enhancement $m^*/m_b \simeq 10$ one obtains a calculated $T_c \simeq 2.9$ K (from the full curve). This value is somewhat larger than the experimental $T_c = 1.8$ K. (After Ref. [180].)

corresponding to $g/\epsilon_\perp = 2$ is required in order to reproduce the experimentally observed mass enhancement within the simplified scheme applied to Eliashberg's equations. With this value a superconducting transition temperature $T_c = 2.9$ K is obtained while the true value is 1.8 K. The dependence of the superconducting T_c on the dimensionless coupling strength $g/\epsilon_\perp$ is shown in Figure V.4. Taking into account that the strong mass anisotropies were derived theoretically *without* adjustable parameters (see Table V.1) it is gratifying to find that when the parameter $I^2 N(E_F)$ (or $g/\epsilon_\perp$) of a simplified model is adjusted so as to reproduce the mass anisotropies, a superconducting transition temperature of the right order of magnitude is obtained.

As regards the experimental situation the superconducting state of UPd$_2$Al$_3$ has been studied in great details for which we refer to the review Ref. [54]. Here we point out merely those experiments which allow for a determination of the symmetry of the order parameter. This applies in particular to the studies of the anisotropic thermal conductivity in an applied magnetic field. As pointed out in Ref. [181] measurements of the thermal conductivity under rotating magnetic field with heat current perpendicular to the rotation plane yield information on

---

[181] K. I. T. Watanabe, Y. Kasahara, Y. Haga, Y. Onuki, P. Thalmeier, K. Maki, and Y. Matsuda, *Phys. Rev. B* **70**, 184502 (2004).

the **k**-space position of gap nodes. Corresponding measurements on single crystals of UPd$_2$Al$_3$[181] and their analysis[182] have indeed shown that $\Delta(\mathbf{p})$ has node lines perpendicular to the c-axis. However these experiments cannot distinguish between a node line at $p_z = \pm\frac{\pi}{2c}$ at the AF zone boundary or at $p_z = 0$ at the zone center, i.e., which one of the theoretically favored gap functions of Eq. (5.23) or Eq. (5.24) is realized. Other evidence for node lines was pointed out by Bernhoeft.[178] He argued that the symmetry property $\Delta(\mathbf{p} \pm \mathbf{Q}) = -\Delta(\mathbf{p})$ is required in order to explain the large intensity of a low energy quasiparticle-like peak in the inelastic neutron scattering spectrum below T_c. This again implies a line of nodes perpendicular to the c axis but cannot distinguish between the two candidates. However already before these results became available a pronounced Knight shift reduction below T_c was found in UPd$_2$Al$_3$.[183] A naive interpretation of this result advocates for the even parity A_{1g} gap function in Eq. (5.23) whose antisymmetric spin function corresponds to the spin singlet state in models with isotropic pairing interaction.

In concluding this section we want to summarize the above findings. It has been shown that in UPd$_2$Al$_3$ superconductivity is due to non-phononic bosons. Intra-atomic excitations of localized $5f$ electrons in U ions, a consequence of strong correlations and described within the dual model provide the glue for the Cooper pairs. We want to stress that their interaction with the itinerant $5f$ electrons (see Eq. (5.18)) is *not* time-reversal invariant. Nevertheless Cooper pairs may form but a sign change of the order parameter along the c-direction is mandatory. Otherwise the interaction would not be pair forming but rather act as a pair breaker. It is likely that a similar pairing mechanism, mediated by the exchange of quadrupolar excitons, is operative in the Pr-skutterudite cage compound.

VI. Charge Ordering

The concept of charge order in electronic systems was introduced by Wigner in the early 1930s.[9] He considered a homogeneous electron gas, i.e., a system in which the positively charged background is distributed uniformly over the sample. Wigner showed that in the limit of low densities the Coulombic repulsion energy between the electrons will always dominate their gain in kinetic energy due to delocalization. Therefore at sufficiently low temperatures electrons will form a lattice. This way they minimize their mutual repulsions.

Following the original proposal of Wigner a number of systems have been discovered which show electronic charge order. The cleanest realization of Wigner

---

[182] P. Thalmeier, T. Watanabe, K. Izawa, and Y. Matsuda, *Phys. Rev. B* **024539**, 72 (2005).
[183] Y. Kitaoka, H. Tou, K. Ishida, N. Kimura, Y. Onuki, E. Yamamoto, Y. Haga, and K. Maezawa, *Physica B* **281–282**, 178 (2000).

crystallization is observed in a classical 2D sheet of electrons generated on a liquid-He surface where the electron density can be varied by an applied electrical field.[184] In constrained geometries like quantum dots clear signatures of Wigner crystallization may be found already for small electron numbers.[185] The most common systems which exhibit this type of electronic charge order have an underlying atomic lattice as an important ingredient in distinction to Wigner's homogeneous positive background. When valence electrons are situated close to an atomic nucleus the overlap with orbitals from neighboring sites is expected to be small. This implies that the associated kinetic energy gain is small when electrons delocalize. Therefore the mutual repulsions between electrons on neighboring sites will dominate the kinetic energy gain at a higher density than it is the case for a uniformly distributed positive background. But a resulting electronic charge order must here be commensurate with the underlying atomic lattice.

The 3d-valence electrons of transition metal compounds are most amenable to this kind of charge order. The latter may occur within a metallic (or at least conducting) state as in the prominent example of magnetite Fe_3O_4 or within an already insulating state as in α'-NaV_2O_5. Since the number of d-electrons in 3d-compounds may commonly be changed by doping, a large variety of commensurate 3d-charge ordered states can be achieved, for example in the cuprate (parent compound $LaCu_2O_4$) and single layer and bilayer manganite families (parent compound $LaMnO_3$ and half-doped compound $LaSr_2Mn_2O_7$, respectively). The amplitude of the total charge order parameter or charge disproportionation on the inequivalent 3d-sites is typically small of the order 0.1 electron per site or less. This is due to the screening of the large 3d-orbital occupation changes by the valence electrons of ligands, as can be clearly seen from LSDA+U calculations. Experimentally the amount of charge disproportionation is extremely difficult to determine. This is mostly done via the empirical valence-bond analysis of X-ray results in the ordered phase, where the change in bond distances is linked to the valence charge disproportionation. In some 3d-oxide compounds like, e.g., the bilayer-manganites one must be aware that 3d-charges are dressed by strong distortions of the surrounding lattice leading to (small) polaron formation. It is then the latter which exhibit the ordering transition.

Charge ordering is less common in $4f$-compounds, because the intersite Coulomb interaction necessary for charge order is well screened and in genuine metallic compounds the hybridization with conduction electrons tends to favor a site-independent $4f$-occupation. This leads to a metallic mixed-valent or heavy fermion state. There is however an important case where $4f$-charge order may occur. In insulating or semimetallic $4f$ compounds which are homogeneous valence fluctuators at high temperatures, the intersite Coulomb interaction may be

---

[184] G. Adams, *Phys. Rev. Lett.* **42**, 795 (1979).
[185] A. V. Filinov, M. Bonitz, and Y. E. Lozovik, *Phys. Rev. Lett.* **17**, 3851 (2001).

strong enough to lead to a $4f$-charge disproportionation, i.e., an inhomogeneous mixed-valent state at low temperatures with different $4f$-orbital occupations on inequivalent sites. Again the amplitude of the total charge order including the effect of ligand screening charges is much smaller than the bare $4f$-charge disproportionation. An important class of compounds where this charge-order (CO) transition occurs are members of the R_4X_3 series (R = rare earth Yb, Sm, Eu and X = As, Bi, P, Sb), notably the semimetal Yb_4As_3 discussed in Section VI.13.

An attractive feature of CO transitions in these $3d$ or $4f$ compounds is the possibility of lowering the effective dimension of the arrangement of magnetic ions, for example to a family of 1D chains, planes of zig-zag chains or ladders and stripes. The effect of low dimensionality has then important consequences for the spin excitation spectrum, e.g., the appearance of a two-spinon continuum in the case of spin chains.

When this type of electronic charge order in $3d$ and $4f$ systems is compared with the original suggestion of Wigner one notices two differences. As pointed out before, the lattice structure of the positive background is very important but also the repulsions between electrons on neighboring sites are not purely Coulombic. Instead they may be modified due to the electrons in inner closed shells which leads to strongly screened Coulomb interactions that usually extend only to nearest and next nearest neighbors.

Another type of charge ordering which does not break spatial symmetries is obtained in systems which can be described by a Hubbard Hamiltonian at half filling. Here it is the on-site Coulomb interaction expressed by an energy U which is competing with electron hopping processes and the associated energy gain. While the repulsion energy U favors single occupancy of sites, and suppresses double occupancies the kinetic energy gain favors a sizeable fraction of sites with double occupations. Without electron correlations 25% of them would be doubly occupied in order to optimize the kinetic energy. For sufficiently large values of U charge order in the form of strongly suppressed on-site charge fluctuations will take place leading to a Mott–Hubbard metal to insulator (M-I) transition. Again, similarities to the previously considered cases are obvious but so are the differences. Repulsions between electrons on different sites are completely neglected in the Hubbard model and it is crucial to have precisely one electron per site in order to obtain charge order. A similar requirement does not exist in the previously considered cases.

Finally, charge order can also occur via formation of a charge density wave (CDW) in metals. In this case the instability is driven by minimizing the kinetic (band) energy of conduction electrons leading to a reconstruction of the Fermi surface. A prerequisite is the presence of nesting properties in the Fermi surface. The generally incommensurate vector $\mathbf{Q}$ which connects the nesting parts determines wavelength and direction of a corresponding CDW. Note that strong electron correlations are not required for a CDW to form. Summarizing we distinguish between the following electronic charge ordering processes:

(a) Wigner crystallization in the homogeneous electron gas or liquid
(b) Charge order due to weak hybridizations and strong intersite interactions
(c) Mott–Hubbard charge order due to strong on-site interaction
(d) Charge density waves due to nesting properties of the Fermi surface
(e) Charge ordering in polaronic systems

In this overview we will not address the genuine Mott–Hubbard M-I transition which is reviewed in existing articles, e.g., Ref. [186]. It demands a considerable technical effort based on the recently developed dynamical mean field theory (DMFT). Low dimensional metallic CDW systems will also not be included, since this topic is well represented in the literature[187] and should be discussed together with spin-density waves and superconductivity which is beyond the scope of this chapter.

12. Wigner Crystallization in Homogeneous 2D Electron Systems

The electron gas may be subject to many different instabilities depending both on the background (lattice) potential and the resulting shape of the Fermi surface as well as the strength and range of the screened Coulomb interactions. A convenient way to characterize the relevant regime is the Brueckner parameter $r_s = \frac{1}{a_0}(\frac{4\pi}{3}n)^{-\frac{1}{3}}$ which is the ratio of average electron distance to the Bohr radius a_0 (n = electron density). In the two limits of small and large r_s the electron system exhibits radically different behavior.

At large density (small r_s) the system is in a metallic state dominated by the kinetic (Fermi) energy and the Coulomb interactions between electrons are well screened. If the Fermi surface has parallel (quasi-1D) sections, 'nested' by a wave vector **Q**, the residual Coulomb interactions may lead to a condensation of electron–hole (Peierls) pairs into a charge- or spin-density wave state with translational and possibly other symmetries broken. Part or all of the nested Fermi surface sheets are then removed by the self-consistent potential in the condensed state. For a conventional density wave spontaneous modulation of the charge- or spin density with period $2\pi/Q$ takes place which may be identified by common methods like X-ray or neutron diffraction. Since real metallic materials have high densities and small or moderate r_s this type of instability is frequently encountered. It often competes or coexists with an alternative condensation mechanism,

---

[186] A. Georges, G. Kotliar, W. Krauth, and M. J. Rosenberg, *Rev. Mod. Phys.* **68**, 13 (1996).
[187] G. Grüner, *Density Waves in Solids*, Addison-Wesley (1994).

namely electron–electron (Cooper) pair formation leading to a superconducting state.

In the opposite limit of small density and large r_s the long range Coulomb interaction is badly screened and it dominates the small kinetic energy gain due to delocalizations. At low enough temperatures the electron liquid condenses in real space, forming a Wigner solid[9] which also breaks translational symmetry. This is complementary to condensation in the **k**-space as found in the previous density wave case. For a two dimensional (2D) homogeneous electron liquid in a uniform positive background the appropriate Hamiltonian $H = T + V$ is

$$H = \sum_{\mathbf{k}\sigma} \epsilon(\mathbf{k}) c^\dagger_{\mathbf{k}\sigma} c_{\mathbf{k}\sigma} + \frac{1}{2\Omega} \sum_{\mathbf{pkq};\sigma\sigma'} v_\mathbf{q} c^\dagger_{\mathbf{p}+\mathbf{q}\sigma} c^\dagger_{\mathbf{k}-\mathbf{q}\sigma'} c_{\mathbf{k}\sigma'} c_{\mathbf{p}\sigma} \qquad (6.1)$$

where $\epsilon(\mathbf{k}) = \mathbf{k}^2/2m$ is the kinetic energy and $v_\mathbf{q} = \frac{4\pi e^2}{\mathbf{q}^2}(1 - \delta_{\mathbf{q}0})$ is the Coulomb interaction. The Kronecker delta $\delta_{\mathbf{q}0}$ ensures that due to the positive background $v_\mathbf{q} = 0$. Furthermore Ω is the volume. For practical calculations on a lattice and for finite systems the Ewald summation technique has to be used to obtain the real space Coulomb potential $V(\mathbf{r})$.

The qualitative shape of the n–T phase boundary for the liquid-solid transition has been given in Ref. [188]. It is derived from the intuitive notion that at the phase transition the average potential and kinetic energies of electrons should be comparable, i.e., $\langle V \rangle / \langle T \rangle \equiv \Gamma_0$. This leads to a parametrically ($z = \exp(-\beta\mu)$) determined n–T phase boundary (μ = chemical potential) shown in Figure VI.1. In the classical limit ($kT \gg \mu$) one has $\langle T \rangle = kT$, furthermore $\langle V \rangle = e^2\sqrt{\pi n}$. Then the melting curve is simply given by $n(T) = (kT\Gamma_0/\sqrt{\pi}e^2)^2$ with an unknown parameter Γ_0. In a homogeneous background the 2D electron liquid solidifies in a trigonal (hexagonal) lattice structure.[189] In the static approximation without kinetic energy the ground-state energy is $E^\triangle_{GS} = -3.921034 e^2 a_c^{-1/2}$ which is lower than for any other of the five 2D Bravais lattices. (Here $a_c = (\sqrt{3}/2)a_0^2$ is the hexagonal cell area with a_0 denoting the lattice constant, the density is $n = 1/a_c$.) This result is simple to understand because of all 2D lattices the trigonal one has the largest lattice spacing for a given density and therefore minimizes the total energy if the latter is dominated by the Coulomb repulsion as in the small density regime. Stability analysis shows that the trigonal lattice is stable under longitudinal and transverse (shear) distortions and the corresponding two phonon branches have real frequencies. In the long wavelength limit they are isotropic, i.e., they depend only on $q = (q_x^2 + q_y^2)^{-1/2}$ and are given by $\omega_1(q) = \omega_p(a_0q)^{1/2}$ and $\omega_2(q) = 0.19\omega_p a_0 q$ for longitudinal and transverse modes respectively. Here

---

[188] P. M. Platzmann and H. Fukuyama, *Phys. Rev. B* **10**, 3150 (1974).
[189] L. Bonsall and A. A. Maradudin, *Phys. Rev. B* **15**, 1959 (1977).

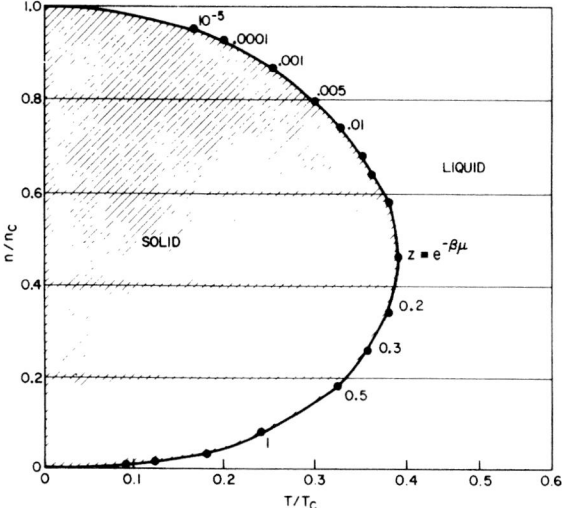

FIG. VI.1. Schematic Fermi liquid–Wigner solid phase diagram of the 2D electron system. Estimates of critical density and temperature are given by $n_c = (4/\pi a_0^2)(1/\Gamma_0^2)$ and $T_c = (2e^4 m/\Gamma_0^2)$. (After Ref. [188].)

$\omega_p = \frac{4\pi e^2}{m^*}(2a_c a_0)^{-1}$ is the 3D plasma frequency of a slab of thickness $2a$. While the Wigner lattice structure and stability is well understood, the melting into the liquid state, both classical melting (as function of T) and quantum melting at $T = 0$ (as function of the control parameter r_s) is much less clear. The situation sketched in Figure VI.1 is certainly oversimplified in one aspect: In both cases presumably an intermediate phase between Wigner solid and Fermi liquid appears which is characterized by the loss of long range translational order but preserves quasi-long range orientational order. In the case of classical melting of 2D trigonal lattices this intermediate phase is known as the 'hexatic phase'.[190] In the case of quantum melting (at $T = 0$) as function of density (r_s) a similar precursor phase to the liquid seems to appear. This can only be investigated by numerical techniques like Quantum Monte Carlo (QMC) simulations. It was known from earlier QMC results[191] that for $r_s > 37 \pm 5$ the Fermi liquid state becomes instable. A more recent investigation of phases as function of r_s was undertaken in Ref. [192] with a fixed-node QMC approach. A variational trial or guiding wave function for the ground state of H (Eq. (6.1)) on a grid (L_x, L_y) was used. It has the following

---

[190] B. I. Halperin and D. R. Nelson, *Phys. Rev. Lett.* **41**, 121 (1978).
[191] B. Tanatar and D. M. Ceperley, *Phys. Rev. B* **39**, 5005 (1989).
[192] H. Falakshahi and X. Waintal, *Phys. Rev. Lett.* **94**, 046801 (2005).

form

$$\Psi(\mathbf{r}_1, \mathbf{r}_2, \ldots \mathbf{r}_N) = \text{Det}\big[\phi_i(\mathbf{r}_j)\big] \prod_{i<j} J\big(|\mathbf{r}_i - \mathbf{r}_j|\big). \tag{6.2}$$

The single particle wave functions $\phi_i(\mathbf{r}_j)$ in the Slater determinant $\text{Det}[\phi_i(\mathbf{r}_j)]$ are taken as plane waves in the liquid state and localized Gaussian orbitals on trigonal sites in the Wigner lattice case. The Jastrow function $J(|\mathbf{r}_i - \mathbf{r}_j|)$ describes correlations and consists of modified Yukawa functions. It takes the Coulomb correlations into account by keeping the electrons apart. The range of correlations scales with the average electron distance $d = 1/\sqrt{\pi n}$ where $n = N/L_x L_y$ denotes the density. With this trial wave function one would obtain indeed the instability of the liquid state at $r_s^* \simeq 40$ since for $r_s > r_s^*$ its energy exceeds that of the Wigner solid (see Figure VI.2). However, the latter with its localized wave functions is still not the most stable state in that range of r_s. This may be seen if one uses instead trigonal lattice Bloch states for the single particle functions $\phi_i(\mathbf{r}_j)$ but with $\mathbf{k}$ constrained to the first BZ. This implies an orientational symmetry breaking in $\mathbf{k}$ space, i.e., a trigonally shaped Fermi surface. But the static density correlation function has still no fully developed Bragg peaks like in the crystal in analogy to the classical melting scenario. This intermediate or hybrid state has indeed a lower energy than the Wigner solid in the range $r_s^* < r < r_s^{**}$ with $r_s^* \simeq 30$ and $r_s^{**} \simeq 80$. For still higher values of r_s the Wigner solid with fully

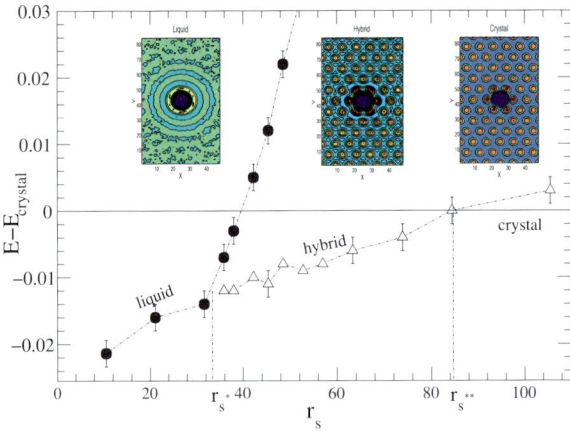

FIG. VI.2. Energy differences in units of $2\pi Nnt$ (t = hopping energy, $n = N/L_x L_y$ = average density) of liquid and intermediate (or hybrid) phases with respect to the Wigner crystal phase. Here $E_{\text{liquid}} - E_{\text{crystal}}$ (circles) and $E_{\text{hybrid}} - E_{\text{crystal}}$ (triangles) are plotted as function of r_s for $N = 72$ electrons on a $L_x = 48 \times L_y = 84$ grid. Critical values are $r_s^* \sim 30$ and $r_s^{**} \sim 80$. Inset: density–density correlation function with reference particle in the center. (After Ref. [192].)

localized functions $\phi_i(\mathbf{r}_j)$ finally becomes stable. There is no symmetry change involved in going from the intermediate to the Wigner solid state. Therefore it is not clear whether there is a real quantum phase transition at r_s^{**}, as there is at r_s^*, or simply a crossover to more pronounced density correlations (see inset of Figure VI.2). The picture of the intermediate phase is also supported by QMC calculations for mesoscopic 2D electron systems in a harmonic trap. There a similar two-step transition is found from liquid to orientational and finally to fully developed lattice correlations.[185] Finally we note that at even larger values of r_s the Wigner lattice becomes spin-polarized due to a ferromagnetic ring exchange in the trigonal lattice.[193]

The experimental realization of the genuine Wigner lattice formation in the 2D electron liquid has been attempted along two alternative approaches. The more successful route is the accumulation of electrons in a monolayer on the surface of liquids such as helium. The accumulation is achieved by applying an electrical pressing field perpendicular to the surface. Its strength allows to vary the electron density in the monolayer over several orders of magnitude. The lower part of the classical melting curve in Figure VI.1 has been determined in Ref. [184] by a rf-resonance method. For the ratio of potential to kinetic energy at the phase boundary a surprisingly large value of $\Gamma_0 \sim 137$ was found in agreement with early MC simulations. Due to a finite surface tension the pressing field deforms the liquid surface around the electrons thereby leading to single electron 'dimples'. This induces an effective attraction between electrons which has to be added to the Coulomb repulsion. It has been proposed that this attraction may lead to a structural phase transition of the Wigner lattice from trigonal to square lattice for sufficiently low density and high pressing fields.[194] The second route to generate 2D electron systems are semiconductor heterojunctions. It has been much less conclusive because the evidence for Wigner lattice formation in transport properties is obscured by the 2D localization effects caused by impurities.[195]

Computational results presented before show that the genuine Wigner lattice formation takes place at r_s values which are more than an order of magnitude larger than those found in real solids. But in real crystals with inhomogeneous electron densities the condition for a Wigner type of lattice formation may be much easier to fulfill. In case that the overlap of atomic wavefunctions of neighboring atoms is small the gain in kinetic energy due to electron delocalization is also small. The mutual Coulomb repulsion of electrons on neighboring sites can become more easily dominant in that case than in a homogeneous electron gas.

---

[193] D. M. Ceperley, *The Electron Liquid Paradigm in Condensed Matter Physics*, IOS Press, Amsterdam (2004), p. 3.

[194] M. Haque, I. Paul, and S. Pankov, *Phys. Rev. B* **68**, 045427 (2003).

[195] E. Abrahams, *Rev. Mod. Phys.* **73**, 251 (2001).

Neighboring rare earth ions have a particularly small overlap of their 4f wavefunctions. Therefore they are particularly good candidates for the formation of charge ordered Wigner like lattices[196] as will be discussed in the next section.

Furthermore in $3d/4f$ compounds the Coulomb interactions are strongly screened whereas in the two-step classical and quantum melting of a 2D Wigner lattice the long range part of the Coulomb interactions is most important. Indeed, charge ordering in compounds usually takes place at a well defined temperature and is mostly of first order. It is driven by a competition between short range next and next nearest neighbor Coulomb interactions and the kinetic energy. Also some compounds are already in a Mott–Hubbard insulating state due to large on-site Coulomb repulsion before charge order caused by the inter-site Coulomb interactions appears. Therefore, for real compounds like vanadates and manganites the extended Hubbard type models are a better starting point than the Hamiltonian in Eq. (6.1) to describe the multitude of charge ordering phenomena in solids. Nevertheless, loosely speaking one may refer to them as a kind of generalized Wigner lattice formation.[196]

13. GENERALIZED WIGNER LATTICE: YB_4AS_3

The intermetallic compound Yb_4As_3 is a perfect example of a system in which charge order takes place, here of $4f$ holes. Yb_4As_3 has a cubic anti-Th_3P_4 structure with a $I\bar{4}3d$ space group. Due to the special lattice structure charge order in this three-dimensional system results in the formation of well separated chains of Yb^{3+} ions. They act like one-dimensional spin chains. The net result is that Yb_4As_3 shows all the signs of a low-carrier-density heavy fermion system which here is due to the properties of the spin chains. This is a good example for the formation of heavy quasiparticles caused by electronic charge order. The Kondo effect plays no role in this compound.

We start out by summarizing some experimental facts and results. Yb_4As_3 and other family members of R_4X_3 were first systematically investigated by Ochiai et al.[43] A compilation of more recent results may be found in Ref. [197]. By counting valence electrons one notices that since As has a valency of -3, three of the Yb ions must have a valency of $+2$ while one ion has a valency of $+3$, i.e., one expects $Yb_4As_3 \to (Yb^{2+})_3(Yb^{3+})(As^{3-})_3$. But Yb^{2+} has a filled $4f$ shell. Thus there is one $4f$ hole per formula unit. The Yb ions occupy four families of

---

[196] P. Fulde, *Ann. Phys. (Leipzig)* **6**, 178 (1997).
[197] B. Schmidt, H. Aoki, T. Cichorek, J. Custers, P. Gegenwart, M. Kohgi, M. Lang, C. Langhammer, A. Ochiai, S. Paschen, et al., *Physica B* **300**, 121 (2001).

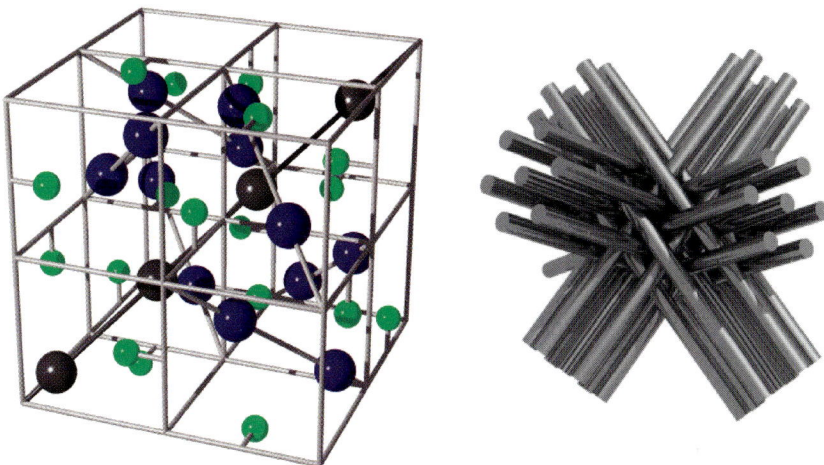

FIG. VI.3. Left panel: Anti-Th$_3$P$_4$ structure of Yb$_4$As$_3$. Large and small spheres symbolize Yb and As ions, respectively. The Yb ions are residing on four interpenetrating families of chains oriented along cubic space diagonals. Right panel: Dense rod packing representation of the Yb-chains. In the CO structure only *one* family of chains carries Yb^{3+} ions with pseudo-spin $S = 1/2$ whereas the other three families are occupied with $S = 0$ Yb^{2+} ions. (After Ref. [197].)

interpenetrating chains which are pointing along the diagonals of a cube (see Figure VI.3). This is often referred to as body-centered cubic rod packing.[198] It is important to notice that the distance between two neighboring Yb ions along a chain is larger than the distance between ions belonging to different chains. This implies that nearest neighbor Yb ions belong to different families of chains. At sufficiently high temperatures, i.e., above 300 K the $4f$ holes move freely between sites and the system is metallic. Measurements of the Hall coefficient R_H confirm that the carrier concentration is approximately one hole per unit cell in that temperature range. The situation is different at low temperatures where the measured Hall coefficient has a value of $R_H = 7 \cdot 10^{18}$ cm^{-3} implying approximately one hole per 10^3 Yb ions (see Figure VI.4). Thus the system changes from a metal to a semimetal as the temperature decreases. This is particularly seen in measurements of the resistivity $\rho(T)$ (see Figure VI.4). While for $T > 300$ K a linear temperature dependence is observed, one notices that near $T_c \simeq 292$ K a first-order phase transition is taking place with a corresponding increase in resistivity. At low temperatures $\rho(T) = \rho_0 + AT^2$ is found, i.e., the semimetal is a Fermi liquid. Despite the low carrier concentration of order $n \simeq 10^{-4}$ per cell obtained from the Hall constant, Yb$_4$As$_3$ shows all the signs of a heavy-quasiparticle system at low temper-

---

[198] Y. M. Li, N. d'Ambrumenil, and P. Fulde, *Phys. Rev. Lett.* **78**, 3386 (1997).

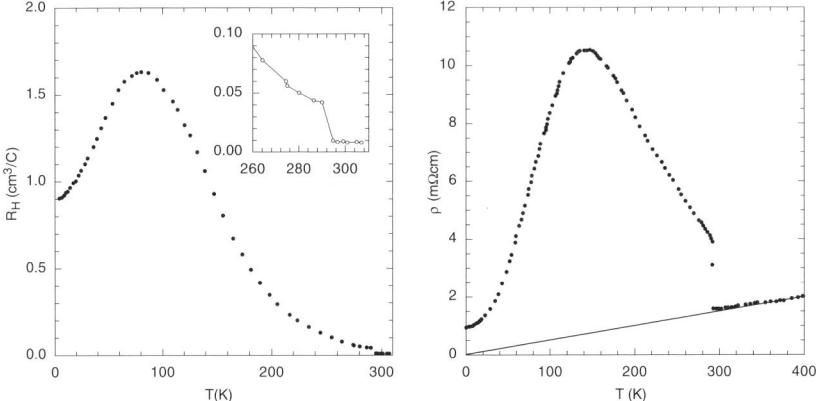

FIG. VI.4. Left panel: Hall coefficient $R_H(T)$ for Yb$_4$As$_3$. The inset shows the change at the phase transition temperature T_c. Right panel: resistivity $\rho(T)$. At $T_c = 295$ K a phase transition due to charge ordering is taking place. Solid line: extrapolation of $\rho(T) \sim T$. (After Ref. [43].)

atures. The γ-coefficient of the specific heat is $\gamma \simeq 200$ mJ/(mol K^2), the Sommerfeld–Wilson ratio is $R_W = 4\pi^2 k_B^2/3(g\mu_B)^2 (\chi/\gamma) \simeq 1$ implying an equally enhanced spin susceptibility and the Kadowaki–Woods ratio A/γ^2 is similar to that of other heavy quasiparticle systems. If one were to postulate an origin of the heavy mass within the Kondo lattice mechanism, this may seem very strange in view of the fact that this semimetal has a very low density of charge carries. The phase transition at $T_c \simeq 300$ K is accompanied by a trigonal distortion and a change of the space group to R3c. This structural transition is volume conserving and is triggered by charge order of the $4f$ holes. The angle between orthogonal axes in the cubic phase changes to $\alpha = 90.8°$ in the trigonal phase for $T \ll T_c$ (Figure VI.5). Associated with this change is a spontaneous elastic strain below T_c which is proportional to the charge order parameter. The structural instability is accompanied by a softening of the c_{44} elastic mode above T_c.[199] This suggests that the trigonal elastic strain $\epsilon_{yz}, \epsilon_{zx}, \epsilon_{xy}$ with Γ_5 symmetry plays a crucial role.

The temperature dependence of the c_{44} mode may be obtained from a Ginzburg–Landau expansion of the free energy in terms of the strains and the charge ordering parameter components (Q_{yz}, Q_{zx}, Q_{xy}).[199] The latter are defined by expanding the charge $\rho = \rho_0 + \Delta\rho$ in the form

$$\Delta\rho = Q_{yz}\rho_{yz}(\Gamma_5) + Q_{zx}\rho_{zx}(\Gamma_5) + Q_{xy}\rho_{xy}(\Gamma_5). \quad (6.3)$$

Here ρ_0 is the part of the charge distribution which remains unchanged by the phase transition while $\rho_{ij}(\Gamma_5)$ are the charge fluctuation modes of Γ_5 symmetry.

---

[199] T. Goto, Y. Nemoto, and A. Ochiai, *Phys. Rev. B* **59**, 269 (1999).

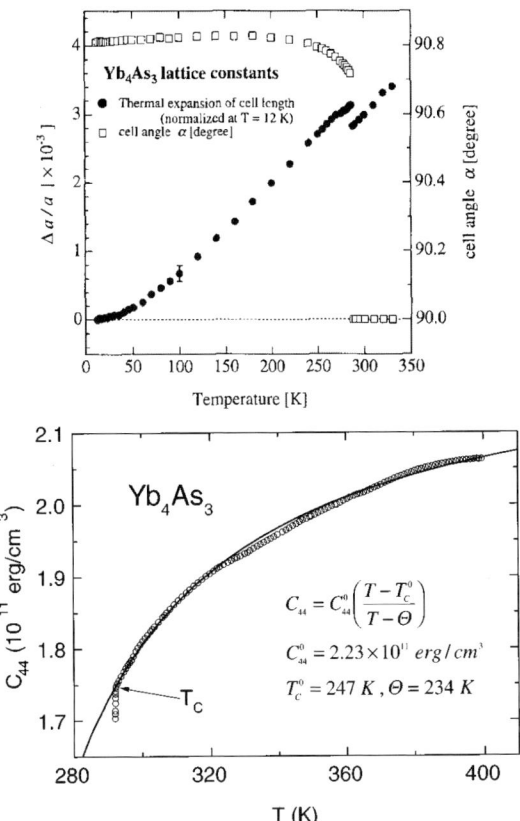

FIG. VI.5. Upper panel: Unit cell angle $\alpha = 90° + \delta$ (open symbols) as function of T showing first order trigonal distortion at the CO transition temperature $T_c \simeq 288$ K together with linear thermal expansion $\Delta a/a$ (full symbols). After Ref. [200]. Lower panel: temperature dependence of the elastic constant $c_{44}(T)$. Above the structural phase transition temperature T_c a strong softening is observed described by Eq. (6.6) and caused by coupling to the Γ_5 type charge order parameter. Due to the first-order nature of the transition the theoretical mean-field transition temperature $T_{c0} = 247$ K is smaller than the actual T_c. (After Ref. [199].)

It is $Q_{ij} = 0$ for $T > T_c$ and $Q_{ij} \neq 0$ for $T < T_c$. Up to a constant the free energy contains three different contributions. One (F_Q) is due to the order parameter, a second one (F_{el}) is due to the elastic energy of the lattice and the third one $(F_{Q\text{-el}})$ describes the interactions of the order parameter with the lattice. For the

---

[200] K. Iwasa, M. Kohgi, N. Nakajima, R. Yoshitake, Y. Hisazaki, H. Osumi, K. Tajima, N. Wakabayashi, Y. Haga, A. Ochiai, et al., *J. Mag. Mag. Mat.* **177–181**, 393 (1998).

c_{44} elastic constant we obtain[201]

$$F_Q = F_0 + \frac{\alpha}{2}(Q_{xy}^2 + Q_{xz}^2 + Q_{yz}^2)$$

$$+ \frac{\beta}{4}\left(Q_{xy}^4 + Q_{xz}^4 + Q_{yz}^4 - \frac{3}{5}(Q_{xy}^2 + Q_{xz}^2 + Q_{yz}^2)^2\right)$$

$$F_{\text{el}} = \frac{c_{44}^0}{4}(\epsilon_{xy}^2 + \epsilon_{xz}^2 + \epsilon_{yz}^2)$$

$$F_{Q\text{-el}} = -g(Q_{xy}\epsilon_{xy} + Q_{xz}\epsilon_{xz} + Q_{yz}\epsilon_{yz}). \tag{6.4}$$

Near a phase transition $\alpha = \alpha_0(T - \Theta)$ changes sign at a characteristic temperature Θ. The fourth-order terms in Q_{ij} stabilize the ordered state of the system. For $T > \Theta$ the softening of the elastic constant is obtained by neglecting the terms $\sim Q_{ij}^4$ in the free energy and minimizing $F = F_Q + F_{\text{el}} + F_{Q\text{-el}}$ by setting $\partial F / \partial Q_{ij} = 0$. For $\beta > 0$ this leads to a trigonal charge order parameter $Q_t = \frac{1}{\sqrt{3}}(Q_{xy}, Q_{xz}, Q_{yz})$. It also leads to a proportionality between Q_{ij} and the strain ϵ_{ij} which may be used to rewrite the free energy in the form

$$F = F_0 + \frac{1}{2}\left(c_{44}^0 - \frac{g^2}{\alpha_0(T-\Theta)}\right)(\epsilon_{xy}^2 + \epsilon_{xz}^2 + \epsilon_{yz}^2). \tag{6.5}$$

Therefore the renormalized elastic constant is

$$c_{44} = c_{44}^0 \left(\frac{T - T_{c0}}{T - \Theta}\right) \quad \text{where } T_{c0} = \Theta + \frac{g^2}{\alpha_0 c_{44}^0} \tag{6.6}$$

denotes the theoretical mean-field transition temperature in the presence of the strain interaction. The explanation of a first-order phase transition at $T_c > T_{c0}$ requires the inclusion of higher order terms in Q_{ij} in (6.4). A detailed group-theoretical analysis of different measured elastic constants is found in Ref. [199]. At temperatures $T > T_c$ the $4f$ holes are equally distributed over the chains. But at low temperatures the system avoids nearest-neighbor Yb^{3+}–Yb^{3+} sites in order to minimize the mutual short-range inter-site repulsion of $4f$ holes. This is accomplished by an accumulation of $4f$ holes in one family of chains, i.e., by charge order. This way nearest-neighbor repulsions of $4f$ holes are reduced. In the idealized case one expects that for $T \to 0$ one family of chains consists of Yb^{3+} sites while in the three remaining families of chains all sites are in a Yb^{2+} configuration. This explains the trigonal distortion (Figure VI.5) which is accompanying charge ordering of the $4f$ holes. Since Yb^{3+} ions have a smaller ionic radius than Yb^{2+} ions the sample is shrinking in the direction of the chains containing the Yb^{3+} ion. In order to keep the volume of the unit cell constant (other-

---

[201] T. Goto and B. Lüthi, *Adv. Phys.* **52**, 67 (2003).

wise a too large amount of elastic energy would be necessary) the remaining three families of chains must expand correspondingly.

It was first pointed out in Ref. [31] that the origin of the heavy quasiparticles are spin excitations (spinons) in the chains of Yb^{3+} ions and that it is not due to the Kondo effect as previously thought. This physical model was beautifully confirmed by inelastic neutron scattering experiments (INS).[32,202] They demonstrated that the magnetic excitations are those of an isotropic antiferromagnetic Heisenberg chain. They agree with the spectrum of two-spinon excitations lying within the lower and upper boundaries $\omega_L(q)$ and $\omega_U(q)$ respectively, given by

$$\omega_L(q) = \frac{\pi}{2} J \sin(q) \quad \text{and} \quad \omega_U(q) = \pi J \sin\left(\frac{1}{2}q\right) \tag{6.7}$$

with q measured in units of inverse lattice constants d^{-1}. The lower boundary $\omega_L(q)$ was calculated long before by des Cloizeaux and Pearson[203] (see Figure VI.6). Note that unlike in the classical spin chain there is no sharp spin wave excitation for a given momentum q but instead a two-spinon continuum. Its corresponding dynamic spin-structure factor (Figure VI.6) is proportional to the INS

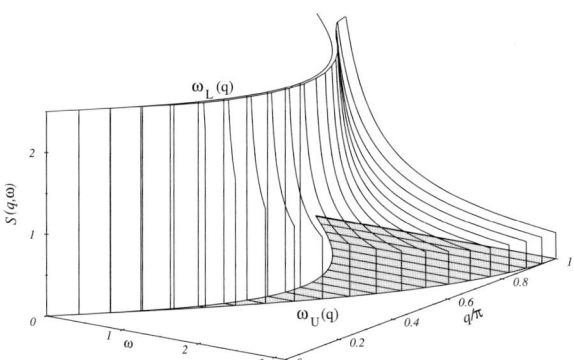

FIG. VI.6. Spectral function $S(q, \omega)$ of the two-spinon excitation spectrum of a Heisenberg chain in the q, ω-plane. It is nonzero in the shaded region above the lower bound $\omega_L(q)$ calculated by Ref. [203] and the below the upper bound $\omega_U(q)$ given by Eq. (6.8). $S(q, \omega)$ diverges at $\omega_L(q)$ approximately with a root singularity. The step-like cutoff at $\omega_U(q)$ is an artefact of the approximate form in Eq. (6.8). The lattice constant is set equal to unity. (After Ref. [204].)

---

[202] M. Kohgi, K. Iwasa, J.-M. Mignot, N. Pyka, A. Ochiai, H. Aoki, and T. Suzuki, *Physica B* **259–261**, 269 (1999).
[203] J. des Cloizeaux and J. J. Pearson, *Phys. Rev.* **128**, 2131 (1962).
[204] M. Karbach, G. Müller, and A. H. Bougourzi, *Phys. Rev. B* **55**, 12510 (1997).

cross-section and may be approximated by[204]

$$S(q,\omega) = \frac{\Theta(\omega - \omega_L(q))\Theta(\omega_U(q) - \omega)}{\sqrt{\omega^2 - \omega_L^2(q)}}. \qquad (6.8)$$

This spectrum diverges at the lower boundary with a square root singularity and therefore has a very asymmetric appearance as function of energy transfer ω (see Figure VI.6). The total integrated intensity, i.e., the frequency integral of $S(q,\omega)$ is linear in q for $q \ll \pi$ and diverges like $[-\ln(1 - q/\pi)]^\alpha$ for $q \to \pi$, where $\alpha = 1$ for the approximation in Eq. (6.8) and $\alpha = 3/2$ is the exact result. Therefore the two-spinon continuum should appear in INS as a spectrum strongly peaked at $\omega_L(q)$ with an asymmetric tail reaching up to $\omega_U(q)$ and a strongly increasing total intensity for $q \to \pi$. This is precisely what has been found in the INS experiments of Kohgi et al. (see Figure VI.7) and constitutes a proof for the 1D character of spin excitations in Yb_4As_3.

That the interacting crystal field ground-state doublets of Yb^{3+} behave like an isotropic Heisenberg system is not immediately obvious. It was shown

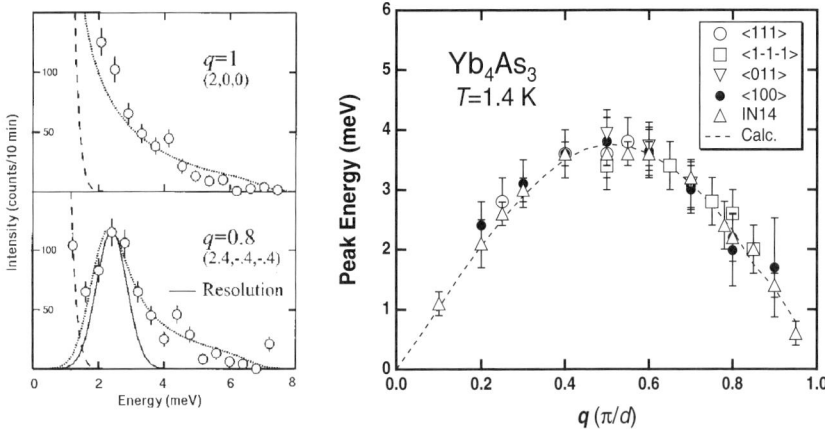

FIG. VI.7. Left panel: Experimental $S(q,\omega)$ from INS[202] for momenta $q = 1$ and 0.8 (here q is given in units of π/d where d is the Yb distance in the chain). The asymmetric shape of the two-spinon spectrum in Figure VI.6 is clearly observed. For $q < 1$ the resolution limited peak position corresponds to the lower boundary $\omega_L(q)$ of the spinon spectrum. Right panel: Dispersion of $\omega_L(q)$ from INS experiments for different directions of the momentum transfer. Here q is the projection of momenta on the $\langle 111 \rangle$ chain direction. All data points fall on the dashed curve which is the theoretical $\omega_L(q) = \frac{\pi}{2} J \sin dq$ with $J/k_B = 25$ K. In addition to the asymmetric shape of $S(q,\omega)$ this proves the 1D character of magnetic excitations in Yb_4As_3. (After Ref. [32,202].)

independently in Refs. [205] and [206]. Thereby the local symmetry of the crystalline electric field was properly accounted for. Spin fluctuations of Heisenberg chains can explain the observed heavy quasiparticle excitations as they appear in the specific heat $C = \gamma T$ and spin susceptibility χ_S. They are given by

$$\gamma = \frac{2}{3} \frac{k_B R}{J} \quad \text{and} \quad \chi_S = \frac{4\mu_{\text{eff}}^2 R}{(\pi^2 J)} \tag{6.9}$$

(see Ref. [207]). Here J is the AF coupling constant of nearest-neighbor sites in the effective $S = \frac{1}{2}$ spin chain described by the 1D Heisenberg Hamiltonian

$$H = J \sum_{\langle ij \rangle} \mathbf{S}_i \mathbf{S}_j. \tag{6.10}$$

Furthermore, R is the gas constant and μ_{eff} is the effective magnetic moment of a Yb^{3+} site. Therefore one finds $R_W = 2$ for the Sommerfeld–Wilson ratio. When the experimental value of $J/k_B = 25$ K (from Figure VI.7) is used in (6.9) the observed size of the γ coefficient is well reproduced.

Having described the underlying physics we discuss a model description for the compound. For that purpose we neglect the electron hopping terms between different chains as it was done, e.g., in the Labbé–Friedel model for A-15 compounds like V_3Si or Nb_3Sn.[208] There one is dealing with three families of intersecting chains while here we have four types of chains. The effective $4f$-model Hamiltonian[31] is then written as

$$H = -t \sum_{\mu} \sum_{\langle ij \rangle \sigma} \left(f_{i\mu\sigma}^+ f_{j\mu\sigma} + h.c. \right) + U \sum_{\mu} \sum_i n_{i\mu\uparrow} n_{i\mu\downarrow}$$

$$+ \epsilon_\Gamma \sum_{\mu} \sum_{i\sigma} \Delta_\mu n_{i\mu\sigma} + \frac{N}{4} c_\Gamma \epsilon_\Gamma^2. \tag{6.11}$$

The first term describes effective $4f$-hole hopping due to hybridization with As $4p$ ligand states within a chain of a family $\mu = 1$–4 from site i to a nearest neighbor site j. From LDA calculations one can deduce that $4t \simeq 0.2$ eV. The second term is due to the on-site Coulomb repulsion of $4f$ holes with $n_{i\mu\sigma} = f_{i\mu\sigma}^+ f_{i\mu\sigma}$ and ensures that in the large U limit Yb^{4+} states with $4f^{12}$ configurations are excluded. The third term describes the volume conserving coupling of the f bands

---

[205] G. Uimin, Y. Kudasov, P. Fulde, and A. Ovchinnikov, *Eur. Phys. J. B* **16**, 241 (2000).
[206] H. Shiba, K. Ueda, and O. Sakai, *J. Phys. Soc. Jpn.* **69**, 1493 (2000).
[207] D. C. Mattis, *The Theory of Magnetism*, vol. 17 of *Springer Series in Solid State Sciences*, Springer-Verlag, Berlin (1981), 2nd ed.
[208] J. Labbé and J. Friedel, *Journ. Phys. (Paris)* **27**, 153, 303 (1966).

to the trigonal strain $\epsilon_\Gamma > 0$ with $\Gamma = \Gamma_5$. It leads to a deformation potential of the form

$$\Delta_\mu = \frac{\Delta}{3}(4\delta_{\mu 1} - 1) \qquad (6.12)$$

for $4f$ holes situated in chains, e.g., in [111] direction denoted by $\mu = 1$. Since the Yb^{3+} ions are smaller than the Yb^{2+} ions, the distance between Yb ions shrinks for $\mu = 1$ while it expands in the other chain directions denoted by $\mu = 2, 3, 4$. As previously pointed out the origin of the deformation potential is the Coulomb repulsion between holes on neighboring sites. It is treated here as an effective attraction V_{eff} between holes on nearest-neighbor sites of a chain. The fourth term in (6.11) is the elastic energy in the presence of a trigonal distortion, where N is the number of sites and c_Γ is the background elastic constant. A reasonable value is $c_\Gamma/\Omega = 4 \cdot 10^{11}$ erg/cm^3 where Ω denotes the volume of a unit cell with a lattice constant of $a_0 = 8.789$ Å.

The Hamiltonians for chains μ with an interaction V_{eff} can be written in the form

$$H_\mu = -t_\mu \sum_{\langle ij \rangle \sigma} f^+_{i\mu\sigma} f_{j\mu\sigma} - V_{\text{eff}} \sum_{\langle ij \rangle} (n_{i\mu} - \bar{n})(n_{j\mu} - \bar{n}) \qquad (6.13)$$

where $\bar{n}$ is the average occupancy of all sites in the system. Within a molecular-field approximation the Hamiltonian reduces to

$$H^{MF}_\mu = -t_\mu \sum_{\langle ij \rangle \sigma} f^+_{i\mu\sigma} f_{j\mu\sigma} - 2V_{\text{eff}}(\bar{n}_\mu - n) \sum_{\langle ij \rangle} (n_{i\mu} - \bar{n}) + \frac{N}{4} V_{\text{eff}}(\bar{n}_\mu - \bar{n})^2. \qquad (6.14)$$

We denote the distinct chains with $\mu = 1$ and note that with the correspondence

$$\frac{4}{3}\epsilon_\Gamma \Delta = -2V_{\text{eff}}(\bar{n}_1 - \bar{n}); \qquad \frac{\Delta^2}{c_\Gamma} = \frac{9}{4} V_{\text{eff}} \qquad (6.15)$$

the Hamiltonians (6.14) and (6.11) become the same. This serves as a justification for the band Jahn–Teller type of description chosen above. When a distortion is taking place the hopping matrix elements also depend on μ, according to

$$t_\mu = t_+ \delta_{\mu 1} + t_-(1 - \delta_{\mu 1})$$
$$t_+ = t e^{\lambda \epsilon_\Gamma}, \qquad t_- = t e^{\lambda \epsilon_\Gamma/3}. \qquad (6.16)$$

But this μ dependence of t is not essential.

The Hamiltonian (6.11) is well suited for describing the lattice distortion caused by charge ordering. The distortion is described here like a collective band Jahn–Teller effect. The four-fold degeneracy of the f bands is lifted by the spontaneous appearance of a trigonal strain which lowers the symmetry. For the purpose of

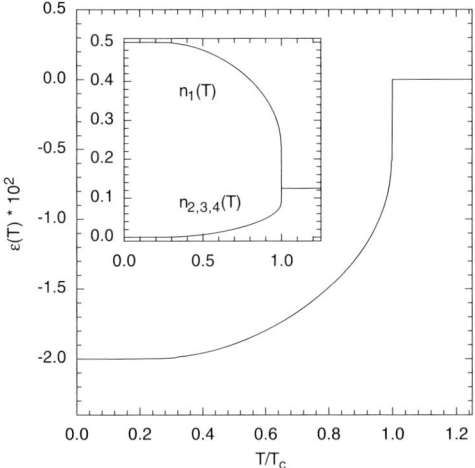

FIG. VI.8. Temperature dependence $\epsilon(T)$ of the (secondary) strain order parameter. Shown in the inset are the corresponding changes in the population of the different families of chains. $Q = n_1 - n_2$ is the primary charge order parameter. (After Ref. [31].)

demonstration we consider first the case of $U = 0$. This neglects the effects of strong on-site correlations on the band Jahn–Teller effect. Near the charge ordering transition where each of the chains contains nearly 25% of Yb^{3+} sites this approximation is justified. However, it is no longer acceptable for low temperatures when nearly all of the sites in the $\mu = 1$ chains are Yb^{3+}. The condition for a collective band Jahn–Teller effect taking place is $\Delta^2/(t\epsilon_\Gamma) > 3$.[31] Choosing $\Delta = 5$ eV results in a transition temperature of $T_c \simeq 250$ K which is close to the experimental value of 300 K. This value of Δ corresponds to a Grüneisen parameter of $\Omega_G \equiv \Delta/(4t) = 25$ which is of a comparable size found in other $4f$-mixed valence systems.

The symmetry strain ϵ_Γ as function of reduced temperature T/T_c is shown in Figure VI.8. As a consequence of a finite strain the four degenerate one-dimensional f bands split into a lower and three upper bands. Their centers of gravity differ by $(4/3)|\epsilon_\Gamma \Delta|$ with the equilibrium strain given by $\epsilon_\Gamma = -\Delta/(2c_\Gamma) \simeq 0.02$. The changes in the population of the chains with $\mu = 1$ and $\mu \neq 1$ with decreasing temperature are shown in the inset of Figure VI.8. The transition is of first order due to the singular DOS of the effective 1D f-bands. Let us reemphasize that those changes are caused by the Coulomb repulsion of holes and drive $\epsilon(T)$ and not vice versa. Band refillings by a band Jahn–Teller transition and crystallization of holes are alternative points of view of the respective descriptions here.

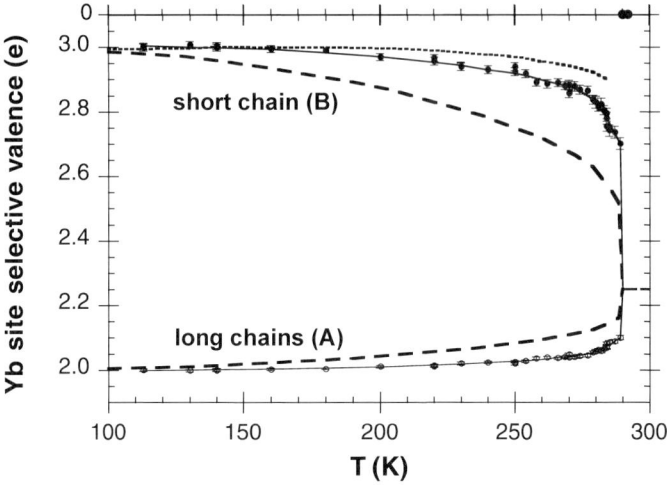

FIG. VI.9. Yb$_4$As$_3$: Site selective values of Yb valence/hole concentration (data points) as obtained from the $30\bar{3}$ reflection in the vicinity of the Yb L$_3$ X-ray absorption edge. Full line: guide to the eye. Dashed line: model calculation from Ref. [198]. Dotted line: from the angular deviation $\delta(T)^{200}$ in Figure VI.5. (After Ref. [209].)

There have been experiments which directly observed one-dimensional charge order.[209] This was achieved, e.g., by resonant X-ray diffraction on the Yb L$_3$ absorption edge. Below T_c forbidden reflections appear in the slightly rhombohedrally distorted cube. From their intensity one can deduce an effective valency for the short and long chains (see Figure VI.9). The discontinuities at the first-order phase transition are even larger than the above model calculation predicts.

We return to the Hamiltonian in Eq. (6.11) and consider the effects of a large on-site interaction U in the charge ordered state for $T \ll T_c$. For this purpose we transform to a t–J Hamiltonian for the chains $\mu = 1$. The exclusion of Yb^{4+} configurations (2 holes) due to U is taken into account by a projector P which projects onto 0 and 1 hole occupations of sites. To lowest order in t/U we obtain

$$H_{\mu=1} = -t \sum_{\langle ij \rangle \sigma} P\left(f_{i1\sigma}^+ f_{j1\sigma} + h.c.\right) P + J \sum_{\langle ij \rangle} \left(\mathbf{S}_i \mathbf{S}_j - \frac{1}{4} n_{i1} n_{j1}\right) \quad (6.17)$$

where $J = 4t^2/U$ and furthermore $\mathbf{S}_i = \sum_{\alpha\beta} f_{i1\alpha}^+ \boldsymbol{\sigma}_{\alpha\beta} f_{i1\beta}$, $n_{i1} = \sum_\sigma n_{i1\sigma}$. For $U = 10$ eV we find $J = 1 \cdot 10^{-3}$ eV. A slight band-narrowing effect due to the

---
[209] U. Staub, B. D. Patterson, C. Schulze-Briese, F. Fauth, M. Shi, L. Soderholm, G. B. M. Vaughan, and A. Ochiai, *Europhys. Lett.* **53**, 72 (2001).

lattice distortion (see Eq. (6.16)) has been neglected here. The Hamiltonian $H_{\mu=1}$ can be treated within the slave-boson approximation. Thereby the projector P is replaced by an auxiliary boson b_j^+ which generates a configuration without a hole, i.e., a Yb^{2+} site. By treating the boson field in mean-field approximation one obtains an effective mass of the quasiparticles

$$\frac{m^*}{m_b} = \frac{t}{t\delta + (3/4)\chi J} \tag{6.18}$$

where m_b is the bare band mass and $\chi = (\frac{2}{\pi})\sin(\frac{\pi}{2}(1-\delta))$. Furthermore, δ is the deviation of the chains $\mu = 1$ from half filling. When $\delta = 0$, i.e., for the half-filled case we find ($J = 4t^2/U$)

$$\frac{m^*}{m_b} = \frac{\pi}{6}\frac{U}{t} = \frac{2\pi}{3}\frac{t}{J} \tag{6.19}$$

which is a large mass enhancement of order 10^2. In this case the heavy quasiparticles are spinons as we are dealing with an antiferromagnetic Heisenberg chain. The low temperature thermodynamics is completely determined by these spin degrees of freedom and we have an example of spin-charge separation here. A de Haas–van Alphen experiment should not yield heavy quasiparticle masses here and the Fermi-liquid relation between the γ coefficient and a renormalized mass of (nearly) free electrons is violated. This becomes obvious when $Yb_4(As_{1-x}P_x)_3$ ($x = 0.3$–0.4)[210,211] is considered. In distinction to Yb_4As_3 this is a charge ordered insulator but it has nevertheless a similar large linear specific heat coefficient γ. The latter is caused by spin excitations in Heisenberg chains as in Yb_4As_3. It is interesting to note that the Hamiltonian (6.11) can be solved exactly by adaptation of Lieb and Wu's solution of the one-band Hubbard model. The solution is based on the Bethe ansatz. This is possible since the different bands are not coupled directly with each other. One interesting finding is that the increase in the strain order parameter just below T_c is much steeper than in Figure VI.8 and therefore in better agreement with experiments. For more details we refer to the original literature.[212]

This brings us to the question: why is Yb_4As_3 at low temperatures still a semimetal and not an insulator? From the present model we would naively expect that at zero temperature Yb_4As_3 is an insulator. Charge order should be complete and a half-filled Hubbard chain is an insulator at sufficiently low temperatures for any positive value of U. In reality this is not the case due to incomplete charge order,

---

[210] H. Aoki, T. Suzuki, and A. Ochiai, unpublished (2005).
[211] A. Ochiai, H. Aoki, T. Suzuki, R. Helfrich, and F. Steglich, *Physica B* **230–232**, 708 (1997).
[212] Y. M. Li, N. d'Ambrumenil, and P. Fulde, *Phys. Rev. B* **57**, R14016 (1998).

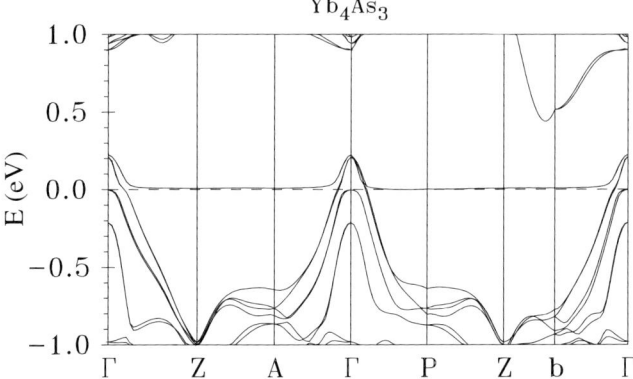

FIG. VI.10. LDA+U energy bands for Yb_4As_3. A small number of As $4p$ holes appears at the Γ-point. The flat band has Yb $4f$-character. (After Ref. [45].)

i.e., the Hubbard chains are not half-filled but doped. Within the model one possible explanation for incomplete charge order would be self-doping. Indeed, as function of U the exact solution of the Hamiltonian (6.11) shows a regime in the c_Γ vs. total electron concentration n_0 plane where charge ordering is incomplete. However, in Yb_4As_3 the true reason for the observed semimetallic behavior is found when bandstructure calculations within the local spin-density approximation plus U approach (LSDA+U) are performed. They show a rigid pinning of the Fermi energy to states close to the top of the As $4p$ band (see Figure VI.10). The filled $4f$ shell of Yb^{2+} is treated as a core shell and the interaction $U = 9.6$ eV between two $4f$ holes was adjusted to obtain the proper number of charge carriers. Because of a large gap between the As $4p$ and Yb $5d$ bands, the $4f$ hole band is pinned to the top of the As $4p$ band. This allows for charge transfer between Yb and As. As a result a small amount of As $4p$ holes is created, which act as mobile charge carriers with a corresponding reduction of Yb $4f$ holes. The calculated cyclotron masses of the almost spherical hole sheet of the As p states are in the range of 0.6 to 0.8 times the free electron mass m_0, in good agreement with the value of $m_{\exp} = 0.72 m_0$ obtained from cyclotron-resonance experiments while a value of $0.275 m_0$ was obtained from Shubnikov–de Haas oscillations.[213] However, the calculated mass should still be renormalized due to electron–phonon interactions. We want to draw attention to the fact that the relation between the large γ coefficient in the specific heat and the mass of the charged quasiparticles responsible for charge transport is lost here. This calls in question one fundamental requirement

---

[213] P. Gegenwart, H. Aoki, T. Cichorek, J. Custers, N. Harrison, M. Jaime, M. Lang, A. Ochiai, and F. Steglich, *Physica B* **312–313**, 315 (2002).

of Landau's Fermi liquid theory. Its basis is a one to one correspondence between the excitations of an interacting electron system and those of a free electron system with renormalized parameters such as the quasiparticle mass. In the case of Yb_4As_3 this correspondence holds only when the charge of the free electrons is renormalized to zero, i.e., when we deal with neutral fermions. This is due to the fact that the low-lying excitations are those of a Heisenberg chain which are of magnetic origin and can be described either by bosons or alternatively by neutral fermions. In addition to the neutral fermionic excitations in the chains which are spinon-like there are charged fermions (the As $4p$-holes), i.e., electrons with a low density of states which provide for charge transport. The large A coefficient in $\rho(T)$ results from scattering of the charged fermions by the neutral ones, while the thermal conductivity is dominated by the neutral ones (spinons). This special feature is amplified when instead of Yb_4As_3 semiconducting mixed crystals $Yb_4(As_{1-x}P_x)_3$ with $x = 0.3$–0.4 are considered. These crystals also exhibit charge order like Yb_4As_3 forming spin chains below the ordering temperature. The Sommerfeld constant is $\gamma = 250$ mJ/(mol K^2) despite the fact that at $T = 0$ there are no charge carriers in the system.[211]

The LSDA+U calculations also show that one must be careful in finding the correct ordered structure by comparing Madelung energies. In competition with the trigonal phase considered here is a cubic $P2_13$ phase with chains of sequence – Yb^{3+}–Yb^{2+}–Yb^{2+}–Yb^{2+}–Yb^{3+}–Yb^{2+}–. The Madelung energy of this structure is slightly lower than the one of the trigonal structure. It does not couple to the Γ_5 strain and therefore cannot explain the observed softening of the elastic constant c_{44}. But when the self-consistent charges are calculated one finds a slight charge disproportionation of 0.05 electrons between the center Yb^{2+} ion and the two adjacent Yb^{2+} ions. This changes the Madelung energy difference in favor of the trigonal structure.

When a magnetic field H is applied to an isotropic antiferromagnetic spin chain ($S = \frac{1}{2}$, $g = 2$), the excitations remain gapless as long as the field remains smaller than $g\mu_B HS = J$ beyond which the system becomes a fully polarized ferromagnet with a Zeeman gap.[214] It was therefore a surprise when it was found that the specific heat of Yb_4As_3 depends strongly on **H**.[215] Experimental results for $\gamma(H)$ are shown in Figure VI.11. Likewise strong anomalies in the thermal expansion were found[215] which is related to the specific heat via the Ehrenfest relation. A number of possible explanations of these observations have been proposed. One is based on magnetic interchain interactions.[216] Another model calcu-

---

[214] J. D. Johnson, *J. Appl. Phys.* **52**, 1991 (1981).
[215] M. Köppen, M. Lang, R. Helfrich, F. Steglich, P. Thalmeier, B. Schmidt, B. Wand, D. Pankert, H. Benner, H. Aoki, et al., *Phys. Rev. Lett.* **82**, 4548 (1999).
[216] B. Schmidt, P. Thalmeier, and P. Fulde, *Europhys. Lett.* **35**, 109 (1996).

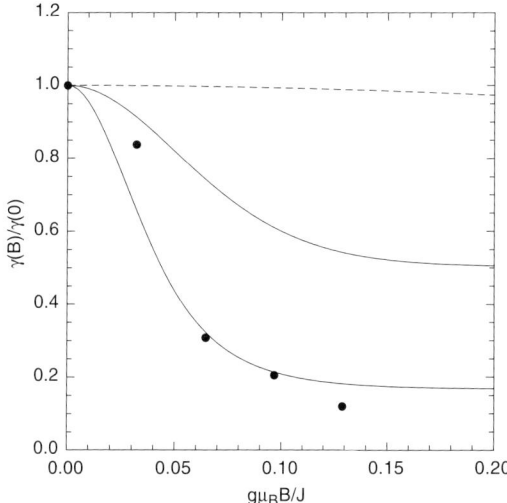

FIG. VI.11. Yb$_4$As$_3$: Experimental field dependence of the specific heat γ coefficient (full circles) at $T \simeq 0.5$ K and theoretical results based on interchain interactions: lower and upper solid lines: $H \perp [111]$ and $H \parallel [111]$. The dashed curve is for $T = 5$ K. A ratio $J'/J = 10^{-4}$ was used. (After Ref. [216].)

lation advocates a staggered field mechanism due to the Dzyaloshinsky–Moriya interaction[206] and a third one suggests that the dipolar interactions within a chain are responsible for the decrease of γ with increasing external field.[205] In all three cases a magnetic field applied perpendicular to the Yb^{3+} chains is opening a gap in the excitation spectrum while the excitations remain gapless when the field is pointing along the chains. But they differ in details, e.g., in the way the gap opens.

When a small interchain coupling J' is assumed between adjacent Yb^{3+} chains a field perpendicular to the chains induces a gap $\Delta \sim (|JJ'|)^{1/2}$. For $J'/J = 10^{-4}$ one obtains a decrease of the γ coefficient with increasing field of the observed size (see Figure VI.11).

The two other models are based on independent Heisenberg chains. The model based on the Dzyaloshinsky–Moriya interaction exploits the fact that the Yb^{3+} sites are not centers of inversion (see Figure VI.12). The local C_3 symmetry together with a glide reflexion and a glide vector parallel to the chains allows for the following features: a uniaxial anisotropy for the symmetric part of the spin–spin interaction and an antisymmetric Dzyaloshinsky–Moriya (DM) interaction. However due to the hidden symmetry they are not independent.[206] A transformation consisting of staggered rotations with angle θ around the chain direction (z) makes this more apparent: The pseudo-spin Hamiltonian for the lowest Yb^{3+}

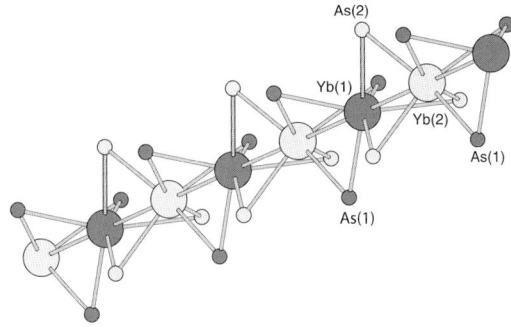

FIG. VI.12. Yb_4As_3: Absence of a center of inversion in Yb^{3+} chains. The distance Yb(1)–As(1) (or Yb(2)–As(2)) is smaller than the distance Yb(1)–As(2) (or Yb(2)–As(1)). Atoms with equal gray scale are equivalent. (After Ref. [206].)

Kramers' doublet, expressed in rotated spin operators, is then given by

$$H_{\text{eff}} = J \sum_{\langle ij \rangle} \mathbf{S}_i \mathbf{S}_j - \sum_i [g_\| S_{iz} H_z + \cos\theta g_\perp (S_{ix} H_x + S_{iy} H_y)$$

$$+ (-1)^i \sin\theta g_\perp (S_{iy} H_x - S_{ix} H_y)]. \quad (6.20)$$

Here the C_3 doublet wavefunction has associated anisotropic g-factors $g_{\|,\perp}$ whose ratio is 2.5.[206] The angle θ is adjustable and in principle determined by the C_3 CEF parameters.

Thus for zero field one has indeed an *isotropic* spin chain ($S = 1/2$) Hamiltonian despite involving strongly anisotropic wave functions of the lowest Kramers' doublet state. This leads to the gapless spinon excitation spectrum discussed previously. When a magnetic field is applied perpendicular to the chain a staggered field perpendicular to both chain and applied field is induced by the DM interaction described by the last term in Eq. (6.20):

$$\mathbf{H}_s = g_\perp \sin\theta [\mathbf{n} \times \mathbf{H}]. \quad (6.21)$$

Here $\mathbf{n}$ is a unit vector in chain direction and $g_\perp = 1.3$ is the corresponding g factor. Due to the staggered field the uniform physical susceptibility $\chi_\perp(0)$ is a mixture of $q = 0$ (uniform) and $q = \pi$ (staggered) susceptibilities χ_{1D} of the ($S = \frac{1}{2}$, $g = 2$) isotropic Heisenberg chain according to

$$\chi_\perp(0) = g_\perp^2 [\cos^2(\theta) \chi_{1D}(0) + \sin^2(\theta) \chi_{1D}(\pi)]. \quad (6.22)$$

Since the latter diverges $\sim 1/T$ at low temperatures this should lead to a Curie-like upturn also in the homogeneous $\chi_\perp(0, T)$. This behavior was indeed found and in principle the value of $\tan^2(\theta) = 0.04$ may be extracted from the analysis of

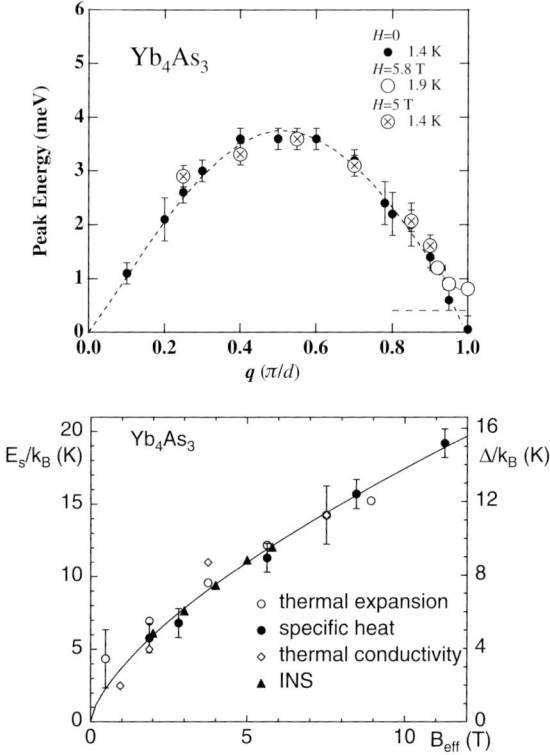

FIG. VI.13. Upper panel: Spin excitations ($\omega_L(q)$) from INS in an Yb^{3+} chain in a magnetic field perpendicular to the chains and without it. Note the gap opening at $q = \pi/d$. After Ref. [217]. Lower panel: Gap as function of applied field. A scaling exponent 2/3 corresponding to quantum sine-Gordon equation is found (full line: $\Delta(H) \sim E_s(H) \sim H^{2/3}$). Here $E_s(H)$ is the gap from thermodynamic and transport quantities and $\Delta(H)$ is the gap obtained in INS. (After Ref. [197].)

the upturn. The most important effect of the staggered field is the appearance of an induced gap Δ in the excitation spectrum with a field dependence

$$\Delta(H) \simeq 1.8 J^{1/3} H_s^{2/3} \sim H^{2/3}. \qquad (6.23)$$

This gap was seen in thermodynamic and transport properties like specific heat, thermal expansion, etc.[215,213] and the field scaling exponent 2/3 was indeed identified (Figure VI.13). A field dependence of this form has also been directly observed by inelastic neutron scattering (INS) on a single crystal at $T = 1.9$ K. A gap

---

[217] M. Kohgi, K. Iwasa, J.-M. Mignot, B. Fåk, P. Gegenwart, M. Lang, A. Ochiai, H. Aoki, and T. Suzuki, *Phys. Rev. Lett.* **86**, 2439 (2001).

was found to open up at $q = \frac{\pi}{d}$ when the field was perpendicular to the short, i.e., Yb^{3+} chains (see Figure VI.13). This supports the above model. In passing we note that the above excitations can also be described by a sine-Gordon equation associated with moving domain walls as applied to Cu-benzoate.[218] But when the magnetic field is parallel to the chain, gapless modes at finite $q = \frac{\pi}{d}(1\pm 2\sigma)$ (σ = magnetization) are expected[219,220] which so far have not been seen and this casts some doubt on the theory.

This brings us to the third model which is based on the weak dipolar interaction of the Yb^{3+} sites. For that purpose the DM interaction is neglected. The dipolar interaction is

$$H_{\text{dip}} = g^2 \mu_B^2 \sum_{i<j} \frac{\mathbf{J}_i \mathbf{J}_j - 3(\mathbf{J}_i \mathbf{e})(\mathbf{J}_j \mathbf{e})}{|\mathbf{R}_i - \mathbf{R}_j|^3} \quad (6.24)$$

where $\mathbf{e} = (\mathbf{R}_i - \mathbf{R}_j)/|\mathbf{R}_i - \mathbf{R}_j|$ and only the ground state doublet of the Yb^{3+} is taken into account. This leads to an effective, slightly anisotropic Heisenberg antiferromagnet in an applied field. In a transverse field a gap opens up.[221] A special feature is that at high fields the gap should close again. Measurements up to $H = 30$ T reveal this effect. Indeed a strong reduction of the gap below the $\Delta(H) \sim H^{2/3}$ scaling curve (Figure VI.13) was observed already above $H = 10$ T.[213] When **H** is parallel to the chains a Zeeman splitting is expected and there should be no incommensurate peaks appearing, in agreement with experiments. In summary it seems that the magnetic field effects are not yet fully understood.

14. CHARGE ORDERING AND 1D SPIN EXCITATIONS IN α'-NAV$_2$O$_5$

In quasi-1D metals such as organic charge transfer salts or the famous KCP chain compound[187] it is known that the ground state exhibits a spontaneous dimerization due to the instability of the 1D Fermi surface. It is signified by a diverging electronic susceptibility at wave number $q = 2k_F$. In real compounds this instability, the 'Peierls-transition' takes place at a finite transition temperature controlled by the interchain-coupling. Amazingly a similar 'spin-Peierls' transition may occur in insulating quasi-1D spin chains with antiferromagnetic nearest neighbor (n.n.) coupling. This is most conveniently understood for an xy-type exchange interaction model which can be exactly mapped by a Jordan–Wigner transformation

---

[218] D. C. Dender, P. H. Hammar, D. N. Reich, C. Broholm, and G. Aeppli, *Phys. Rev. Lett.* **79**, 1750 (1997).
[219] G. Müller, H. Thomas, H. Beck, and J. C. Bonner, *Phys. Rev. B* **24**, 1429 (1981).
[220] H. Shiba, K. Ueda, O. Sakai, and S. Qin, *Physica B* **312–313**, 309 (2002).
[221] D. V. Dmitriev, V. Y. Krivnov, and A. A. Ovchinnikov, *Phys. Rev. B* **65**, 172409 (2002).

(JW) to a model of free spinless 1D fermions at half filling.[222] Naturally this leads to a chain dimerization via the same 'electron'–phonon coupling as in the case of real conduction electrons. Adding the z-term of the Heisenberg exchange interaction one obtains an interacting fermion model after the JW transformation but the basic mechanism of dimerization is unchanged. The spin-Peierls transition has originally been found in a number of organic spin chain compounds and has been theoretically investigated first in Ref. [223] within a mean-field treatment and later in Ref. [224] using the Heisenberg exchange and its coupling to the lattice in the fermionic (JW) representation. Surprisingly the spin-Peierls transition rarely occurs in inorganic 1D-spin chain compounds, presumably because magnetic order due to interchain–exchange is preferred in most cases. The only such spin-Peierls compound known until recently was $CuGeO_3$[225] where the $S = \frac{1}{2}$ spins of Cu-chains undergo dimerization at $T_{SP} = 14$ K. The corresponding dimerization of the exchange integral along the chain then creates an isotropic spin excitation gap. The ensuing isotropic drop in the spin susceptibility below the transition temperature is therefore the most direct method to identify the spin-Peierls mechanism. Actually this is not unambiguous because the presence of n.n.n exchange interactions J' with a ratio $J'/J > 0.24$ where J is the n.n. exchange also leads to a spin gap and this issue is still controversial in $CuGeO_3$.

Therefore the discovery of a structural phase transition in the layered perovskite insulator α'-NaV_2O_5 at $T_c \simeq 34$ K with a corresponding isotropic spin-gap formation[226] has created enormous interest and activity, both experimental and theoretical. This compound was seen as a second candidate for an inorganic spin-Peierls system which seemed to be consistent with the original crystal structure determination[227] that suggested the existence of $S = 1/2$ V^{4+} spin chains along the crystal b-axis (every second V chain in Figure VI.14), isolated by intervening nonmagnetic V^{5+} ($S = 0$) chains. Both belong to V-V ladders formed by the oxygen pyramids, where neighboring ladders are shifted by $b/2$. In the ab-plane this leads to a Trellis lattice structure shown in Figure VI.14. Exchange-dimerization of the $S = 1/2$ V chains due to the spin-Peierls mechanism below T_c would then lead to the isotropic spin gap. This interpretation was however much too naive as became clear subsequently. It turned out that the structural phase transition is not of the simple chain dimerization type but leads to a very low symmetry (monoclinic) structure that can only be understood if in addition a charge

---

[222] G. Beni and P. Pincus, *J. Chem. Phys.* **57**, 3531 (1972).
[223] E. Pytte, *Phys. Rev. B* **10**, 4637 (1974).
[224] M. C. Cross and D. S. Fisher, *Phys. Rev. B.* **19**, 402 (1979).
[225] M. Hase, I. Terasaki, and K. Uchinokura, *Phys. Rev. Lett.* **70**, 3651 (1993).
[226] M. Isobe and Y. Ueda, *Phys. Soc. Jpn.* **65**, 1178 (1996).
[227] A. Carpy and J. Galy, *Acta Crystallogr. B* **31**, 1481 (1975).

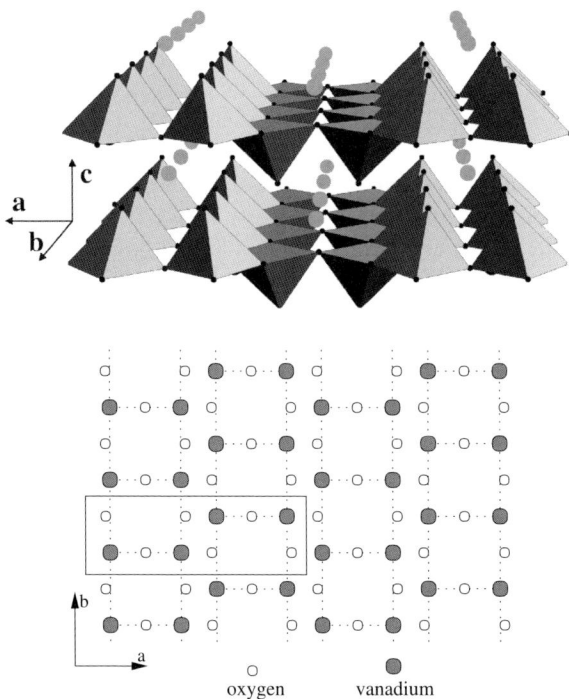

FIG. VI.14. Upper panel: Layered perovskite structure of α'-NaV$_2$O$_5$ consisting of chains of oxygen pyramids that contain the V atoms aligned along the crystal b axis. The layers are stacked along c axis. Na atoms (grey spheres) are centered above the ladder plaquettes. Lower panel: ab-plane Trellis lattice structure of V-V ladders alternatingly shifted along b by half a lattice constant. This leads to a quasi-'triangular' structure for V-V rung units. Orthorhombic high temperature unit cell is indicated.

ordering transition of V^{4+}/V^{5+} is present, starting from a homogeneous high temperature insulating mixed valence (V$^{4.5+}$) state, in contrast to the first assumption in Ref. [227]. Nevertheless above T_c the susceptibility exhibits the typical Bonner–Fisher maximum of 1D spin chains.[226,228] This compound therefore is another example of an intricate relation between charge ordering and quasi-1D spin excitations which in the present case are gapped.

In this section we first discuss the result of the increasingly more detailed X-ray analysis of the high and low temperature structure of α'-NaV$_2$O$_5$, nevertheless no final agreement on the low-T structure has been reached. Electronic structure calculations are essential to construct an effective Hamiltonian for the charge ordering. We will argue that this transition and the ensuing spin

---

[228] P. Horsch and F. Mack, *Eur. Phys. J. B* **5**, 367 (1998).

gap formation requires the inclusion of lattice degrees of freedom which lead to the appearance of two inequivalent V-V ladder types in the perovskite layers. The gapped spin excitation spectrum in the distorted CO phase will be discussed within a simple dimer-RPA approach and a comparison with INS results is given which also advocates the inequivalent ladder model. Finally we briefly mention the stacking of CO layers along c and the destruction of CO by Na deficiency doping or substitution of Na by Li or Ca and in connection with the nature of the insulating state in α'-NaV$_2$O$_5$. In this section we will not review the various alternative theoretical models that have been proposed for α'-NaV$_2$O$_5$. We also will exclude the discussion of most optical experiments, since this is not the ideal method to investigate spin excitations.

The determination of the crystal structure of α'-NaV$_2$O$_5$, both above and below T_c, is fundamental for constructing a microscopic model for the phase transition and spin gap mechanism. This was surprisingly controversial and has led to much confusion, mostly for the low- but even for the high-temperature structure. The latter was originally thought to have non-centrosymmetric symmetry[227] already above T_c leading to the assumption of V^{5+}/V^{4+} in-line charge disproportionation on each V-V rung of the ladders from the outset. However later it was discovered[229] that the proper high temperature structure belongs to the centrosymmetric Pmmn space group. Due to the reflection plane containing y both V-sites in the rungs must be equivalent, i.e., they are in a mixed valence state $V^{4.5+}$. The structure of the low temperature phase ($T \ll T_c$) proved even more controversial. In Refs. [230–232] an orthorhombic space group Fmm2 with a $2a \times 2b \times 4c$ supercell was proposed. In the low temperature phase the modulation (atomic shifts of V, O and Na atoms with respect to high temperature structure) strongly differs for neighboring ladders A and B. Loosely speaking the atomic positions on A are modulated while on B they are not (see Figure VI.17). Subsequently it was shown[233] that the symmetry is even lower, characterized by the monoclinic space group A112 which does not contain an inversion symmetry. This implies the existence of a spontaneous dipole moment (induced ferroelectric order parameter) below the phase transition which has indeed been observed by measuring the dielectric function. The two space groups differ in the number of inequivalent atoms

---

[229] H. G. von Schnering, Y. Grin, M. Knaupp, M. Somer, R. K. Kremer, O. Jepsen, T. Chatterji, and M. Weiden, *Z. Kristallographie* **213**, 246 (1998).

[230] J. Lüdecke, A. Jobst, S. van Smaalen, E. Morré, C. Geibel, and H. G. Krane, *Phys. Rev. Lett.* **82**, 3633 (1999).

[231] S. van Smaalen and J. Lüdecke, *Europhys. Lett.* **49**, 250 (2000).

[232] J. L. deBoer, S. A. Meetsma, J. Baas, and T. T. M. Palstra, *Phys. Rev. Lett.* **84**, 3962 (2000).

[233] H. Sawa, E. Ninomiya, T. Ohama, H. Nakao, K. Ohwada, Y. Murakami, Y. Fuji, Y. Noda, M. Isobe, and Y. Ueda, *J. Phys. Soc. Jpn.* **71**, 385 (2002).

per unit cell, e.g., 6 V sites in two inequivalent layers a, b for Fmm2 and 8 V sites again in two layers for A112. Likewise there are 6 vs. 8 inequivalent Na atoms and 16 vs. 20 inequivalent O atoms in both cases respectively. Nevertheless, the full refinement of the modulated crystal structure with A112 and Fmm2 space group leads to almost identical atomic positions.[233]

This raises the question how the valences, most importantly of V-atoms should be assigned in the low temperature structure. Unfortunately the modulated structure has not yet been investigated within LDA+U calculations. Therefore one has to resort to the empirical valence-bond method.[234] In this approach every atom at an inequivalent site is assigned a formal valence V_i (or oxidation number) which may be thought of as the number of electrons it contributes to all the bonds connected to ligand atoms according to the prescription

$$|V_i| = \sum_j v_{ij} \quad \text{with } v_{ij} = \exp\left(\frac{R_0 - R_{ij}}{B}\right) \quad (6.25)$$

where v_{ij} is the valence of a given central atom-ligand bond which depends on the bond length. Here R_0 and B are empirical parameters characteristic of a given type of chemical bond, e.g., V–O bond, independent of the material. The optimal parameters R_0 and B may however be weakly temperature dependent. Using the sum rule in Eq. (6.25) and identifying valence with the number of electrons contributing to bonds the valence of constituents may be determined. In α'-NaV$_2$O$_5$ we are primarily interested in the number of electrons in the V-d_{xy} orbital because the $3d_{xy}$-bands are well separated from all other bands as seen below. Its orbital occupation is then given by $n_{d_{xy}} = 5 - V_i$, assuming that bonding takes place mainly with the lower lying O3p orbitals. This quantity is the most interesting to study in view of possible charge ordering.

We now discuss a few basic observations that prove the connection between charge ordering and spin gap formation at the structural phase transition. It has been proposed early from thermal expansion measurements[235] that actually there are two separate but close by phase transitions at $T_{c1} \simeq 33$ K and $T_{c2} \simeq 32.7$ K which are of first and second order respectively. These values depend considerably on sample quality and may be higher (see Figure VI.15) than those given above. Detailed investigation of NMR frequencies and Knight shifts have shown[236] that T_{c1} and T_{c2} may be associated with charge ordering on V-sites and a spin-gap formation respectively (Figure VI.15). In the following we will denote the whole

---

[234] J. D. Brown, *Acta Cryst. B* **48**, 553 (1992).
[235] M. Köppen, D. Pankert, R. Hauptmann, M. Lang, M. Weiden, C. Geibel, and F. Steglich, *Phys. Rev. B* **57**, 8466 (1998).
[236] Y. Fagot-Revurat, M. Mehring, and R. K. Kremer, 84, 4176 (2000).

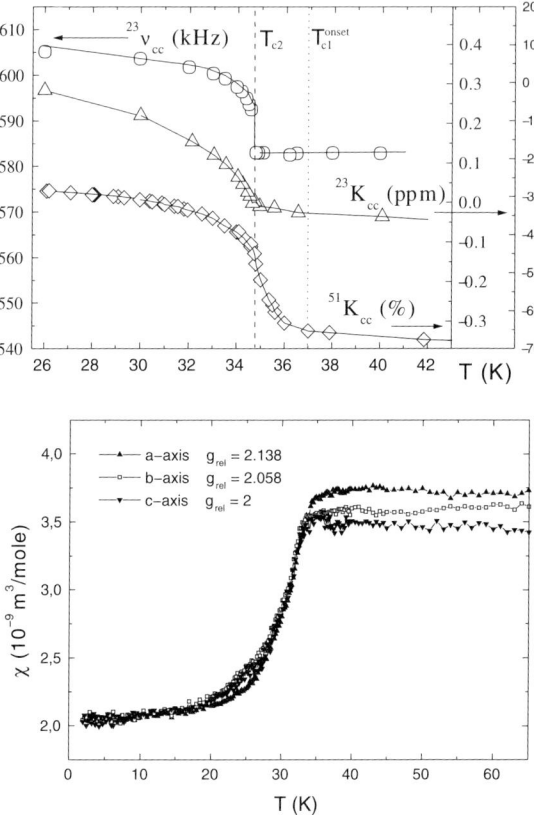

FIG. VI.15. Upper panel; V-Knight shift $^{51}K_{cc}$ and Na-Knight shift $^{23}K_{cc}$ as well as quadrupolar frequency $^{23}\nu_{cc}$ of α'-NaV$_2$O$_5$. The first reflects charge order, the second the opening of the spin gap and the third the lattice distortion around Na-sites. Since $^{23}K_{cc}$ and $^{23}\nu_{cc}$ behave synchronously below T_{c2} this suggests that spin gap formation is connected with exchange dimerization due to Na-shifts. After Ref. [236]. Lower panel: Magnetic susceptibility for single crystalline α'-NaV$_2$O$_5$ showing the isotropic spin gap opening below T_{c2}. (After Ref. [237].)

region of these close transitions simply by 'T_c'. At the lower transition the critical exponent of the spin-gap opening $\Delta_{SP} \sim (1 - T/T_c)^{\beta_\Delta}$ is $\beta_\Delta \sim 0.34$ close to the theoretical value 0.33 obtained from $\Delta_{SP} \sim \delta^{\frac{2}{3}}$ (see Ref. [224]) and a mean-field chain and exchange dimerization behavior $\delta \sim (1 - T/T_c)^{\frac{1}{2}}$. This suggests indeed a spin-Peierls dimerization at the lower transition. The first transition on

---

[237] M. Weiden, R. Hauptmann, C. Geibel, F. Steglich, M. Fischer, P. Lemmens, and G. Güntherod, Z. Phys. B **103**, 1 (1997).

the other hand has a 2D-Ising character, as indicated by a logarithmic peak in the specific heat superposed by the jump of the second transition. Furthermore, according to temperature dependence of X-ray satellites the total lattice distortion has a temperature exponent $\beta \sim 0.2$ closer to the Ising value $1/8$ than to the mean-field exponent. In summary, these findings suggest that the T_{c1}-transition leads to charge ordering on the V-sites and has partly 2D Ising character. Its field dependence is smaller than for an Ising transition but larger than for a pure structural transition. At T_{c1} the main lattice distortion takes place. At the slightly lower T_{c2} the spin gap opens with a spin-Peierls like exponent and connected with an additional exchange dimerization along the charge ordered chains.

Before discussing microscopic models for these peculiar transitions results of the electronic structure calculations for α'-NaV$_2$O$_5$ have to be summarized. So far they were only done for the undistorted structure. Both LDA calculations[238] and spin polarized LDA+U calculations[239] for the charge ordered (but undistorted) case have been performed. In both cases it is found that the planar V d_{xy} bands are well separated from hybridized Vd-Op-bands which are fully occupied or empty as shown in Figure VI.16. In the LDA calculation the Fermi level is centered in the d_{xy} band predicting a metallic state, contrary to the fact that even without CO above T_c α'-NaV$_2$O$_5$ is an insulator. As in the undoped cuprates this is due to a neglect of on-site Coulomb correlations in LDA. In the LDA+U treatment they are simulated by breaking the orbital symmetry, i.e., by using an orbital-occupation dependent one-electron potential

$$V_\sigma^{LDA+U} = V_\sigma^{LSDA} + \sum_\alpha (U - J)\left(\frac{1}{2} - n_{\alpha\sigma}\right)|\alpha\sigma\rangle\langle\alpha\sigma| \qquad (6.26)$$

where U, J are the on-site Hubbard and exchange energy and α denotes the 3d-orbitals. In a charge ordered state the different orbital occupations of inequivalent V-sites cause relative shifts of their orbital energies. As shown in Figure VI.16 this leads to a d_{xy} subband splitting with a charge transfer gap Δ_{CT}. It moves the Fermi level to the top of the lower subband, thus creating a charge transfer insulator in the CO phase. The size of the gap is $\Delta_{CT} \simeq 0.5$–1 eV. A gap of this size was indeed found in optical conductivity measurements.[240] Calculation of the total LDA+U energy for various CO structure favors the zig-zag structure in one ladder,[239] however so far it cannot say anything on the arrangement of adjacent CO ladders or whether only every second ladder exhibits CO as proposed in

---

[238] H. Smolinsky, C. Gros, W. Weber, U. Peuchert, G. Roth, M. Weiden, and C. Geibel, *Phys. Rev. Lett.* **80**, 5164 (1998).

[239] A. N. Yaresko, V. N. Antonov, H. Eschrig, P. Thalmeier, and P. Fulde, *Phys. Rev. B* **62**, 15538 (2000).

[240] C. Presura, D. van der Marel, M. Dischner, C. Geibel, and R. K. Kremer, *Phys. Rev. B* **62**, 16522 (2000).

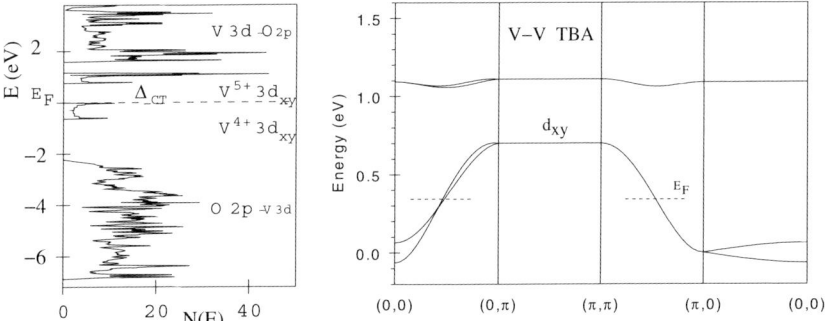

FIG. VI.16. Left panel: LDA+U DOS of α'-NaV$_2$O$_5$ with $U = 4.1$ eV and $J = 1.1$ eV exhibits *two* isolated $3d_{xy}$ subbands due to enforced charge order with a charge transfer gap $\Delta_{CT} \sim 0.5$–1 eV between them. Right panel: Tight binding (TB) fit to LDA bands ($\Delta_{CT} = 0$) with full V-O basis corresponding to effective hopping elements $t_R = 0.380$ eV, $t_L \simeq t_D = 0.085$ eV for the effective V-V TB model with only d_{xy}-states included. Note that within LDA a metallic state is predicted with E_F lying in the lower d_{xy} band. (After Ref. [239].)

the structure model of Ref. [230] and others. The nature of the insulating phase above T_c is an unsolved problem since every rung is only singly occupied, i.e., one has quarter filled V-ladders. Even without CO when Δ_{CT} vanishes there is an excitation gap for double occupancy of V-V rungs. An interpretation as a simple Mott–Hubbard insulator in an effective d_{xy}-one band model based on the molecular orbital (bonding) state of a rung seems inadequate. An interpretation of the insulating state in terms of an extended Hubbard model has been proposed in Ref. [241]. Such microscopic models require as an input the effective hopping parameters obtained from a mapping of LDA band structure to a tight binding model. It was shown in Ref. [239] that an adequate mapping requires a basis of both V d_{xy}- and O p_x, O p_y orbitals. The result is shown in Figure VI.16 for the full basis. Nevertheless in a crude approximation this may be further simplified by mapping to an effective d_{xy} model containing only three hopping parameters t_R, t_L and t_D along the rung, leg and diagonal of a single ladder (inset of Figure VI.17). Their values are given in Figure VI.16. It is found that $t_L \simeq t_D$. This leads to the essentially flat upper band in Figure VI.16 (right) and therefore t_D cannot be neglected. It is also essential to obtain the proper exchange Hamiltonian for the zig-zag CO structure [111] which will be discussed now.

The description of the coupled CO and exchange dimerization in α'-NaV$_2$O$_5$ starts from an extended Hubbard model containing hopping terms discussed before, the on-site effective Coulomb energy $U_{\text{eff}} \simeq 3$ eV and unknown intersite-Coulomb energies V_R, V_L and V_D for the same bonds as the hopping integrals

---

[241] A. Bernert, *Ph.D. thesis*, TU Dresden (2001).

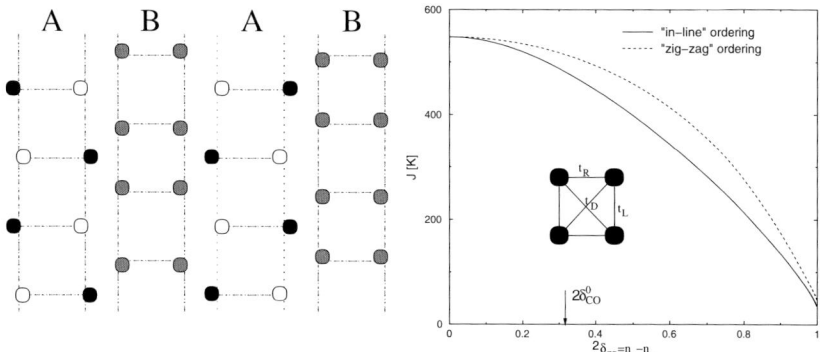

FIG. VI.17. Charge order and dimerization in α'-NaV$_2$O$_5$. Left panel: Shift of V-positions in distorted a-layer below T_{c1}. Stacking along c is of $aaa'a'$ type. In a'-layers A shifts have the same phase and B shifts are moved by one lattice constant along b (ladder direction). Each layer has three inequivalent V (full, open and gray circles), leading to 6 inequivalent V. Charge order (zig-zag) happens mainly on A, ladder dimerization on B. Below T_{c2} the A ladders also dimerize and V sites on B become inequivalent leading to 8 inequivalent V sites altogether. Right panel: Variation of effective spin exchange along the ladder (b-direction) with V charge order parameter $2\delta_{CO}$ for zig-zag CO (A). δ_{CO}^0 is the actual charge order parameter for α'-NaV$_2$O$_5$ on A-ladders corresponding to $J_A = 440$ K (37.9 meV). Inset shows TBA hopping elements t_R (along rung $\parallel a$), t_L (along leg $\parallel b$) and t_D (along diagonal) on a ladder plaquette. The same convention for inter-site Coulomb interactions V_R, V_L and V_D is used. (After Ref. [242].)

(inset of Figure VI.17). For an investigation of CO, this model may be reduced to a much simpler one taking for granted an insulating state with double occupancies of rungs prohibited. Then the charge degrees of freedom are described by a pseudo spin $T = \frac{1}{2}$ where $T_z = \pm\frac{1}{2}$ describes a d_{xy} electron that occupies the left $(-\frac{1}{2})$ or right $(+\frac{1}{2})$ V-atom of a rung.[243] The projected Hamiltonian in the charge sector then has the form[244]

$$H_{\text{ITF}} = \sum_{\ll i,j \gg_L} K_{Lz}^{ij} T_i^z T_j^z + \sum_{\langle i,j \rangle_{IL}} K_{IL}^{ij} T_i^z T_j^z + \sum_i 2\tilde{t}_R^i T_i^x \quad (6.27)$$

which is the Ising model in a transverse field (ITF) whose role is plaid by the renormalized intra-rung hopping $\tilde{t}_R^i$. In the spin sector one has a Heisenberg Hamiltonian with exchange between the spins on neighboring rungs and legs (Eq. (6.30)). Like the Ising interaction 'constants' K_{Lz}^{ij} (intra-ladder) and K_{IL}^{ij}

---

[242] A. Bernert, P. Thalmeier, and P. Fulde, *Phys. Rev. B* **66**, 165108 (2002).
[243] P. Thalmeier and P. Fulde, *Europhys. Lett.* **44**, 242 (1998).
[244] A. Bernert, T. Chatterji, P. Thalmeier, and P. Fulde, *Eur. Phys. J. B* **21**, 535 (2001).

(inter-ladder) $\tilde{t}_R^i$ is related to the original Hubbard parameters which are renormalized by terms that depend on the spin configuration:[242]

$$K_{Lz}^{ij} = 2V_L + \delta K_{Lz}(\mathbf{S}_i, \mathbf{S}_j)$$
$$K_{IL}^{ij} = -V_{IL} + \delta K_{IL}(\mathbf{S}_i, \mathbf{S}_j)$$
$$\tilde{t}_R = t_R + \sum_{\langle ij \rangle_L} \delta \tilde{t}_R(\mathbf{S}_i, \mathbf{S}_j). \qquad (6.28)$$

In this model the charge (**T**) degrees of freedom and spin (**S**) degrees of freedom are coupled through the renormalized interaction constants of Eq. (6.28) that depend on the spin configuration. As a consequence, even above T_c the optical conductivity which probes the (rung) charge excitations also shows a signature of coupled spin excitations.[245] For the moment, considering only the possibility of charge ordering we may freeze the spins in an AF or FM configuration along or between the ladders according to the sign of the spin exchange obtained in LDA+U calculations.[239] This leads to values $\tilde{t}_R = -0.19$ eV and $K_{Lz} \simeq -K_{IL} = 0.68$ eV.[242] One might expect that the Hamiltonian in Eq. (6.27) on the rigid Trellis lattice describes charge order in a natural way, however this is not so obvious. It is true that the ITF on a *single* ladder (first and last term in Eq. (6.27)) has a (doubly degenerate) ground state with staggered 'AF' ($K_{Lz} > 0$) pseudo-spins corresponding to zig-zag charge order along the ladder if the magnitude of the 'transverse field' $\tilde{t}_R$ is smaller than a critical value $4\tilde{t}_R^c = K_{Lz}$. This value defines the quantum critical point $\lambda_c = 1$ of the ITF model[93] with the dimensionless control parameter $\lambda = K_{Lz}/4\tilde{t}_R$. For $\lambda > \lambda_c$, and assuming an infinitesimal staggered field to lift the twofold ground-state degeneracy, the order parameter of zig-zag CO is given by the exact solution of the 1D ITF:

$$\langle T_i^z \rangle = (-1)^i \frac{1}{2} \left[1 - \left(\frac{\lambda_c}{\lambda}\right)^2 \right]^{\frac{1}{8}} = (-1)^i \delta_{CO}(\lambda). \qquad (6.29)$$

However due to the Trellis lattice structure (Figure VI.14) the rungs of a ladder form a trigonal covering lattice and therefore the *inter*-ladder coupling K_{IL} frustrates the zig-zag charge ordering on a given ladder. Therefore the critical $\lambda_c(K_{IL})$ increases monotonously with the inter-ladder coupling which defines a quantum critical line for zig-zag CO at $T = 0$ in the K_{Lz}–K_{IL} plane. At finite temperature the presence of frustration prevents zig-zag CO to occur. If K_{IL} becomes equal to the intra-ladder K_{Lz}, CO melts and for some range of K_{IL} even at $T = 0$ a disordered state appears. Finally if K_{IL} increases even further one obtains again a CO state, but now with in-line order with parallel alignment of pseudo spins

---

[245] M. V. Mostovoy, D. I. Khomskii, and J. Knoester, *Phys. Rev. B* **65**, 064412 (2002).

along a ladder, i.e., V^{4+} configurations are on one side of the ladder and V^{5+} on the other side. These conclusions have been drawn from an exact diagonalization study of the 2D model with finite K_{IL}.[246]

The LDA+U results for α'-NaV$_2$O$_5$ (see Ref. [239]) imply a relation $2V_L - V_{IL} \simeq 0.027$ eV meaning $K_{Lz} \simeq K_{IL}$. Therefore α'-NaV$_2$O$_5$ is indeed close to the quantum critical line for zig-zag charge order and purely Coulombic interactions cannot lead to a phase transition at finite temperature in this compound due to geometric frustration. This is rather similar to the charge ordering in the pyrochlore- or spinel lattices (or their 2D analogue, the checkerboard lattice) where charge ordering is also prohibited by the inherent geometric frustration due to corner sharing tetrahedra of V- or other 3d ions of different valencies (Section VII). In such structures CO requires the lifting of macroscopic degeneracy of the charge configurations by a lattice distortion as is the case in AlV$_2$O$_4$ (see Section VII.17). Something similar happens in α'-NaV$_2$O$_5$. The driving mechanism for the lattice distortion is here the spin superexchange energy between singly occupied rungs along a ladder. Due to the quarter-filled ladders the superexchange does not only contain terms coming from intermediate states with doubly occupied sites but also contributions from rungs with two singly occupied sites. The latter depend on the pseudospin configurations, which is the complementary effect as compared to Eq. (6.28). Therefore the effective spin exchange constants will depend on the degree of charge order. One obtains

$$H_{\text{ex}} = \sum_{i,j} J_{ij} \mathbf{S}_{ij} \mathbf{S}_{i+1j}$$

$$J_{ij} = \left(1 + \sum_\alpha \mathbf{u}^\alpha_{i+\frac{1}{2},j} \nabla^\alpha_{i+\frac{1}{2},j}\right) J\left(T^z_{ij}, T^z_{i+1j}\right)$$

$$J\left(T^z_{ij}, T^z_{i+1j}\right) = J^{ij}_0 \left(1 + f\left(T^z_{ij}, T^z_{i+1j}\right)\right)$$

$$\mathbf{u}_{ij} = \sum_{\lambda q} \frac{1}{(mN)^{1/2}} \exp(i\mathbf{q}\mathbf{R}_{ij})\left(b^{\lambda\dagger}_{qj} + b^\lambda_{qj}\right). \quad (6.30)$$

The summation runs over the atomic shifts $\mathbf{u}_{ij}$ of neighboring atoms ($\alpha = $ O, Na) with respect to V which modulate the superexchange. In addition the latter depends on the charge configuration. Therefore spin ($\mathbf{S}_i$), charge ($\mathbf{T}_i$) and lattice (b^λ_{qj}) degrees of freedom are now coupled via Eq. (6.30). The spin exchange in Eq. (6.30) has to be added to Eq. (6.27) to get the total Hamiltonian H_{ISSP} of the Ising-spin Peierls model for α'-NaV$_2$O$_5$. To simplify the model the $J(T^z_{ij}, T^z_{i+1j})$

---

[246] A. Langari, M. A. Martin-Delgado, and P. Thalmeier, *Phys. Rev. B* **63**, 144420 (2001).

dependence is approximated by $J_{ij} = J_0^{ij}(1 - 4\langle T_j^z \rangle^2)$. This reduction of superexchange (Figure VI.17) with increasing charge order was explained before. The remaining spin-lattice part is treated within the Cross–Fisher theory.[224] The combined spin-Peierls transition and charge ordering takes place at T_{c1}. At that temperature the renormalized frequency of the VO-bond dimerization mode with wave vector $q_0 = \pi/b$ along the ladder given by

$$\tilde{\omega}_{q_0}^2(T) = \omega_{q_0}^2 - 0.26|g_{VO}(q_0)|^2 T^{-1} - 4g_{Is}^2 \chi_{q_0}(T) \tag{6.31}$$

becomes soft, i.e., when $\tilde{\omega}_{q_0}^2(T_{c1}) = 0$. Here the last term is determined by the pseudo-spin susceptibility $\chi_{q_0}(T)$ of H_{ITF} where g_{Is} is a coupling constant that describes the change of the transverse pseudo-spin field with lattice distortion. It may be shown that inclusion of a spin-Zeeman term leads to the proper field dependence of $T_{c1}(H)$ observed in experiment. It is found to be much weaker than for a pure spin Peierls transition. The choice of the precise order parameter and distortion pattern between the ladders below T_{c1} is determined by the minimization of the total free energy. On a given ladder there is a competition between the exchange dimerization energy gain of the spin Peierls distortion and the zig-zag charge order because the former is proportional to J and the latter reduces J as just discussed (Figure VI.17). This can be avoided by generating two *inequivalent* types of ladders A and B below T_{c1} where one (A) is mainly charge ordered and the other (B) mainly dimerized.[242] At the same time this reduces the symmetry of the Trellis lattice in such a way that geometric frustration is reduced and long range charge order is stabilized at a finite temperature. Indeed, X-ray results[230,233] show that the distortion of B-type ladders is much stronger as of A-type ladders, independent of whether orthorhombic Fmm2 or monoclinic A112 space groups are used for the structure refinement (Figure VI.17, left panel). At this stage, slightly below T_{c1} one has a large spin gap on B ladders but none on the zig-zag spin chains on the A ladders. However the rapidly growing charge order $\delta_{CO}(T)$ on A-type ladders below T_{c1} modulates the hopping integrals and orbital energies via Na- and rung O- shifts. This occurs in such a way that an additional but smaller exchange dimerization is also induced on the A-type ladders. It grows with $\delta_{CO}(T)$. Once it is big enough it leads to a further dimerization below T_{c2} on the A-type ladders which is now of the pure spin-Peierls type and opens the spin gap seen in experiment (Figure VI.15). This interpretation is supported by the appearance of additional Na-NMR splittings caused by the shifts of Na atoms on top of the A-ladders for $T < T_{c2}$. They correspond to 8 instead of 6 inequivalent Na positions below T_{c2} which is indeed compatible with the monoclinic A112 structure from X-ray analysis.[233]

Although the mechanism for the phase transitions is intricate, involving charge, spin and lattice degrees of freedom and their competing coupling effects, the low temperature ($T \ll T_c$) spin dynamics is quite simple again. With charge order and lattice distortion saturated only the pure spin part in Eq. (6.30) of the original ISSP Hamiltonian remains. Thereby the existence of two inequivalent A,B-type ladders and their low temperature ($T \ll T_{c1}$) distortions in the parameterization of exchange constants has to be taken into account. Since the B ladders are strongly exchange dimerized with a $\delta_B \sim 0.25$ and $J = J_B(1 \pm \delta_B)$ they have a large spin-excitation gap $\Delta_B = 38$ meV.[242] It is not visible in the inelastic neutron scattering (INS) results which show a minimum excitation energy around 10 meV. It must therefore result from excitations on the more weakly dimerized ($\delta_A \ll \delta_B$) A-type CO ladders. Consequently the low temperature spin Hamiltonian comprises only slightly dimerized 1D zig-zag $S = 1/2$ spin chains (on A) with an intra-chain exchange $J_A(1 \pm \delta_A)$ along the b-direction. Since they are separated by intervening B-ladders the A-ladders are only weakly coupled with $J'_a \ll J_A$ in the transverse a-direction. One therefore would expect gapped quasi-1D spin excitations in α'-NaV$_2$O$_5$. Indeed it was found that their dispersion is much stronger along b than along a.[247,248] As mentioned earlier the stacking of layers along c is of the $aaa'a'$-type.[249] Therefore the zig-zag chains on A are in-phase on aa (and $a'a'$) bilayers and out-of-phase between aa'. In the c-direction one may therefore assume an exchange J_c within the bilayers and neglect coupling between them.

Using this effective exchange model with parameter set (J_A, δ_A, J'_a, J_c), where $J'_a = J_a - 4J_D$ is an effective intra-chain exchange along a, the magnetic excitations have been calculated within a local dimer approach. It is applicable here since $J_A \gg J'_a, J_c$. The susceptibility of an isolated dimer pair in the bilayer is given by

$$u^\pm(\omega) = \frac{2(J_A(1+\delta_A) \mp J_c)}{(J_A(1+\delta_A) \mp J_c)^2 - \omega^2}. \tag{6.32}$$

The collective susceptibility of the bilayers in RPA is then expressed as

$$\chi^\pm(\vec{q},\omega) = \left[1 - J(\vec{q})u^\pm(\omega)\right]^{-1} u^\pm(\omega). \tag{6.33}$$

The dispersion of spin excitations in the ab plane may be obtained from the poles of $\chi^\pm(\vec{q},\omega)$ as

$$\omega_\pm^2(q_x, q_y) = \left[J_A(1+\delta_A) \mp J_c\right]^2 - \left[J_A(1+\delta_A) \mp J_c\right] \\ \times \left[J_A(1-\delta_A)\cos 2q_y \pm \left(J_a \cos(q_x - q_y)\right)\right. \\ \left. - 4J_D \cos q_x \cos q_y\right)\right]. \tag{6.34}$$

---

[247] B. Grenier, L. P. Regnault, J. E. Lorenzo, T. Ziman, J. P. Boucher, A. Hiess, T. Chatterji, J. Jegoudez, and A. Revcolevschi, *Phys. Rev. Lett.* **86**, 5966 (2001).

[248] T. Yosihama, M. Nishi, K. Nakajima, K. Kakurai, Y. Fuji, M. Isobe, C. Kagami, and Y. Ueda, *J. Phys. Soc. Jpn.* **67**, 744 (1998).

[249] K. Ohwada, Y. Fuji, K. Y. J. Muraoka, H. Nakao, Y. Murakami, H. Sawa, E. Ninomiya, M. Isobe, and Y. Ueda, *Phys. Rev. Lett.* **94**, 106401 (2005).

The comparison of mode dispersions with experimental results is shown in Figure VI.18. Using the low temperature $J_A = 440$ K (see Ref. [248]) which corresponds to the charge order parameter $2\delta_{CO} = 0.32$ in Figure VI.17 the remaining parameters may be determined from three of the four observed gaps $\omega^\pm(0, \frac{1}{2})$ and $\omega^\pm(0, 1)$ in Figure VI.18. The fourth (lowest) gap is then correctly calculated as $\omega^-(0, 1) = 8.14$ meV. The exchange dimerization obtained is $\delta_A = 0.03 \ll \delta_B$. The model calculation explains a number of observations from INS:

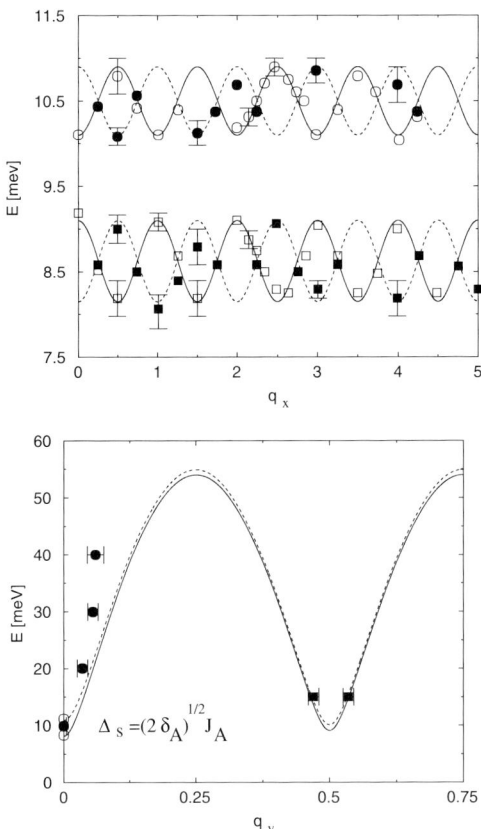

FIG. VI.18. Dispersion of spin excitations in α'-NaV$_2$O$_5$ ($q_x \parallel a$ and $q_y \parallel b$ given in r.l.u. $2\pi/a$ and $2\pi/b$ respectively). Comparison of theoretical fit after Eq. (6.34) with experimental results from Ref. [247]. Full lines: $q_y^0 = \frac{1}{2}$, broken lines $q_y^0 = 1$. Upper panel: $\omega^+(q_x)$-dispersion (top) and $\omega^-(q_x)$-dispersion (bottom). From extremal values at $q_x = 0$ model parameters $\delta_A = 0.03$, $J'_a = 0.21$ meV and $J_c = 0.43$ meV are obtained. Lower panel: Dispersion along b with exchange $J_A = 37.9$ meV (440 K) along zig-zag chain on A. (After Ref. [242].)

(i) The dispersion along b is much larger than along a because $J'_a \ll J_A$ due to weak coupling of A-ladders through intervening B-ladders. (ii) The a-dispersion of $\omega^+(q_x, q_y^0)$ is considerably smaller than for $\omega^-(q_x, q_y^0)$ with $q_y^0 = \frac{1}{2}$ or 1. (iii) Contrary to earlier results[248] the high resolution experiments[247] presented in Figure VI.18 show a finite gap between lower and upper mode $\Delta_{+-} \simeq 1$ meV which is determined by a combination of J_A and J_c. It provides strong evidence for the inequivalent A-B ladder model. If only equivalent A-ladders were present, the gap should vanish, i.e., $\omega^+(q_x, q_y^0)$ and $\omega^-(q_x, q_y^0)$ would touch with their minima and maxima and the dispersion would have twice the observed period along a.[250] We conclude that α'-NaV$_2$O$_5$ presents another example where charge ordering may lead to pronounced 1D character of spin excitations below T_c, here they are gapped due to exchange dimerization caused by the complicated nature of the associated lattice distortions.

It has become clear that CO in α'-NaV$_2$O$_5$ is severely inhibited by effects of geometric frustration. In fact its comparatively low $T_c \simeq 33$ K indicates that it is close to the quantum critical point of the ITF model, where the CO is of essentially 1D Ising type. The true 2D ordered state at finite T is then established by a staggered longitudinal pseudo spin field set up by the distortions of the neighboring ladders. In this scenario it is suggestive that even small perturbations of the 1D Ising spin correlations along the ladder might suppress the CO state in α'-NaV$_2$O$_5$. This can be achieved by reducing the filling factor n of the 1D ladders below $\frac{1}{4}$ by doping with holes. This introduces "empty" rungs into the ladder which cut the Ising bonds. The ensuing destruction of long-range 1D correlations should then strongly reduce T_c as function of the hole concentration δ_h ($n = \frac{1}{4} - \delta_h$). This has indeed been found by introducing holes into the ladders through Na-deficiency doping[251] where a few per cent holes are sufficient to destroy the CO state and the associated spin Peierls transition. Rapid suppression of charge order has also been found in various other doping series, i.e., replacing Na by Li and K (isoelectronic) or Ca (electron-doping).[252] This may be observed directly by specific heat measurements[252] which show a progressive suppression of $\Delta C(T_c)$ with increasing doping. It is also seen in the susceptibility[251] which exhibits a closing of the spin gap associated with CO. Most importantly, in this doping range α'-Na$_x$V$_2$O$_5$ remains an insulator. This is not easy to understand within a Hubbard like model for the quarter filled ladder.[244,242] Possibly 1D localization and polaronic effects play a role, indeed the conductivity was found to exhibit variable-range hopping behavior[251] for hole doping. It was shown in

---

[250] C. Gros and R. Valenti, *Phys. Rev. Lett.* **82**, 976 (1999).
[251] M. Isobe and Y. Ueda, *J. Alloys Compounds* **261–263**, 180 (1997).
[252] M. Dischner, C. Geibel, E. Morré, and G. Sparn, *J. Magn. Magn. Mat.* **226–230**, 405 (2001).

Ref. [253] that even for the (hole-) doped case the CO problem may be treated with an extended pseudo-spin model. In order to incorporate the possibility of empty rungs on a ladder a pseudo spin $T = 1$ is introduced where the $|T_z = \pm 1\rangle$ states describe occupation of the right or left $V d_{xy}$ orbital of a rung respectively, and in addition $|T_z = 0\rangle$ the empty (hole-doped) rung. The total effective $T = 1$ Hamiltonian of a single ladder is then given by

$$H = \sum_i \left[(\epsilon - \mu) T_{zi}^2 + t_R \left(T_{xi}^2 - T_{yi}^2 \right) - h_{si}^0 T_{zi} \right]$$
$$+ 2t_L \sum_{\langle ij \rangle} \left[O_{zy}^i O_{zy}^{j\dagger} + O_{zy}^{i\dagger} O_{zy}^j \right]$$
$$+ V_- \sum_{\langle ij \rangle} T_{zi} T_{zj} + V_+ \sum_{\langle ij \rangle} T_{zi}^2 T_{zj}^2. \quad (6.35)$$

Here ϵ is the on-site orbital energy and $h_s^0(i) = h_s^0(-1)^i$ is a longitudinal staggered pseudo-spin field that simulates the effect of distortion connected with CO and the coupling to neighboring ladders, $\langle ij \rangle$ denotes n.n. rungs along the ladder. Here $O_{zy} = i T_z T_y$ may be interpreted as a quadrupolar operator in the $T = 1$ pseudo-spin space. Furthermore ϵ is the d_{xy} orbital energy and μ, a 'chemical potential' to fix the number of holes $\delta = 1 - n$ (n = number of d-electrons per rung) at a value determined by the doping. Interaction parameters are defined as $V_\pm = \frac{1}{2}(V_L \pm V_D)$. If the ladder diagonal term V_D is neglected then $V_\pm = \frac{1}{2} V_L$ leads to a control parameter $\lambda = 2t_R/V_L$ (to stay in accordance with Ref. [253] the inverse value of the previous definition for λ is used here). In isospin language the (staggered) CO parameter δ_{CO} and the hole doping δ_h are given by

$$\delta_{CO} = \langle T_{z1} \rangle = -\langle T_{z2} \rangle, \qquad \delta_h = 1 - \langle T_z^2 \rangle. \quad (6.36)$$

Here $i = 1, 2$ denote the two 1D sublattices along the ladder direction b. The selfconsistent mean-field solution of the model is shown in Figure VI.19 (lower panel). Its usefulness relies on the Ising type nature of the ordered state. It shows that close to the QCP $\lambda_c = 1$ the charge order parameter is rapidly suppressed with increasing doping which is also evident from the δ_h-dependence of the slope shown in the inset. This behavior corresponds qualitatively to the rapid reduction of CO in Na-deficiency doped α'-Na$_x$V$_2$O$_5$ where the spin-gap and hence the associated charge order is suppressed by a few per cent Na-deficiency $1 - x$. For small doping one may assume that the latter is equal to the average hole concentration δ_h in the V-rungs.

---

[253] P. Thalmeier, *Physica B* **334**, 60 (2003).

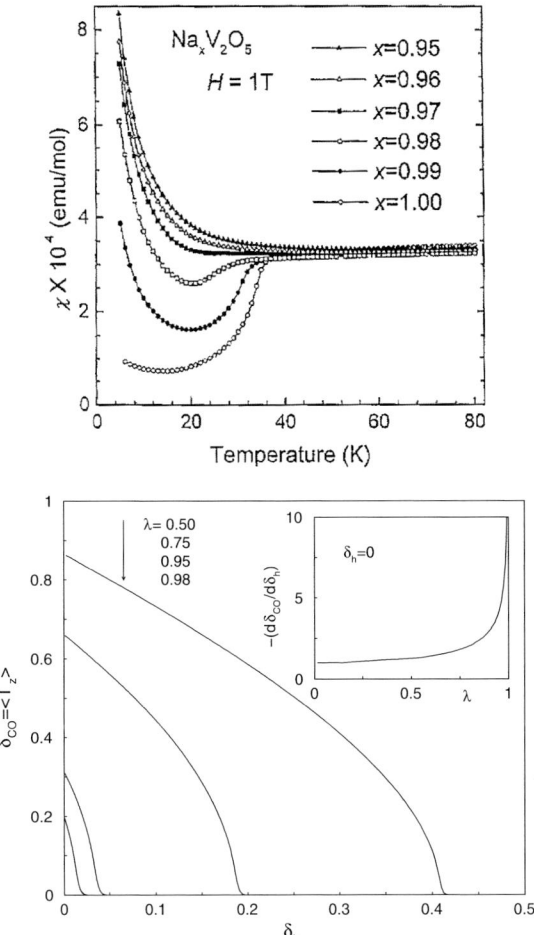

FIG. VI.19. Upper panel: Melting of charge order by Na-deficiency $(1-x)$ (hole) doping seen from the rapid closing of the spin gap in $\chi(T)$ with increasing $1-x$ (corresponding nominally to δ_h). (After Ref. [251].) Lower panel: Calculated melting of CO with hole doping δ_h for various $\lambda = 2t_R/V_L$ given in decreasing order. The absolute slope value $-(d\delta_{CO}/d\delta_h)$ increases strongly when approaching the quantum critical point $\lambda_c = 1$ of CO; this is shown in the inset. (After Ref. [253].)

Finally we briefly discuss the β-vanadium bronzes β-Na$_{0.33}$V$_2$O$_5$ with large but stoichiometric $(1-x = 2/3)$ Na deficiency doping. Their crystal structure is different but it still contains the Trellis lattice layers. For high temperatures these compounds are 1D metals (only in the stoichiometric case) along the b-axis and exhibit a CDW-instability at $T_{CDW} = 136$ K into an insulating state.[254] While the Wigner-lattice type CO transition discussed for α'-NaV$_2$O$_5$ is due mostly to intersite Coulomb interaction energies of localized 3d electrons, the CDW Fermi-surface instability in β-Na$_{0.33}$V$_2$O$_5$ is driven by the kinetic energy of 1D conduction electrons. Under pressure the CDW transition may be suppressed again and around $p_c \simeq 8$ GPa the $T_{CDW}(p)$-line ends in a quantum critical point with an associated superconducting dome around a maximum superconducting T_c of about 10 K.

15. REENTRANT CHARGE ORDERING AND POLARON FORMATION IN DOUBLE EXCHANGE BILAYER MANGANITES La$_{2-2x}$Sr$_{1+2x}$Mn$_2$O$_7$

The layered perovskite manganites have been at the center of $3d$-oxide research since the discovery of the colossal magnetoresistance effect (CMR).[255] Its signature is a change in the resistivity over several orders of magnitude under comparatively small magnetic field changes for doped metallic manganites close to the ferromagnetic phase transition. The investigation of CMR has led to a global survey of doped manganite compounds. Their structures consist of MnO$_6$ octahedra corner-linked to layers that may be stacked in different fashions. As in the cuprates, the parent compounds are AF Mott–Hubbard insulators and hole doping destroys the AF order. However, the metallic state is not superconducting but ferromagnetic. In the manganites the doping with holes only reduces the Mn-moments, rather than creating non-magnetic Zhang–Rice singlets as in the cuprates. The Mn moments then order ferromagnetically via the double exchange (DE) mechanism[256,257] that lowers kinetic energy of e_g conduction electrons when their spins are aligned with spins of localized Mn-t_{2g} spins (see Figure VI.25).

The physics of manganites, especially those derived from the infinite layer parent compound LaMnO$_3$, has been reviewed in many articles, e.g., Ref. [22]. In this chapter we shall focus exclusively on aspects of the bilayer manganites with half-doped insulating compound LaSr$_2$Mn$_2$O$_7$ that are related to charge ordering and possible polaronic effects. In addition we discuss magnetic excitations in the doped metallic ferromagnetic bilayer compounds which give evidence for the double exchange mechanism that is central to the physics of magnetic phases and magnetotransport. This topic has also been more extensively reviewed in

---

[254] K. Okazaki, A. Fujimori, T. Yamauchi, and Y. Ueda, *Phys. Rev. B* **69**, 045103 (2004).
[255] Y. Moritomo, Z. Asamitsu, and Y. Tokura, *Nature* **380**, 141 (1996).
[256] C. Zener, *Phys. Rev. B* **82**, 403 (1951).
[257] P. W. Anderson and H. Hasegawa, *Phys. Rev. B* **100**, 675 (1955).

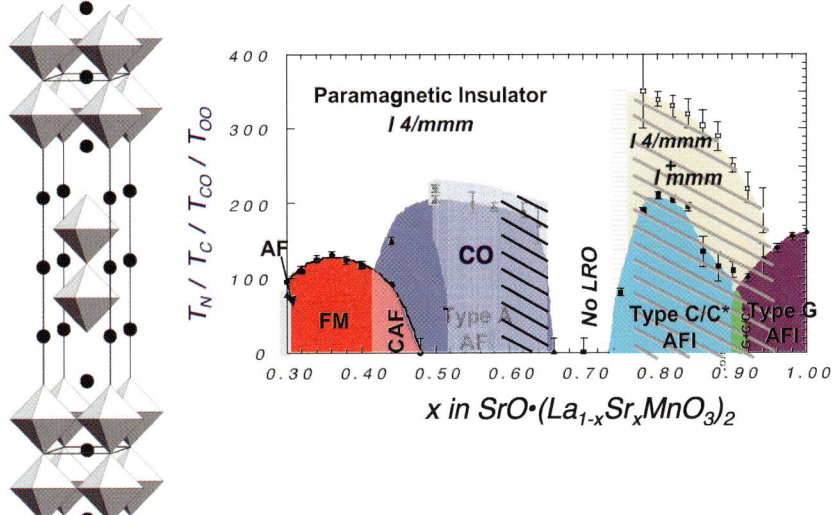

FIG. VI.20. Left panel: Crystal structure of LaSr$_2$Mn$_2$O$_7$ and La$_{2-2x}$Sr$_{1+2x}$Mn$_2$O$_7$ consisting of MnO$_6$ octahedra and (La,Sr)-cations (circles). Lattice constants are ($x = 0.5$) $a = 3.874$ Å, $c = 19,972$ Å. The space group is *I4/mmm*. Right panel: Structural and magnetic phase diagram of bilayer La$_{2-2x}$Sr$_{1+2x}$Mn$_2$O$_7$ in the doping range $0.3 < x < 1.0$. Charge order (CO) appears for $0.5 \leqslant x \leqslant 0.65$. Around $x = 0.7$ magnetic order is absent. (After Ref. [258].)

Ref. [259]. We first give a brief summary of structural properties. The manganites belong to the Ruddlesden–Popper phases which may be described by intergrowth of rock-salt like MnO slabs and n slabs of the perovskite LaMnO$_3$. For $n = \infty$ one has the infinite layer LaMnO$_3$ perovskite, while $n = 2$ corresponds to the bilayer manganite La$_{2-2x}$Sr$_{1+2x}$Mn$_2$O$_7$ considered here. Its structure is shown in Figure VI.20. In the (hypothetical) compound with $x = 0$ La and Sr are tri- and divalent cations which implies a Mn^{3+} ($S = 2$) state for the magnetic cations. Replacing La by Sr according to the chemical formula La$_{2-2x}$Sr$_{1+2x}$Mn$_2$O$_7$ is equivalent to hole doping and creates Mn^{4+} ($S = 3/2$) with a nominal concentration of x holes/Mn site. For $x = 0.5$ one has a stoichiometric mixed-valent compound LaSr$_2$Mn$_2$O$_7$ with a 1:1 ratio of Mn^{3+}/Mn^{4+} ions. This suggests the possibility of charge (and orbital) ordering leading to an insulating state. In the possible concentration range $0.2 < x < 1$ the crystal structure remains the same although the Mn–O bond lengths depend on x due to the Jahn–Teller distortion of

---

[258] X. Qiu, S. J. L. Billinge, C. R. Kmety, and J. F. Mitchell, cond-mat/0307652 (2003).

[259] T. Chatterji, G. Jackeli, and N. Shannon, in *Colossal Magnetoresistance Manganites*, Kluwer Academic Publishers (2004), chap. 8, p. 321.

the octahedrons containing Mn^{3+}. The AF ($x = 0.5$) and FM ($x < 0.4$) structures are shown later in Figure VI.25.

Electronic structure calculations for $LaSr_2Mn_2O_7$ within LSDA+U (see Ref. [260]) lead to a quasi-2D band structure that is close to that of a half metal. The gap between majority and minority spin bands is $\Delta_{\uparrow\downarrow} = 2.7$ eV. The influence of possible charge ordering has been neglected in this calculation. Due to the CEF splitting of $3d$ states into t_{2g} and e_g states, the lower lying t_{2g} bands, which are ~ 1 eV below E_F are almost dispersionless while the bands crossing E_F have mainly e_g character. This allows one to use a simple model for the electronic structure: While the e_g electrons are described in a nearest neighbor (n.n.) tight-binding (TB) approximation, the t_{2g} electrons are treated as localized with the intra-atomic exchange aligning their spins to a total spin $\mathbf{S}$ ($S = 3/2$) (see Figure VI.25). Furthermore, there is a Hund's rule coupling of strength J_H which tries to align localized t_{2g} and itinerant e_g spins. Finally on-site (U) and inter-site (V) Coulomb interaction terms for the e_g electrons have to be added. Then one obtains the total Hamiltonian

$$H = \sum_{ij\sigma} t_{ij}\left(c_{i\sigma}^\dagger c_{j\sigma} + h.c.\right) + U \sum_i n_{i\uparrow}n_{i\downarrow} + V \sum_{\langle ij \rangle} n_i n_j$$
$$- J_H \sum_i \mathbf{S}_i \mathbf{s}_i + J \sum_{ij} \mathbf{S}_i \mathbf{S}_j. \qquad (6.37)$$

Here the first three terms describe e_g conduction electrons ($c_{i\sigma}$) where $n_{i\sigma} = c_{i\sigma}^\dagger c_{i\sigma}$ and $n_i = n_{i\uparrow} + n_{i\downarrow}$, the fourth term describes FM Hund's rule coupling ($J_H > 0$) between e_g-spins ($\mathbf{s}$) and t_{2g} spins ($\mathbf{S}$) and the last one a superexchange between the localized t_{2g} spins. Note that $i = (l, \lambda)$ where $\lambda = 1, 2$ is the bilayer index and l the site within a layer. The model is able to describe both the charge ordering and magnetism in the manganites. The orbital degree $\alpha = 1, 2$ of e_g electrons is still missing. It is too complex to be solved in full generality, we therefore treat CO and magnetic aspects separately and disregard the possibility of simultaneous orbital order.

To investigate charge order as function of the e_g band filling n we take into account only the first three terms which constitute an extended Hubbard model (EHM) for the e_g electrons. Assuming identical CO in both layers the problem reduces to the EHM on a 2D square lattice. Magnetic order will be suppressed and two-sublattice (A,B) charge order is assumed. In the limit $U \gg t > V$ this model may be treated[261] by a combination of Hartree–Fock approximation for the inter-site term (V) and a CPA approximation for the on-site Hubbard term (U).

---

[260] P. K. de Boer and R. A. de Groot, *Phys. Rev. B* **60**, 10758 (1999).
[261] A. T. Hoang and P. Thalmeier, *J. Phys. Condens. Matter* **14**, 6639 (2002).

The latter rests on the alloy-analogy, i.e., the EHM is replaced by a single-particle Hamiltonian with diagonal (sites 1,2 within each layer) disorder of e_g orbital energies $E_{A/B\sigma}^{(1,2)}$ and corresponding probabilities $p_{A/B\sigma}^{(1,2)}$ on each sublattice. Accordingly,

$$E_{A/B\sigma}^{(1)} = zVn_{A/B} \quad \text{with} \quad p_{A/B\sigma}^{(1)} = 1 - n_{A/B-\sigma}$$
$$E_{A/B\sigma}^{(2)} = zVn_{A/B} + U \quad \text{with} \quad p_{A/B\sigma}^{(2)} = n_{A/B-\sigma}. \quad (6.38)$$

The Green's functions of the equivalent single-particle Hamiltonian is then configuration averaged within CPA, i.e., requiring the average T-matrix of the system to vanish. This leads to averaged Green's functions

$$\bar{G}_{A/B}(\mathbf{k},\omega) = \left[\omega - \Sigma_{A/B}(\omega) - \frac{t_\mathbf{k}^2}{\omega - \Sigma_{B/A}(\omega)}\right] \quad (6.39)$$

where $t_\mathbf{k} = (t/2)(\cos k_x + \cos k_y)$. The corresponding self-energies are given by

$$\Sigma_{A/B}(\omega) = \bar{E}_{A/B} - \left[zVn_{B/A} - \Sigma_{A/B}(\omega)\right]\bar{G}_{A/B}(\omega)\left[zVN_{B/A} + U - \Sigma_{A/B}(\omega)\right]. \quad (6.40)$$

Here $\bar{E}_{A/B} = zVn_{B/A} + \frac{1}{2}Un_{A/B}$ are the effective orbital energies on A/B sublattices. On the bipartite lattice CO has to be symmetric which leads to the restriction $n_{A/B} = n \pm n_{CO}$. This requires only a single averaged Green's function defined by $G(\pm n_{CO}, \omega) = \bar{G}_{A/B}(\omega)$. From the above equations the self energies may be eliminated. One obtains a cubic equation for G from which the order parameter n_{CO} may be calculated as function of filling n and Coulomb interaction parameters U, V by requiring charge conservation. Setting $n_{CO} = 0$ the phase boundary between CO and homogeneous phase is obtained as a surface in (n, U, V) space. It is obtained as an implicit solution of the equations

$$n = -\frac{2}{\pi}\int d\omega\, f(\omega)\,\text{Im}\,G(0,\omega) \quad \text{and}$$
$$1 = -\frac{2}{\pi}\int d\omega\, f(\omega)\,\text{Im}\,G'(0,\omega) \quad (6.41)$$

where the derivative is defined as $G'(\omega) = \partial G(n_{CO},\omega)/\partial n_{CO}|_{n_{CO}=0}$ and $f(\omega)$ is the Fermi function. The self-consistent solution for charge order has been determined for the 2D square lattice ($z = 4$ and $W = 4t =$ half bandwidth). The resulting n–V and V–T phase diagrams for CO are shown in the left and right part of Figure VI.21 respectively. In the n–V phase diagram we notice two regions. For $n < n^* \sim 0.67$ the CO boundary $V_c(n)$ is almost independent of U while for $n > n^*$ when the half filled case $n = 1$ is approached CO is strongly suppressed with increasing U. This is due to the fact that in a Mott–Hubbard in-

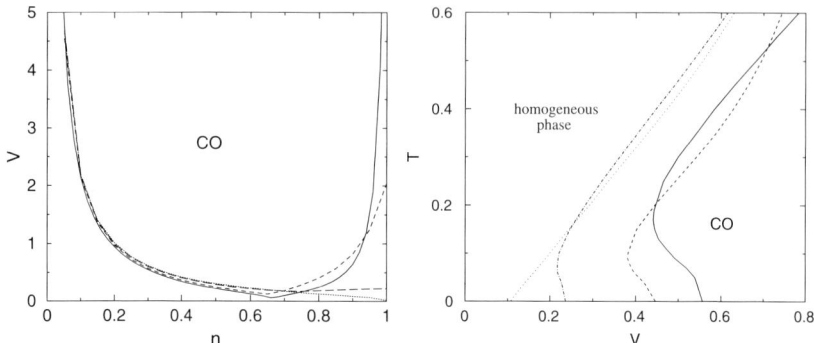

FIG. VI.21. Left: n–V phase diagram of charge order (CO) for the 2D EHM ($T = 0$, $W = 4t \equiv 1$) and $U = 0, 0.5, 1.5, \infty$ corresponding to dotted, long-dashed, dashed and solid lines respectively. Right: V–T phase diagram of CO for $U = 2$ and various band filling $n = 0.3, 0.5, 0.65$ and 0.8 corresponding to solid, dash-dotted, dotted and dashed curves respectively. The CO regime is to the right of the boundary for each value of n. Energies and temperature are given in units of t. (After Ref. [261].)

sulator charge fluctuations are already strongly suppressed and additional spatial symmetry breaking CO then needs a larger threshold value V_c. On the other hand for $n < n^*$, e.g., quarter filling $n = 1/2$ the $V_c \sim 0.25W = t$ is determined essentially by the hopping t. The minimum V_c where CO is most easily achieved is obtained around $n = n^*$.

The V–T phase diagram shows an interesting aspect: For each value of U a range of V values exists for which the CO transition is reentrant as function of temperature. For such a V value the ground state is homogeneous while CO appears in an intermediate temperature range. This behavior cannot be obtained by treating the EHM in Hartree–Fock approximation and thus it is a genuine correlation effect. However, as we shall see, another mechanism based on polaron formation may also lead to reentrant CO. In the present model the absolute value of transition temperatures is unrealistically large because only the n.n. interaction V is assumed. This may be improved by including also a competing n.n.n. Coulomb repulsion which reduces the CO temperature as shown later.

The predicted phase diagram of charge order qualitatively agrees with experimental observations, keeping in mind that due to electron–hole symmetry the calculated phase diagram can be used for the hole doped case of $La_{2-2x}Sr_{1+2x}Mn_2O_7$ ($x = 1 - n$). One must note however that there is no unanimous agreement on doping range and temperature behavior of CO in this compound. Both depend considerably on the experimental method used to detect CO, e.g., X-ray, electron diffraction or transport measurements. In the latter case one

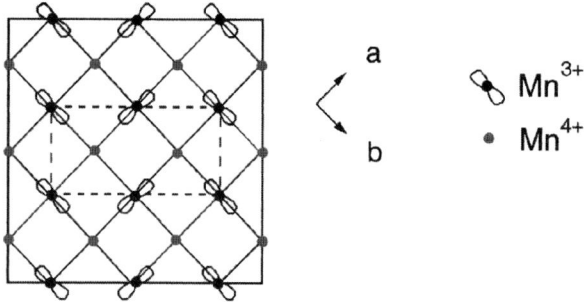

FIG. VI.22. Structure of charge and orbital order in stoichiometric ($x = 0.5$) LaSr$_2$Mn$_2$O$_7$ corresponding to wave vector $\mathbf{Q} = (\frac{1}{4}, \frac{1}{4}, 0)$. (After Ref. [262,263].)

finds CO in a broad range of doping ($0.44 \leqslant x \leqslant 0.8$) at $T_{CO} \sim 200$ K.[264] Another observation is the collapse of CO or reentrance of the homogeneous phase at lower temperatures. Surprisingly a second reentrance of CO appears below 50 K as concluded from an upturn in the resistivity. From diffraction experiments the doping range of CO is somewhat smaller ($0.5 \leqslant x \leqslant 0.65$).[258] For the stoichiometric case $x = 0.5$ (Mn^{3+}/Mn^{4+} = 1) the CO has first been observed in Ref. [262] by X-ray diffraction. If one had only CO one would expect a simple two-sublattice structure with Mn^{3+}/Mn^{4+} ordering corresponding to a commensurate superstructure with $\mathbf{Q} = (\frac{1}{2}, \frac{1}{2}, 0)$. However one rather observes $\mathbf{Q} = (\frac{1}{4}, \frac{1}{4}, 0)$. This is due to additional orbital ordering of Mn^{3+} e_g orbitals in a staggered fashion ($d_{3x^2-r^2}/d_{3y^2-r^2}$) on top of CO (see Figure VI.22). The $\mathbf{Q} = (\frac{1}{4}, \frac{1}{4}, 0)$ superlattice reflexions are then due to the associated JT distortion of the crystal structure. The additional orbital order is not contained in the above single-orbital model. Below 100 K the superstructure reflections vanish, indicating melting of both CO and orbital order as conjectured from transport.[264] Later X-ray experiments on LaSr$_2$Mn$_2$O$_7$ ($x = 0.5$) have shown that there is also evidence for a CO reentrance below 50 K (see Ref. [265]) in agreement with Ref. [264]. This evidence for the second (CO) reentrance is seen in Figure VI.23. The intensity of reflexions in the reentrant CO region below 50 K is however much smaller, indicating much weaker CO. In fact, it seems sample dependent since it was not observed in other experiments. The possible existence of the second CO reentrance cannot be

---

[262] T. Kimura, R. Kumai, Y. Tokura, J. Q. Li, and Y. Matsui, *Phys. Rev. B* **58**, 11081 (1998).

[263] Y. Wakabayashi, Y. Murakami, I. Koyama, T. Kimura, Y. Tokura, Y. Moritomo, Y. Endoh, and K. Hirota, *J. Phys. Soc. Jpn.* **72**, 618 (2003).

[264] J. Dho, W. S. Kim, H. S. Choi, E. O. Chi, and N. H. Hur, *J. Phys. Condens. Matter* **13**, 3655 (2001).

[265] T. Chatterji, G. J. McIntyre, W. Caliebe, R. Suryanarayanan, G. Dhalenne, and A. Revcolevschi, *Phys. Rev. B* **61**, 570 (2000).

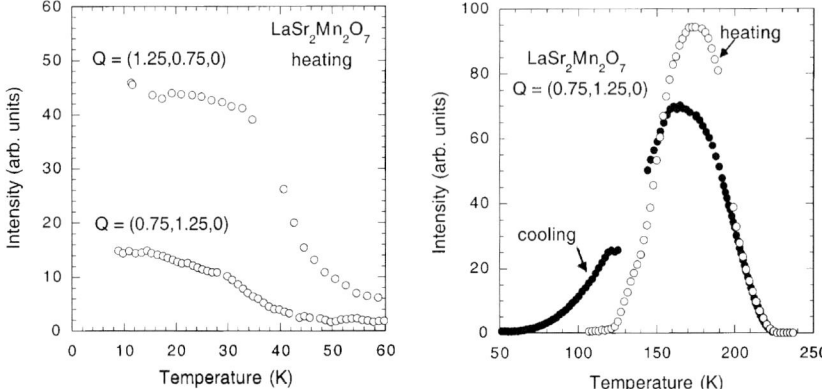

FIG. VI.23. Temperature variation of superlattice reflexion intensities in $LaSr_2Mn_2O_7$ corresponding to wave vector $\mathbf{Q} = (\frac{1}{4}, \frac{1}{4}, 0)$ (cf. inset of Figure VI.24). Right: CO and first reentrance of homogeneous state. Left: second possible reentrance of CO below 50 K. (After Ref. [265].)

explained by the purely electronic EHM model. In fact around $x = 1 - n^*$ there is not even the first reentrance found to the homogeneous phase. Instead the charge order parameter increases monotonously with decreasing temperature as may be inferred from the $V-T$ phase diagram ($n = 0.65$) in Figure VI.21.

We note that Mn^{3+}/Mn^{4+} charge order has also been observed in the infinite-layer manganites, typically around the half-doped ($x = 0.5$) compounds, e.g., in $La_{0.5}Ca_{0.5}MnO_3$ and $R_{0.5}Sr_{0.5}MnO_3$ (R = Pr, Nd). For the Pr-compound CO was observed in a large doping range ($0.3 \leqslant x \leqslant 0.7$) similar as in the bilayer $La_{2-2x}Sr_{1+2x}Mn_2O_7$. The reentrance into the homogeneous phase was also observed in $Pr_{0.5}Sr_{0.5}MnO_3$, again only away from half doping ($x = 0.5$). It becomes especially pronounced in the presence of magnetic fields up to 10 T.[266] There is no evidence for a second CO reentrance in the infinite layer compounds.

To explain these phenomena in the stoichiometric bilayer $LaSr_2Mn_2O_7$ a more extended but still single-orbital based theory is apparently needed. It was proposed in Ref. [267] that in this case CO is profoundly affected by polaron formation caused by a strong coupling to the lattice. This is suggestive since polaron formation changes the ratio of kinetic vs. intersite Coulomb energy of the holes and thus affects the conditions of CO. The electron–lattice coupling is due to the modulation of pd-hybridization along the in-plane bond directions caused by vibrations of oxygen atoms that form corner-sharing octahedra around the Mn ions.

---

[266] Y. Tomioka, A. Asamitsu, H. Kuwahara, Y. Moritomo, and Y. Tokura, *Phys. Rev. B* **53**, R1689 (1996).
[267] Q. Yuan and P. Thalmeier, *Phys. Rev. Lett.* **83**, 3502 (1999).

Effectively this leads to a coupling of the Mn^{3+}-e_g level shift to a bond-stretching vibration of the oxygen. The latter may be assumed dispersionless with a frequency ω_0. We note that this refers to conventional Holstein-type polarons. They differ from the Jahn–Teller (JT) type polarons where the e_g level is split by coupling to the JT symmetry distortions of the whole oxygen octahedron. The latter have been proposed for the slightly doped infinite layer $La_{1-x}Sr_xMnO_3$ in the region $0.1 \leqslant x \leqslant 0.2$.[268] The present Holstein-type model for the $x \simeq 0.5$ bilayer $LaSr_2Mn_2O_7$ is described by

$$H_{e-ph} = g \sum_{i,\delta}(b_{i,\delta} + b^\dagger_{i,\delta})(n_{i+\delta} - n_i) + \omega_0 \sum_{i,\delta} b^\dagger_{i,\delta} b_{i,\delta}. \quad (6.42)$$

Here $b^\dagger_{i,\delta}$ is the local vibration at oxygen site δ associated with Mn site i and g is the coupling constant. This has to be added to the EHM part, i.e., to the first three terms of Eq. (6.37). The latter is also generalized by including both n.n. interactions (V_1) and n.n.n. interactions (V_2) to achieve a realistic T_{CO} which is controlled by the ratio $(V_1 - V_2)/t$. In the limit of strong coupling ($\alpha \equiv g^2/\omega_0^2 \sim 1$) the phonon coordinates may be eliminated by a combined Lang–Firsov (LF) transformation

$$U_1 = \exp\left[-\sum_{i,\delta} g/\omega_0 (b_{i,\delta} - b^\dagger_{i,\delta})(n_{i+\delta} - n_i)\right]$$

(see Ref. [269]), and squeezing transformation

$$U_2 = \exp\left[\gamma \sum_{i,\delta}(b_{i,\delta}b_{i,\delta} - b^\dagger_{i,\delta}b^\dagger_{i,\delta})\right].$$

Here $\gamma > 0$ is a variational parameter[270] determined by minimization of the ground-state energy. This leads again to an effective electronic model, but with renormalized hopping and interaction parameters. Furthermore, the on-site correlation problem is simplified by using the limit $U \to \infty$ where the layers are fully spin polarized and we may assume spinless fermions. For $x = 0.5$ this implies the spinless half-filled band is realized in $LaSr_2Mn_2O_7$. Then, after applying U_2U_1 the effective Hamiltonian reads, up to a constant,

$$H_{\text{eff}} = -\tilde{t} \sum_{i,\delta}(c^\dagger_i c_{i+\delta} + h.c.) + (V_1 + 2\alpha\omega_0) \sum_{i,\delta} n_i n_{i+\delta}$$

$$+ V_2 \sum_{i,\eta} n_i n_{i+\eta}. \quad (6.43)$$

---

[268] Y. Yamada, O. Hino, S. Nohdo, and R. Kanao, *Phys. Rev. Lett.* **77**, 904 (1996).
[269] I. G. Lang and Y. A. Firsov, *Sov. Phys. JETP* **16**, 1301 (1963).
[270] H. Zheng, *Phys. Rev. B* **37**, 7419 (1988).

The first term describes hopping of small polarons. The essential point is that the effective hopping element $\tilde{t} = t \exp[-5\alpha\tau \coth(\omega_0/2T)]$ with $\tau = \exp(-4\gamma)$ is strongly temperature dependent. It is small compared to the bare hopping t when $T \gg \omega_0$ and increases for $T \ll \omega_0$ because the occupation of the n-phonon modes decreases with T eventually until only the effect of zero-point fluctuations in $\tilde{t}$ is left. This T-dependent renormalization of $\tilde{t}$ is very important for charge ordering to take place, since the latter is determined by the balance of kinetic and intersite Coulomb energies. Treating the intersite Coulomb terms for two sublattices A,B in Hartree–Fock approximation as before and choosing $\langle n_i \rangle = 1/2 \pm n_{CO}$, $i \in A, B$, the condition for the CO instability line is

$$1 = \frac{2V}{\pi^2} \int_0^1 \frac{\tanh[2\sqrt{(\tilde{t}z)^2 + (Vn_{CO})^2}/T]}{\sqrt{(\tilde{t}z)^2 + (Vn_{CO})^2}} K(\sqrt{1-z^2})\, dz \quad (6.44)$$

where K is the complete elliptic function of the first kind and

$$z = \sqrt{\varepsilon^2 - (4Vn_{CO})^2}/4\tilde{t}$$

with $V = V_{12} + 2\alpha\omega$, $V_{12} = V_1 - V_2$. For $T \to 0$ this equation has always a solution with $n_{CO} > 0$. For a reasonable parameter set (t, ω_0, V_{12}) its solution leads to the α–T phase diagram shown in Figure VI.24. Depending on the size of α, both the non-reentrant and double-reentrant scenarios are possible. There is an intermediate region for $\alpha \sim 0.65$ where on lowering the temperature one obtains

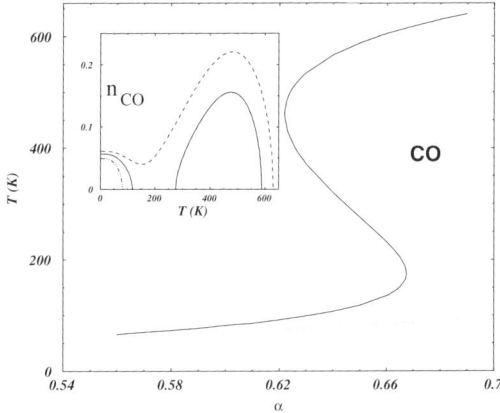

FIG. VI.24. α–T phase diagram for the spinless (half-filled) case of the Holstein–EHM model with parameters $V_{12} = 0.1$, $\omega_0 = 0.05$ in units of $t = 4 \cdot 10^3$ K. For comparison, the thin line corresponds to a $\tilde{t}$ which is kept constant (cf. Figure VI.23). The inset shows $n_{CO}(T)$. The dotted, solid and dashed lines correspond to the three possible types of diagrams with $\alpha = 0.60, 0.65$ and 0.68 respectively. (After Ref. [267].)

a CO phase, then a first reentrance into the homogeneous phase and subsequently a second reentrance into the CO phase. The corresponding variation of the order parameter $n_{CO}(T)$ is shown in the inset of Figure VI.24 (full line). The other possible cases are also illustrated. Physically the behavior around $\alpha \sim 0.65$ may be explained as follows. First the effective hopping $\tilde{t}$ is reduced from the bare value and polaronic CO appears at relatively high temperatures. With decreasing temperature the kinetic energy ($\sim \tilde{t}$) increases until CO can no longer be maintained and the homogeneous state is stable again. For even lower temperatures CO always reappears because the n.n. tight binding Fermi surface has a nesting instability to CO for the half filled (spinless) case for arbitrary small values of V. The reentrance behavior discussed here is entirely due to the temperature dependence of the effective polaron hopping $\tilde{t} = t \exp[-5\alpha\tau \coth(\omega/2T)]$. Without it only the lowest CO transition does occur. This agrees with the previous study where no reentrance was observed for quarter filling including spin or spinless half filling. In the discussion of the reentrance behavior in $La_{2-2x}Sr_{1+2x}Mn_2O_7$ we have so far invoked electronic correlation effects and polaron formation as origin but neglected the influence of coexisting magnetic order. A treatment of the full problem of CO would require the inclusion of spin degrees of freedom for finite values of U together with the last two exchange terms in Eq. (6.37) and the electron–lattice coupling. This is still an open problem.

In the last part of this chapter we therefore focus on a rather complementary case: When the doping is large enough Coulomb correlation effects are less prominent and CO is absent for all temperatures. Then only the exchange terms in Eq. (6.37) are important. Double exchange (DE) illustrated in Figure VI.25 favors FM order within and between layers by optimizing kinetic energy gain for parallel e_g ($s = 1/2$) conduction and t_{2g} ($S = 3/2$) localized spins. In contrast superexchange (last term in Eq. (6.37)) favors AF orientation of moments between layers leading to the AF order in Figure VI.25 for $x = 0.5$. For $0.4 < x < 0.5$ its competition with FM double exchange along the c-axis leads to a canting of moments out of the plane[271] (CAF region in Figure VI.20) which may be interpreted as a superposition of an AF and FM structure. For $x \leqslant 0.4$ the e_g conduction bands become increasingly two-dimensional with $d_{3z^2-r^2}$ character. Therefore double exchange along c increases significantly as discussed below. Consequently only the FM structure with FM bilayers remains and the last term in Eq. (6.37) may be neglected. The less than half-doped ($x \leqslant 0.4$) metallic $La_{2-2x}Sr_{1+2x}Mn_2O_7$ compounds are then described by the ferromagnetic ($J_H > 0$) Kondo-lattice Hamil-

---

[271] K. Hirota, S. Ishihara, H. Fujioka, M. Kubota, H. Yoshizawa, Y. Moritomo, Y. Endoh, and S. Maekawa, *Phys. Rev. B* **65**, 64414 (1998).

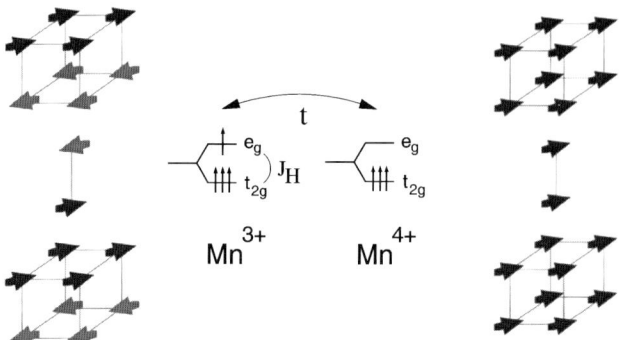

FIG. VI.25. Left: AFM structure of stoichiometric ($x = 0.5$) CO insulator LaSr$_2$Mn$_2$O$_7$ consisting of AF stacked FM layers along c. Right: FM structure of metallic ($x = 0.4$) La$_{2-2x}$Sr$_{1+2x}$Mn$_2$O$_7$. Moments are parallel to [110] in both cases. Center: Illustration of FM double exchange mechanism in doped manganites. FM polarization of itinerant e_g spins ($s = 1/2$) leads to kinetic energy gain ($\sim t$) due to FM Hund's rule coupling ($\sim J_H$) to localized t_{2g} spins ($S = 3/2$).

tonian

$$H_K = -t \sum_{\langle ij \rangle \lambda \alpha} c^\dagger_{i\lambda\alpha} c_{j\lambda\alpha} - t_\perp \sum_{i\alpha} \{c^\dagger_{i1\alpha} c_{i2\alpha} + h.c.\}$$

$$- J_H \sum_{i\lambda\alpha\beta} \mathbf{S}_{i\lambda} \cdot c^\dagger_{i\lambda\alpha} \mathbf{s}_{\alpha\beta} c_{i\lambda\beta}. \quad (6.45)$$

Here $t, t_\perp > 0$ are the nearest neighbor hopping parameters within a layer and between the two partners ($\lambda = 1, 2$) of a bilayer along c respectively. Furthermore $\mathbf{S}$ and $\mathbf{s}$ are localized and conduction electron spins respectively with $\alpha, \beta = \uparrow, \downarrow$. In La$_{2-2x}Sr_{1+2x}Mn_2O_7$ the lattice constants ($x = 0.4$) are $a = 3.87$ Å and $c = 20.14$ Å. The intra-bilayer splitting $d \simeq a$ is much smaller than the distance $D = 6.2$ Å between adjacent bilayers (Figure VI.20). Therefore inter-bilayer hopping or exchange has been neglected above. Diagonalization of the hopping term leads to bonding and antibonding tight binding bands split by $t_\perp$ with identical in-plane dispersion and a DOS given by

$$\epsilon_{\sigma\pm}(\mathbf{k}) = -\left[\pm t_\perp + t(\cos k_x + \cos k_y)\right]$$
$$N_\pm(\epsilon) = N(\epsilon \pm t_\perp);$$

$$N(\epsilon) = \frac{2}{\pi^2} \frac{1}{W} K\left(\left[1 - \left(\frac{\epsilon}{W}\right)^2\right]^{\frac{1}{2}}\right). \quad (6.46)$$

The justification for this model and the size of its parameters can be obtained by considering the spin wave excitations below the Curie temperature T_C. They have been investigated by inelastic neutron scattering by various groups[271–274] and analyzed in Refs. [275,276] both in the classical limit and with quantum corrections. In the manganites the condition $J_H \gg t, t_\perp$ is fulfilled because the Hund's rule coupling $J_H \sim 2$ eV is quite large. This greatly simplifies spin wave calculations because firstly the e_g bands $\epsilon_{\sigma\pm}(\mathbf{k})$ will be spin-split such that only majority bands are occupied, i.e., $n_{\uparrow\pm} = 1$ and $n_{\downarrow\pm} = 0$ and secondly the large $S = 3/2$ local t_{2g} spins allow for a $1/S$ expansion. This approach has been first applied to the cubic manganites[277] and later used for the bilayer manganites.[275,276] It turns out that in the limit $J_H \to \infty$, and to order $1/S$ one obtains classical spin waves of an effective Heisenberg model for t_{2g} spins. The effective exchange constants are determined by the e_g conduction band dispersion and filling. One obtains

$$\omega_A(\mathbf{q}) = zJ^{DE}S[1 - \gamma_q]$$
$$\omega_O(\mathbf{q}) = zJ^{DE}S[1 - \gamma_q] + 2J_\perp^{DE}S. \quad (6.47)$$

In two dimensions $\gamma_q = \frac{1}{2}(\cos q_x + \cos q_y)$ and $\omega_A(\mathbf{q})$ and $\omega_O(\mathbf{q})$ are the dispersion of acoustic (A) and optical (O) spin wave branches respectively. They have equal dispersion in the ab-plane and are split by the A-O gap $\Delta_{AO} = 2SJ_\perp^{DE}$. In this approximation there is no spin-space anisotropy and therefore the A branch has no gap at $\mathbf{k} = 0$. The effective Heisenberg exchange constants J^{DE}, $J_\perp^{DE}$ for t_{2g} spins are given by

$$J^{DE} = -\frac{1}{2S^2}\frac{1}{2z}(\epsilon_0 + \epsilon_\pi) \quad \text{with } \epsilon_{0,\pi} = \int_{-W}^{\epsilon_F \pm t_\perp} N(\epsilon)\epsilon\, d\epsilon$$

$$J_\perp^{DE} = \frac{1}{2S^2}\frac{t_\perp}{2}(n_0 - n_\pi) \quad \text{with } n_{0,\pi} = \int_{-W}^{\epsilon_F \pm t_\perp} N(\epsilon)\, d\epsilon. \quad (6.48)$$

The effective t_{2g} exchange parameters J^{DE}, $J_\perp^{DE}$ are therefore completely determined by the e_g band parameters in the limit $J_H \to \infty$. This surprisingly simple result for the spin waves in the double exchange ferromagnet is in good agreement with inelastic neutron scattering experiments on the $x = 0.4$

---

[272] T. Chatterji, P. Thalmeier, G. J. McIntyre, R. van de Kamp, R. Suryanarayanan, G. Dhalenne, and A. Revcolevschi, *Europhys. Lett.* **46**, 801 (1999).
[273] T. Chatterji, L. P. Regnault, P. Thalmeier, R. Suryanarayanan, G. Dhalenne, and A. Revcolevschi, *Phys. Rev. B* **60**, R6965 (1999).
[274] T. G. Perring, D. T. Adroja, G. Chaboussant, G. Aeppli, T. Kimura, and Y. Tokura, *Phys. Rev. Lett.* **87**, 217201 (2001).
[275] T. Chatterji, L. P. Regnault, P. Thalmeier, R. van de Kamp, W. Schmidt, A. Hiess, P. Vorderwisch, R. Suryanarayanan, G. Dhalenne, and A. Revcolevschi, *J. Alloys Compounds* **326**, 15 (2001).
[276] N. Shannon, T. Chatterji, F. Ouchni, and P. Thalmeier, *Eur. Phys. J. B* **27**, 287 (2002).
[277] N. Furukawa, *J. Phys. Soc. Jpn.* **65**, 1174 (1996).

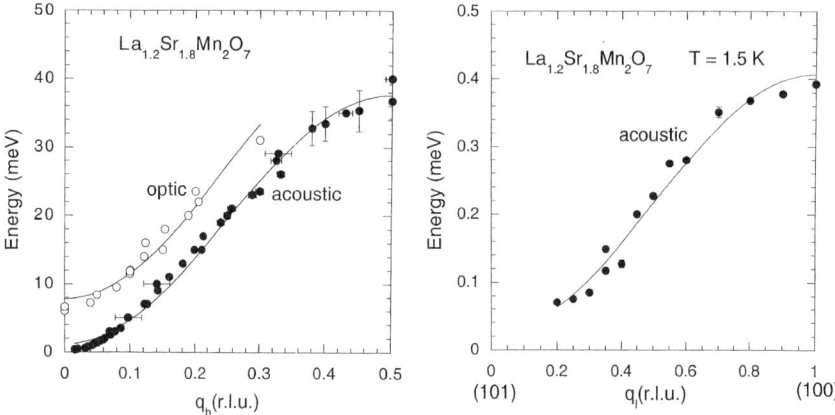

FIG. VI.26. Left panel: Acoustic and optic spin wave branches along [100] direction with a splitting $\Delta_{AO} = 6$ meV and overall dispersion of 40 meV. Full lines are obtained from the classical dispersion expression in Eq. (6.47). Right panel: Acoustic spin wave dispersion along [001]. An extrapolated anisotropy gap $\Delta_A(0) = 0.04$ meV is obtained. (After Ref. [275].)

$La_{2-2x}Sr_{1+2x}Mn_2O_7$. This is seen in the left panel of Figure VI.26. Two parallel A and O modes are observed with a maximum zone-boundary energy of $\omega_A(q_x = \frac{\pi}{a}) \sim 40$ meV and an A-O splitting of $\Delta_{AO} = 6$ meV. From these values one obtains $SJ^{DE} = 10$ meV and $SJ_\perp^{DE} = 3$ meV. Using Eqs. (6.48) this leads to e_g band parameters $t = 0.175$ meV and $t_\perp = 0.1$ eV.

In the present theory there is no double exchange between adjacent bilayers and hence no spinwave dispersion along the c-axis. But experimentally a small dispersion along c was found (Figure VI.26, right panel) although it is two orders of magnitude smaller than the one along the a-axis. This requires an inter-bilayer DE constant $J_\perp'^{DE}$ with $J_\perp'^{DE}/J_\perp^{DE} \simeq 1.5 \cdot 10^{-2}$ (see Ref. [272]) and shows that $La_{2-2x}Sr_{1+2x}Mn_2O_7$ is an almost ideal two dimensional double exchange ferromagnet. This conclusion was supported by an analysis of diffuse neutron scattering which exhibits long range FM in-plane correlations far above the Curie temperature ($\sim 2.3T_C$).[278] Finally, Figure VI.26 shows that a small extrapolated anisotropy gap $\Delta_A(0) \sim 0.04$ meV for the acoustic mode exists at the zone center whose microscopic origin is not clear.

The double exchange model based on the FM Kondo lattice Hamiltonian is able to describe ferromagnetism and basic properties of spin wave excitations quite well. However it has its limits. Firstly quantum corrections of order $1/S^2$ can-

---

[278] T. Chatterji, R. Schneider, J. U. Hoffmann, D. Hohlwein, R. Suryanarayanan, G. Dhalenne, and A. Revcolevschi, *Phys. Rev. B* **65**, 134440 (2002).

not be completely neglected as discussed in Refs. [259,276] and references cited therein. They lead to two effects: (i) reduction of the overall dispersion of spin waves. In the effective Heisenberg model this would necessitate a rescaling of the parameters J^{DE}, $J_{\perp}^{DE}$ or possibly an inclusion of more parameters; (ii) the local spin moment is coupled to density fluctuations in the itinerant system which leads to damping effects not present in the classical $(1/S)$ approximation. Comparison with experiment shows that $(1/S^2)$ corrections provide still insufficient damping and also cannot explain the observed deviations from the classical **q**-dependence of spin waves.[276] It has also been proposed that a crossing with a phonon branch might be involved in these anomalies. Another shortcoming of the model is the neglect of orbital degeneracy of e_g states. This has indeed dramatic effects on the doping dependence of the A-O spin wave splitting and the resulting effective exchange constants in the FM regime $x = 0.3$–0.4 as shown in Ref. [274]. When x decreases from the present value $x = 0.4$ the effective $J_{\perp}^{DE}$ strongly increases while J^{DE} stays almost constant. In the above double exchange model the anisotropy ratio is given by

$$\frac{J_{\perp}^{DE}}{J^{DE}} = -\left(\frac{t_{\perp}}{t}\right)\frac{W(n_0 - n_\pi)}{\epsilon_0 + \epsilon_\pi}. \qquad (6.49)$$

Assuming a doping independent $t_{\perp}/t$ the above ratio changes at most by $\sim 10\%$ in the doping range $x = 0.3$–0.4, which is much too small to explain experimental observations. Therefore another mechanism must be invoked. It has been found[279] that for decreasing x the MnO_6 octahedra elongate significantly along the c axis due to the JT effect on e_g orbitals which leads to a lower energy and hence larger occupation for $d_{3z^2-r^2}$ orbitals as compared to $d_{x^2-y^2}$ orbitals. Since the former have larger overlap along the c axis, the effective $t_{\perp}$ strongly increases with decreasing x. Thus it is really the prefactor $(t_{\perp}/t)$ in the above equation which leads to the dramatic increase of $J_{\perp}^{DE}$ and the A-O spin wave splitting with decreasing x. This effect was described phenomenologically in Ref. [276] and within a microscopic model in Ref. [280].

In our discussion of bilayer manganites we have focused on models for charge ordering which can be described as simple periodic superstructures caused by inter-site but short range Coulomb interactions supplemented by small polaron formation. We have also discussed the importance of the double exchange mechanism for explaining the spin excitations in the metallic ferromagnet away from half-doping. We have mostly neglected the complications of orbital order, JT distortions and the effect of longer range Coulomb interactions in the low doping

---

[279] M. Kubota, H. Fujioka, K. Hirota, K. Ohoyama, Y. Moritomo, H. Yoshizawa, and Y. Endoh, *J. Phys. Soc. Jpn.* **69**, 1606 (2000).
[280] G. Jackeli and N. B. Perkins, *Phys. Rev. B* **65**, 212402 (2002).

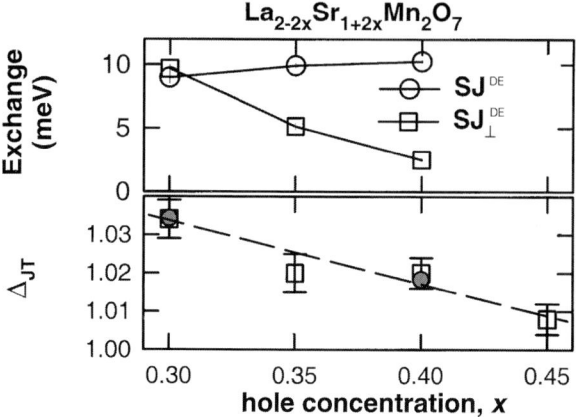

FIG. VI.27. Upper panel: Doping dependence of effective exchange parameters. Note that $J_{\perp}^{DE}$ is proportional to A-O spin wave splitting and J^{DE} to the overall dispersion. Lower panel: Doping dependence of JT distortion Δ_{JT} (elongation along c) of MnO_6 octahedra. (After Ref. [274].)

regime. This may lead to the important possibility of an inhomogeneous state due to phase separation of ferromagnetic metallic and charge ordered insulating regions. Such states may consist of metallic droplets or stripes of holes in an insulating environment. This possibility has been reviewed in Ref. [281]. These aspects may also be of great importance for explaining the giant magnetoresistance of the manganites.[22]

VII. Geometrically Frustrated Lattices

Usually the concept of frustration is used in connection with magnetic systems. When Ising spins with an antiferromagnetic interaction are placed onto certain lattices like a triangular one, the pair-wise interactions cannot be satisfied simultaneously and therefore are frustrated. Here we will associate the concept of frustration with lattice structures. We call a lattice geometrically frustrated when in case that its sites are occupied by antiferromagnetically coupled Ising spins the interactions are frustrated. Examples are in two dimensions the just mentioned triangular, the checkerboard or the Kagomé lattice. In three dimensions the pyrochlore lattice (see Figure VII.1) is the one most frequently investigated. For those lattices we want to study charge degrees of freedom, i.e., when the electron number at a lattice site is fluctuating. A frustrated

---

[281] A. Moreo, S. Yunoki, and E. Dagotto, *Science* **283**, 2034 (1999).

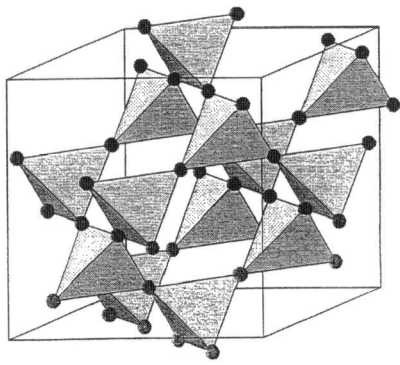

FIG. VII.1. Pyrochlore lattice consisting of corner-sharing tetrahedra.

lattice structure can have a degenerate ground state for special band fillings when electron correlations are strong. In fact, in the limit of large on-site and nearest-neighbor electron repulsions there exists an exponentially large number of configurations with minimal potential energy. This is particularly so when the number of electrons equals half the number of lattice sites. It is not surprising that this special feature which is closely related to a frustrated geometry leads to new theoretical models and special effects when the electrons are strongly correlated. In the center of our attention will be the above mentioned pyrochlore lattice and a two-dimensional projection of it, the checkerboard lattice.

A pyrochlore lattice is a substructure of the spinels which have the composition AB_2O_4. They can be considered as face-centered cubes of O^{2-} ions. The B ions are surrounded by an octahedron of O^{2-} ions, i.e., BO_6 and are positioned on corner-sharing tetrahedra which define the pyrochlore lattice. Here we want to consider metallic spinels in which the electrons are itinerant. They may undergo metal to insulator transitions which are usually accompanied by a structural distortion. The most studied example has been magnetite Fe_3O_4. A transition to an insulator already indicates that electron correlations may be strong in spinels but an unambiguous proof is the observed heavy-fermion behavior of LiV_2O_4 at low temperatures.[282] Another interesting case is AlV_2O_4. This material is either a poor metal or a semiconductor at low temperatures. It undergoes a structural phase transition at lower temperatures which apparently is caused by strong electron correlations. Finally, $LiTi_2O_4$ is a metallic spinel which becomes supercon-

---

[282] S. Kondo, D. C. Johnston, and L. L. Miller, *Phys. Rev.* **59**, 2609 (1999).

TABLE VII.1. SPINELS WITH A HALF-INTEGER VALENCY OF d IONS

M =	Ti	V	V(Cr)	Mn
Li(Al)M$_2$O$_4$	LiTi$_2$O$_4$	LiV$_2$O$_4$	AlV$_2$O$_4$ (LiCr$_2$O$_4$)	LiMn$_2$O$_4$
average d-electron count per M-atom	$d^{0.5}$	$d^{1.5}$	$d^{2.5}$	$d^{3.5}$

ducting at a relatively high transition temperature of $T_c = 13.7$ K.[283,284] Table VII.1 summarizes these materials with half-integer valency of the cations.

In the following we start out with a reminder on Fe$_3$O$_4$ for which a huge literature does exist. For references see, e.g., Ref. [285]. The purpose is to merely recall some of the basic facts in order to understand better the special features of strong electron correlations in LiV$_2$O$_4$, AlV$_2$O$_4$ and other spinels. This is followed by a discussion of fractional charges. They are found when a model Hamiltonian describing strongly correlated electrons with strong on-site and nearest-neighbor repulsions is used and applied to a frustrated lattice.

Magnetite Fe$_3$O$_4$ has been much investigated because of the important role it has played in the development of magnetism and magnetic materials. It is a spinel of the form AB$_2$O$_4$ with A = Fe^{3+} and B = Fe$^{2.5+}$ sites. We assume that 50% of the B sites are in a Fe^{2+} and Fe^{3+} configuration each, i.e., that electron correlations are so strong that Fe$^+$ or Fe^{4+} configurations are suppressed. Magnetite undergoes at 120 K a phase transition from a metallic high-temperature phase to an insulating low-temperature phase. This transition was first observed by Verwey and Haayman[8] and is usually referred to as Verwey transition. Verwey presented also a model for its description in terms of an order-disorder transition, which is entropy driven. The implicit assumption regarding B sites is thereby that the repulsions of electrons on neighboring sites are so strong that the kinetic energy term of the electrons plays only a minor role and may be neglected when the phase transition is considered. With an average valence of the B sites of +2.5 this implies that two neighboring Fe^{2+} sites and also two Fe^{3+} sites repel each other, while Fe^{2+}–Fe^{3+} sites attract. Let us denote by $V_{\alpha\beta}$ the interaction between two neighbors Fe$^{\alpha+}$–Fe$^{\beta+}$. An order-disorder phase transition will take place when

$$\delta V = V_{33} + V_{22} - 2V_{23} > 0. \qquad (7.1)$$

Verwey suggested a particular charge ordering for the insulating low-temperature phase in which the Fe^{2+} and Fe^{3+} sites of the pyrochlore lattice order in form of

---

[283] D. C. Johnston, *J. Low Temp. Phys.* **25**, 145 (1976).
[284] R. W. McCallum, D. C. Johnston, C. A. Luengo, and M. B. Maple, *J. Low Temp. Phys.* **25**, 177 (1976).
[285] M. Isoda and S. Mori, *J. Phys. Soc. Jpn.* **69**, 1509 (2000).

two families of chains pointing in the [110] and [1$\bar{1}$0] direction, respectively. However, the situation is more complicated than that. First one should realize that in the absence of electron hopping the ground state is highly degenerate. In order to minimize the Coulomb interactions $V_{\alpha\beta}$, two of the sites of each tetrahedron of the pyrochlore structure must be occupied by a Fe^{2+} and two by a Fe^{3+} ion (tetrahedron rule). There is an exponentially large number of different configurations which satisfy this rule.[286] When a small hopping of electrons is taken into account this degeneracy is partially lifted. How that is taking place remains an unsolved problem. It is possible that the electronic ground state would remain disordered, i.e., liquid like as long as the lattice is unchanged. In that case charge order could result from a structural distortion which is accompanying the metal-insulator transition. Indeed, the experimental determination of the electronic low-temperature phase has been a challenging and controversial subject as has been a proper theoretical interpretation. A recent review[287] describes the development of different theoretical models starting from a Hubbard-like Hamiltonian including nearest-neighboring repulsions[288] up to inclusion of electron–phonon interactions together with the tetrahedron rule.[289] Calculations in the frame of density functional theory based on LDA+U and the observed low temperature lattice structure produce charge order which agrees with estimates based on a valence bond analyses.[290] They also conclude that the tetrahedron rule is not strictly fulfilled. This is due to the kinetic energy, i.e., electronic hopping terms which lead to violations of that rule. Nevertheless, for a simple reason the tetrahedron rule must be satisfied to a high degree. Without considerable short-range order it would be difficult to understand the relatively low transition temperature of the Verwey transition. For a conventional order–disorder phase transition the transition temperature is given by $T_V = 2\delta V/k_B$, where k_B is Boltzmann's constant. This would imply a T_V of several thousands of Kelvin.[286]

16. METALLIC SPINELS: LiV_2O_4 — A METAL WITH HEAVY QUASIPARTICLES

In order for a spinel oxide to be conducting, the electron count of the B ions in AB_2O_4 should differ from an integer number. The compounds listed in Ta-

---

[286] P. W. Anderson, *Phys. Rev.* **102**, 1008 (1956).
[287] F. Walz, *J. Phys. Cond. Matt.* **14**, R285 (2002).
[288] J. R. Cullen and E. Callen, *J. Appl. Phys.* **41**, 879 (1970).
[289] D. Ihle and B. Lorenz, *Phil. Mag.* **42**, 337 (1980).
[290] I. Leonov, A. N. Yaresko, V. N. Antonov, M. A. Korotin, and V. I. Anisimov, *Phys. Rev. Lett.* **93**, 146404 (2004).

ble VII.1 are therefore of special interest. As mentioned before LiTi$_2$O$_4$ with $d^{0.5}$ per Ti ion is a superconductor with a transition temperature of $T_c = 13.7$ K. LiV$_2$O$_4$ with $d^{1.5}$ per V ion is a metal with heavy quasiparticle excitations. At ambient pressure no spin- or charge order has been observed down to the lowest temperature. The compound LiCr$_2$O$_4$ is not stable and therefore AlV$_2$O$_4$ with $d^{2.5}$ per V ion is particularly interesting. This system becomes a charge ordered insulator by so-called valence skipping (see the next section). Finally LiMn$_2$O$_4$ with $d^{3.5}$ per Mn ion is an antiferromagnetic insulator with a Néel temperature of $T_N = 280$ K. Charge ordering is taking place in that material which has been used in batteries. In the following we will discuss LiV$_2$O$_4$ in more detail.

As pointed out above LiV$_2$O$_4$ shows at low temperatures heavy-fermion behavior, i.e., it supports heavy quasiparticle excitations.[282,291] It has been the first system where the heavy quasiparticles originate from d electrons. Experiments show that the γ coefficient of the low temperature specific heat $C = \gamma T$ is strongly enhanced and of order $\gamma \simeq 0.4$ J mol^{-1} K^{-2}. The spin susceptibility is equally enhanced at low T. Over a large temperature regime it shows a behavior

$$\chi_S = \chi_0 + \frac{C}{T + \Theta}, \quad \Theta = 63 \text{ K} \tag{7.2}$$

i.e., of Curie–Weiss type. The Sommerfeld–Wilson ratio $R_W = \pi k_B^2 \chi_S(T = 0)/(3\mu_B^2 \gamma)$ is found to be $R_W = 1.7$. The temperature independent term $\chi_0 = 0.4 \cdot 10^{-4}$ cm^3/mol and the Curie constant is $C = 0.47$ cm^3 K/(mol V). The sign of Θ indicates antiferromagnetic interactions between V sites but no magnetic ordering was found down to 4.2 K. In view of the frustrated lattice this is understandable. The resistivity is found to be $\rho(T) = \rho_0 + AT^2$ with a large coefficient $A = 2\mu\Omega$ cm K^{-2}. The Kadowaki–Woods ratio A/γ^2,[292] a hallmark of heavy quasiparticles is in the range of other heavy-quasiparticle materials.[293] These findings are typical signatures of heavy fermion systems. From the specific heat data one may determine the entropy $S(T)$. One finds that $S(T = 60$ K$) - S(T = 2$ K$) = 10$ J/(mol · K) which is close to $2R \ln 2$ where R is the gas constant. The implication is that at 60 K the system has almost one excitation per V ion. This is inconceivable with a conventional band description of the d electrons. According to Pauli's principle only a small fraction of them is participating in the excitations

---

[291] S. Kondo, D. C. Johnston, C. A. Swenson, F. Borsa, A. V. Mahajan, L. L. Miller, T. Gu, A. I. Goldman, M. B. Maple, D. A. Gajewski, *Phys. Rev. Lett.* **78**, 3729 (1997).
[292] K. Kadowaki and S. B. Woods, *Solid State Comm.* **58**, 507 (1986).
[293] C. Urano, M. Nohara, S. Kondo, F. Sakai, H. Takagi, T. Shiraki, and T. Okubo, *Phys. Rev. Lett.* **85**, 1052 (2000).

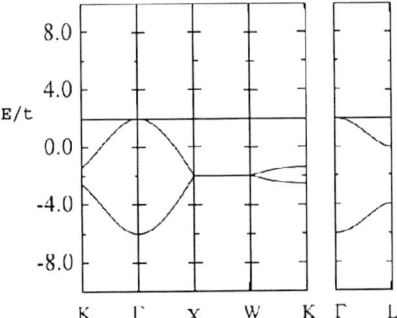

FIG. VII.2. Energy bands for electrons in a pyrochlore lattice with one orbital per site in the presence of nearest neighbor hopping. The upper flat band is two-fold degenerate. (After Ref. [285].)

when a one-electron picture is applied. Calculations based on the LDA show that the electrons near the Fermi energy have t_{2g} character.[294–296] These states are well separated from the e_g states as well as from the oxygen states (see Figure VII.3). The width of the t_{2g} bands is of order 2 eV and therefore at 60 K only a small fraction of the electrons in t_{2g} states would contribute to the excitations. In fact, the calculated density of states must be multiplied by a factor of 25 in order to account for the large γ value. This provides convincing evidence for strong electron correlations in LiV_2O_4. Further support is given by the observation that the material undergoes a phase transition into a charge ordered state at approximately 6 GPa.[297,298] Presumably this metal–insulator transition is again accompanied by a structural distortion as in Fe_3O_4.

It is worth pointing out that the band structure of a pyrochlore lattice has interesting features, which are simple to derive when only nearest-neighbor hopping is taken into account. The Hamiltonian of noninteracting electrons with nearest-neighbor hopping t is given in diagonal form by

$$H_0 = \sum_{k\alpha\sigma} [\epsilon_\alpha(\mathbf{k}) - \mu] a^\dagger_{\mathbf{k}\alpha\sigma} a_{\mathbf{k}\alpha\sigma} \qquad (7.3)$$

---

[294] J. Matsuno, A. Fujimori, and L. F. Mattheiss, *Phys. Rev. B* **60**, 1607 (1999).
[295] D. J. Singh, P. Blaha, K. Schwarz, and I. I. Mazin, *Phys. Rev. B* **16**, 359 (1999).
[296] V. Eyert, K.-H. Höck, S. Horn, A. Loidl, and P. S. Riseborough, *Europhys. Lett.* **46**, 762 (1999).
[297] K. Takeda, H. Hidaka, H. Kotegawa, T. C. Kobayashi, K. Shimizu, H. Harima, K. Fujiwara, K. Miyoshi, J. Takeuchi, Y. Ohishi, et al., *Physica B* **359–361**, 1312 (2005).
[298] K. Fujiwara, K. Miyoshi, J. Takeuchi, Y. Shimaoka, and T. Kobayashi, *J. Phys.: Condens Matter* **16**, S615 (2004).

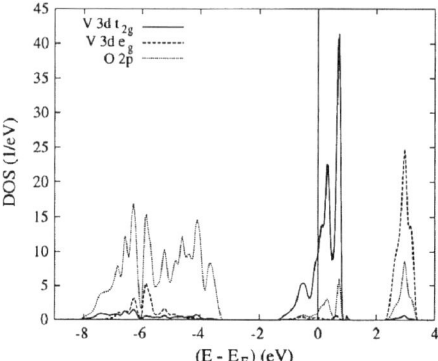

FIG. VII.3. Partial densities of states (DOS) for LiV$_2$O$_4$ calculated in LDA. The flat band of Figure VII.2 corresponds to the spike at $\simeq 1$ eV. (After Ref. [296].)

where $\alpha = 1, \ldots, 4$ is a band index due to 4 atoms/unit cell and one orbital per site is assumed. The band energies are

$$\epsilon_\alpha(\mathbf{k}) = \begin{cases} 2t, & \alpha = 3, 4 \\ -2t\left[1 \pm (1+\eta_\mathbf{k})^{\frac{1}{2}}\right] & \alpha = 1, 2 \end{cases}$$

$$\eta_\mathbf{k} = \cos(2k_x)\cos(2k_y) + \cos(2k_y)\cos(2k_z) + \cos(2k_z)\cos(2k_x) \quad (7.4)$$

where k_ν is given in reciprocal lattice units $\frac{2\pi}{a}$. The bandstructure is shown in Figure VII.2. One notices a two-fold degenerate flat band which is unoccupied provided $t > 0$. Hopping processes beyond nearest neighbors do not give the flat band a dispersion. This is only the case when the hopping matrix elements differ for different orbitals. The essential features of the simplified bands are still visible in the bands calculated by LDA (see Figure VII.3).

As regards heavy quasiparticles the crucial question is which degrees of freedom are associated with their formation. Usually one relates spin degrees of freedom with the low energy excitations giving rise to the heavy quasiparticles. This has been discussed at length, e.g., in Sections III, V and VI. However, in a frustrated lattice one might also think of charge degrees of freedom giving rise to a large number of low energy excitations. The high degeneracy of the ground state in the absence of a kinetic energy term in the Hamiltonian is lifted when hopping processes are included and the entropy can be released over a small temperature range. This is discussed in one of the next sections for the strong correlation limit. Nonetheless, the finding that the entropy at 60 K is close to $2R \ln 2$ suggests that here too spin degrees of freedom are responsible for the large low temperature specific heat coefficient γ.

In the following we give an estimate which shows that spin degrees of freedom in the pyrochlore structure are indeed able to explain the size of γ in LiV$_2$O$_4$. In setting up the Hamiltonian we include repulsive interactions between electrons on neighboring sites. In view of the observed charge ordering under pressure, the following Hamiltonian seems appropriate

$$H = -\sum_{\langle ij\rangle\nu} t_\nu \left(c^\dagger_{i\nu\sigma} c_{j\nu\sigma} + h.c.\right) + U \sum_{i\nu} n_{i\nu\uparrow} n_{i\nu\downarrow} + U \sum_{i;\nu>\mu} n_{i\nu} n_{i\mu}$$
$$+ \tilde{J} \sum_{i\nu\mu} \mathbf{s}_{i\nu} \mathbf{s}_{i\mu} + V \sum_{\langle ij\rangle} n_i n_j + \sum_{\langle ij\rangle} J_{ij}(S_i, S_j) \mathbf{S}_i \mathbf{S}_j. \quad (7.5)$$

Here i is a site and ν is an orbital index ($\nu = 1, 2, 3$) denoting the different t_{2g} orbitals. The first term is the kinetic energy or electronic hopping term. For the purpose of the intended estimate for γ we will later neglect it. The following three terms describe the intra-atomic Coulomb repulsions and spin interactions. For simplicity we have neglected the differences in the repulsions when different orbitals at site i are involved. Otherwise we would have to introduce an additional parameter U' (compare with Eq. (7.7) below). Finally, the last two terms are due to the Coulomb repulsions and antiferromagnetic spin-spin interactions between neighboring sites. Hereby $\mathbf{S}_i = \sum_\nu \mathbf{s}_{i\nu}$. For an estimate of the spin contributions to the γ coefficient we assume that the t_ν are very small so that they do not play a role. Then the d^2 configurations have spin $S = 1$. The Coulomb repulsions are minimized if on each tetrahedron two sites are in a d^1 configuration with $S = 1/2$ and two are in a d^2 configuration with $S = 1$. Let us pick out one of the exponentially large number of degenerate ground-state configurations (see Figure VII.4). One notices that all $S = 1/2$ sites form chains and rings and the same holds true for the sites with spin 1. The smallest rings consist of six sites. These features are independent of the chosen configuration. By means of constrained LDA+U calculations one can determine the nearest-neighbor spin coupling constants $J_{ij}(S_i S_j)$.[299] One finds for $J(1/2, 1/2) = 3$ meV and $J(1, 1) = 24$ meV implying that the spin 1 sites are much stronger coupled to each other than the spin 1/2 sites. Note that spin 1 chains have a gap in the expectation spectrum, i.e., the Haldane gap Δ_H.[300,301] Therefore spin 1/2 chains and rings are virtually uncoupled from each other. They can be coupled only via spin 1 chains. But the coupling between the two is frustrated and it takes a considerable energy $\Delta_H \simeq 0.41 J(1, 1)$ to excite the spin 1 chains and rings. Therefore the spin 1/2 chains remain essentially uncoupled and

---

[299] P. Fulde, A. N. Yaresko, A. A. Zvyagin, and Y. Grin, *Europhys. Lett.* **54**, 779 (2001).
[300] F. D. M. Haldane, *Phys. Lett. A* **93**, 464 (1983).
[301] F. D. M. Haldane, *Phys. Rev. Lett.* **50**, 1153 (1983).

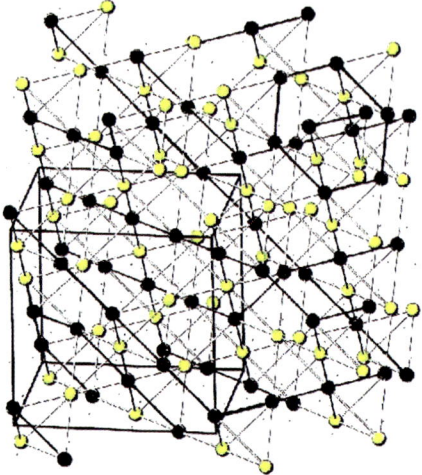

FIG. VII.4. Pyrochlore lattice: Example of a configuration satisfying the tetrahedron rule. Occupied sites with $S = 1$ (black dots) are connected by thick solid lines which form chains or rings. The same may be done for occupied sites with $S = 1/2$ (yellow dots).

we can determine directly the γ coefficient of the specific heat and the susceptibility from the relations.[302]

$$\gamma = \frac{2}{3}\frac{k_B R}{J(1/2, 1/2)}, \qquad \chi_s = \frac{4\mu_{\text{eff}}^2 R}{\pi^2 J(1/2, 1/2)}. \qquad (7.6)$$

Note that the Sommerfeld–Wilson ratio is $R_W = 2$. An experimental fit of γ_{exp} would require $J(1/2, 1/2) = 1.2$ meV instead of the calculated 3 meV. It is known that spin interactions are overestimated by a LDA+U calculation. But in any case, the improvement which is needed by applying the above localized electron picture to determine the γ coefficient is much less than the factor of 25 which is missing when a band approach is used. The above estimate suggests that a description of electrons in LiV_2O_4 should start from the localized limit instead from the band limit because in the former case the required corrections by including $t_v \neq 0$ are much less.

There have been also a number of attempt to explain the heavy quasiparticle in LiV_2O_4 with an on-site Hubbard interaction U only, i.e., without including Coulomb repulsions between neighboring sites. Thereby one of the t_{2g} electrons is kept as localized while the remaining 0.5 electron per V site is treated as delocalized. The following argument is used to justify this distinction. Due to a slight

---

[302] J. C. Bonner and M. E. Fisher, *Phys. Rev. A* **135**, 640 (1964).

distortion of the oxygen octahedra surrounding the V sites the t_{2g} states split into a lower a_{1g} and two e'_g states. The splitting is much smaller than the corresponding bandwidths. Indeed, a LDA bandstructure calculation[294–296] finds the total occupancies $n(e'_g) = 1.1$ and $n(a_{1g}) = 0.4$ implying a similar population of the different t_{2g} orbitals. But when instead a LDA+U calculation is performed[303] the a_{1g} state is singly occupied while the remaining 0.5 electrons per V site are of e'_g character. However, this seems to be a typical mean-field result. From an atomic point of view there is no reason why an a_{1g} electron should not be able to hop to a neighboring site like an e'_g type electron. The following Hamiltonian based on the LDA+U findings has been used and investigated beyond mean-field approximation[304]

$$H = -\sum_{\langle ij \rangle} t_{12}\left(c^\dagger_{i1\sigma} c_{j2\sigma} + h.c.\right) + U\sum_{i\alpha} n_{i\alpha\uparrow} n_{i\alpha\downarrow} + U'\sum_i n_{i1} n_{i2}$$
$$- \tilde{J}\sum_i \mathbf{S}_i(\sigma_{i1} + \sigma_{i2}) + J\sum_{\langle ij \rangle} \mathbf{S}_i \mathbf{S}_j. \tag{7.7}$$

The indices 1 and 2 refer to the two e'_g orbitals. Due to Hund's rule the coupling at site i between a localized a_{1g} electron with spin $\mathbf{S}_i$ and an e'_g electron with spin $\frac{1}{2}\sigma_{i\alpha}$ ($\alpha = 1, 2$) is ferromagnetic. The spin–spin interactions between neighboring sites i and j are antiferromagnetic. Otherwise the system would order ferromagnetically. The on-site Coulomb repulsion of e'_g electrons is chosen to be different when the electrons are in the same orbital and when they are not. The spin–spin interaction between the e'_g electrons is neglected. A strong Hund's rule coupling is assumed between the a_{1g} and the e'_g electrons by taking the limit $\tilde{J} \to \infty$. The nearest neighbor spin correlations between the localized a_{1g} electrons imply that the effective hopping matrix element $t_{12}(S)$ of the e'_g electrons depends on the relative spin orientation of the a_{1g} electrons, i.e.,

$$t_{12}(S) = t_{12}\sqrt{\frac{1 + \langle \mathbf{S}_i \mathbf{S}_j \rangle}{2S^2}}. \tag{7.8}$$

The determination of $\langle \mathbf{S}_i \mathbf{S}_j \rangle$ takes into account the frustrated lattice, here the pyrochlore structure [305] With the above simplification the Hamiltonian (7.7) reduces to

$$H = -\sum_{\langle ij \rangle} t_{12}(S)\left(c^\dagger_{i1} c_{j2} + c^\dagger_{i2} c_{j1} + h.c.\right) + U'\sum_i n_{i1} n_{i2}. \tag{7.9}$$

---

[303] V. I. Anisimov, M. A. Korotin, M. Zölfl, T. Pruschke, K. L. Hur, and T. M. Rice, *Phys. Rev. Lett.* **83**, 364 (1999).
[304] M. S. Laad, L. Craco, and E. Müller-Hartmann, *Phys. Rev. B* **67**, 033105 (2003).
[305] B. Canals and C. Lacroix, *Phys. Rev. Lett.* **80**, 2933 (1998).

The c_i operators have only one additional index which takes the values 1 and 2 and acts like a pseudospin. Therefore Eq. (7.9) has the form of a Hubbard Hamiltonian with a spin dependent hopping matrix element. This Hamiltonian has been treated for the 1/4 filled case by iterated perturbation theory.[304] When U' is increased a Kondo-like sharp resonance is obtained at the Fermi surface resulting in heavy quasiparticles at low temperatures. Thermodynamic as well as transport properties can be expressed in terms of a cross-over temperature T^*, going from a heavy Fermi liquid to a spin liquid at $T > T^*$.

There have been also a number of other attempts to explain the heavy quasiparticles which we want to mention. One approach starts from a Hamiltonian similar to (7.7) but replacing the two e'_g orbitals by a single one.[306] Nearest-neighbor spin correlations are treated by a mean-field ansatz $\langle \mathbf{S}_i \mathbf{S}_j \rangle = -\frac{3}{2}\Gamma^2$ and so are Hund's rule correlations $\langle \mathbf{S}_i \boldsymbol{\sigma}_i \rangle = u^2$. The mean field u is determined from a pseudo-hybridization between the a_{1g} electron and the itinerant, i.e., e'_g electron. A subsidiary condition ensures that there is one a_{1g} electron per site. Two temperatures characterize that approach. Above $T = T_{\text{mag}}$ the quantity $\Gamma = 0$ and the susceptibility is Curie–Weiss like because the intersite correlations have vanished. Similarly, a vanishing mean field u marks the second temperature T_{HF} ($\simeq 20$ K) below which a narrow a_{1g} band at E_F appears giving rise to heavy quasiparticles. More details are found in Ref. [306].

The mean-field approach has been generalized by treating also the a_{1g} electron as itinerant. The on-site Coulomb repulsion between the e'_g and a_{1g} electrons is taken into account in a slave boson mean-field approximation.[307] In effect the a_{1g} bandwidth is strongly renormalized and gives rise to a sharp resonance at E_F.

There have been also weak coupling approaches to explain the heavy quasiparticles. One suggestion is based on multicomponent fluctuations due to the t_{2g} and spin degrees of freedom.[308] One of the consequences of the large orbital contributions to the γ coefficient of the specific heat is a small Sommerfeld–Wilson ratio of $R_W \simeq 0.1$. Another approach treats the pyrochlore structure of the V ions as a network of Hubbard chains.[309] Due to this one-dimensional feature electron correlations have strong effects on electron-hole excitations and hence on the self-energy.

---

[306] C. Lacroix, in *Proceed. HFM 2000, Canad. J. Phys.* **79**, 135 (2000).
[307] H. Kusunose, S. Yotsuhashi, and K. Miyake, *Phys. Rev. B* **62**, 4403 (2000).
[308] Y. Yamashita and K. Ueda, *Phys. Rev. B* **67**, 195107 (2003).
[309] S. Fujimoto, *Phys. Rev. B* **65**, 155108 (2002).

17. STRUCTURAL TRANSITION AND CHARGE DISPROPORTIONATION: AlV_2O_4

The spinel AlV_2O_4 is of interest because the average d-electron number per V ion is 2.5 and therefore the configurations are expected to fluctuate between $3d^2$ and $3d^3$. However, what actually happens is that the system undergoes a phase transition at approximately $T_c = 700$ K to a charge ordered state. It is associated with a change of the lattice structure from pyrochlore to alternating Kagomé and triangular planes. This is shown in Figure VII.5. In the low temperature rhombohedral phase the [111] axis is elongated while the perpendicular axes are shortened in order to keep the volume of the unit cell nearly constant. Experimental results are shown in FigureVII.6. The figure contains also a plot of the observed changes of the angle between two unit vectors of the rhombohedral lattice as the temperature

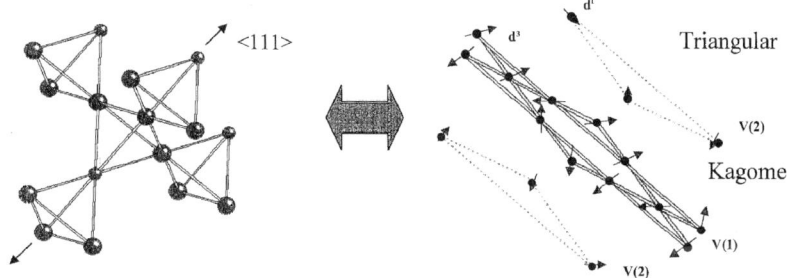

FIG. VII.5. Distortion of a pyrochlore lattice by elongation along the [111] axis. The resulting rhombohedral lattice consists of Kagomé and triangular planes.

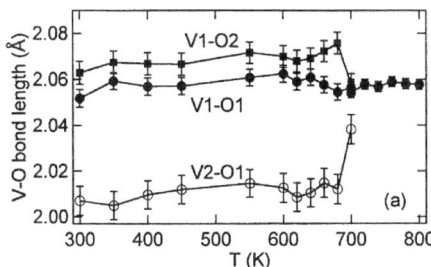

FIG. VII.6. AlV_2O_4: Dependence of the V–O bond lengths on temperature. V1 and V2 refer to V^{2+} and V^{4+} ions while O1 and O2 are different oxygen ions. (After Ref. [310].)

---

[310] J. Matsuno, T. Katsufuji, S. Mori, Y. Moritomo, A. Machida, E. Nishibori, M. Takata, M. Sakata, N. Yamamoto, and H. Takagi, *J. Phys. Soc. Jpn.* **70**, 1456 (2001).

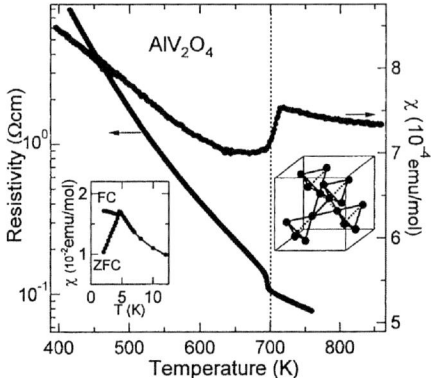

FIG. VII.7. Resistivity and magnetic susceptibility at 1 Tesla as function of temperature for AlV_2O_4. The left inset shows the low temperature field cooled (FC) and zero field cooled (ZFC) susceptibility in a field of 10^{-2} Tesla. (After Ref. [310].)

is lowered below the phase transition temperature. Note that there are three times as many sites on the Kagomé lattice than there are on the triangular one.

In a simplified description the V ions have a valency of V^{4+} on the triangular and V^{2+} on the Kagomé sites corresponding to d^1 and d^3 configurations, respectively. The charge disproportionation is often called valence skipping. Of course, this is a simplified view since in reality the disproportionation is considerably less than the separation $4 \times d^{2.5} \rightarrow 3 \times d^3 + 1 \times d^1$ would suggest. The system avoids frustration by distorting. Unfortunately the experimental results are still sparse. It is not even clear whether AlV_2O_4 is a poor metal or a small gap semiconductor at low temperatures. Resistivity measurements as well as those of the susceptibility are shown in Figure VII.7. The structural phase transition shows up in both quantities, i.e., by a small but steep increase in the resistivity and a pronounced decrease of the magnetic susceptibility.

For a description of the phase transition one must set up a model Hamiltonian. It should also allow for explaining differences of charge ordering in AlV_2O_4 and in LiV_2O_4 under pressure. While AlV_2O_4 is a semiconductor or a metal in the charge ordered state, LiV_2O_4 becomes an insulator.[297] Electronic structure calculations based on LDA+U provide a realistic description of magnetic insulators but are not suitable for strongly correlated paramagnetic metals. Therefore we exclude them here and start from a microscopic model Hamiltonian at the price of having to introduce adjustable parameters. Nevertheless, we can uncover this way the processes which lead to the observed structural transition and the accompanying charge order. From standard band-structure calculations it is known that only t_{2g} states are near the Fermi energy (compare with Figure VII.3). All other bands are well above or below E_F. Therefore it suffices to include only t_{2g} electrons in the model Hamiltonian. We write it in the form

$$H = H_0 + H_{\text{int}} + H_{\text{e-p}}, \quad \text{with}$$

$$H_0 = \sum_{\langle l\mu, l'\mu'\rangle} t^{\nu\nu'}_{\mu\mu'}(l,l') c^{\dagger}_{l\mu\nu\sigma} c_{l'\mu'\nu'\sigma}$$

$$H_{\text{int}} = \sum_{l\mu} \left\{ (U+2J) \sum_{\nu} n_{l\mu\nu\uparrow} n_{l\mu\nu\downarrow} + U \sum_{\nu > \nu'} n_{l\mu\nu\sigma} n_{l\mu\nu'\bar{\sigma}} \right.$$

$$\left. + (U-J) \sum_{\nu > \nu'} n_{l\mu\nu\sigma} n_{l\mu\nu'\sigma} \right\} + \frac{V}{2} \sum_{\langle l\mu, l'\mu'\rangle \nu\nu'\sigma\sigma'} n_{l\mu\nu\sigma} n_{l'\mu'\nu'\sigma'}$$

$$H_{\text{e-p}} = \epsilon \Delta \sum_{l\nu\sigma} \sum_{\mu} \left(\delta_{\mu,1} - \frac{1}{3}(1-\delta_{\mu,1}) \right) n_{l\mu\nu\sigma} + K \sum_{l} \epsilon_l^2. \quad (7.10)$$

Here H_0 describes the kinetic energy. The electron creation and annihilation operators are specified by four indices, i.e., for the unit cell (denoted by l), the sublattice ($\mu = 1-4$), the t_{2g} orbital ($\nu = d_{xy}, d_{yz}, d_{zx}$), and the spin ($\sigma = \uparrow, \downarrow$). The brackets $\langle \ldots \rangle$ indicate a summation over nearest-neighbor sites. The term H_{int} describes the on-site Coulomb and exchange interactions U and J among the t_{2g} electrons. The last term contains the Coulomb repulsion V of an electron with those on the six neighboring sites. Finally $H_{\text{e-p}}$ describes the coupling to lattice distortions. The deformation potential is denoted by Δ. It is due to a shift in the orbital energies of the V sites caused by relative changes in the oxygen positions. While the energy shift is positive for the $V(1)$ sites it is negative for the $V(2)$ sites.

The elastic constant K refers to the c_{44} mode and describes the energy due to the rhombohedral lattice deformation. It is reasonable to assume that like in Fe_3O_4[311] and Yb_4As_3[199] only the c_{44} mode is strongly coupled with the charge disproportionation. One can give at least approximate values for all parameters except for V and Δ. Their ratio will be fixed by the charge-ordering transition temperature while keeping the constraint $V \ll U$. For the on-site Coulomb- and exchange integrals we set $U = 3.0$ eV and $J = 1.0$ eV which are values commonly used for vanadium oxides.[312] Band structure calculations which we have performed demonstrate that the hopping matrix elements $t^{\nu\nu'}_{\mu\mu'}$ between different orbitals $\nu \neq \nu'$ are negligible. For simplicity we can therefore omit ν, ν'. Then

---

[311] H. Schwenk, S. Bareiter, C. Hinkel, B. Lüthi, Z. Kakol, A. Koslowski, and J. M. Honig, *Eur. Phys. J. B* **13**, 491 (2000).
[312] A. Fujimori, K. Kawakami, and N. Tsuda, *Phys. Rev. B* **38**, R7889 (1988).

$t_{\mu\mu'}(l, l') = -t$ when l, l' are nearest neighbors. We will take into account the orbital dependent hopping matrix elements coming from a tight-binding fit of LDA calculations. Furthermore the c_{44} elastic constant is neither known for AlV$_2$O$_4$ nor for LiV$_2$O$_4$, while computational methods for its ab initio calculation in the case of materials with strong electronic correlations are not mature enough. Therefore a representative value $c_{44}^{(0)}/\Omega = 6.1 \cdot 10^{11}$ erg/cm^3 is used for AlV$_2$O$_4$ where Ω is the volume of the cubic unit cell with a lattice constant of $a = 5.844$ Å. This value is close to the experimental value for Fe$_3$O$_4$ which has also the spinel structure. This leads to $K \simeq 1.1 \cdot 10^2$ eV. The deformation potential Δ is not known but is commonly of the order of the band width. For convenience we introduce the dimensionless coupling constant $\lambda = \Delta^2/Kt$ and lattice distortion $\delta_L = \epsilon\Delta/t$. From LDA calculations the bandwidth is $8t = 2.7$ eV, and therefore a reasonable value is $\lambda t = \Delta^2/K = 1$ eV. This means $\Delta = 10.5$ eV which is twice the value of Yb$_4$As$_3$.[31]

Not contained in the Hamiltonian (7.10) is a spin–spin interaction term between the V ions. This might turn out a shortcoming since bandstructure calculations based on a local spin-density approximation (LSDA) to density functional theory find effective exchange constants which are strongly enhanced in the low temperature phase. Estimates are $J_{kk} = J_{kt} = 202$ K for the high temperature phase where the subscripts k and t refer to Kagomé and triangular sites. For the low temperature phase the corresponding estimates are $J_{kk} = 360$ K and $J_{kt} = 167$ K.[313]

Let us first consider H_0 which is easily diagonalized. With four V ions per unit cell and three t_{2g} orbitals there are altogether 24 bands, assuming that the spin symmetry is broken. Of those twelve are dispersionless and degenerate. Furthermore, there are two sixfold degenerate dispersive bands (compare with Figure VII.2). The Fermi energy is in a region of high density of states. More details are found in Ref. [285].

The next term we discuss is H_{e-p} which describes the deformation potential coupling. When it is sufficiently strong it leads to charge ordering. There is no opening of a gap though, but only a sharp decrease of the density of states near E_F, i.e., the system remains a metal. This feature does not change when the interactions U and V are included in mean-field approximation, provided we deal with a paramagnetic state. Generally U suppresses charge ordering while V enhances it. Again, no gap opens at E_F but the density of states decreases in its neighborhood. In mean-field approximation the relation between the homogeneous lattice distortion δ_L along [111] and charge disproportionation is given by

$$\delta_L = \frac{\lambda(n_2 - n_1)}{2} \qquad (7.11)$$

---

[313] A. Yaresko, I. Leonov, and P. Fulde, *Physica B*, in press.

where $n_1 = \sum_{v\sigma} \langle n_{l1v\sigma} \rangle$ is the occupational number of the triangular sites while $n_2 = n_3 = n_4$ with $n_2 = \sum_{v\sigma} \langle n_{l2v\sigma} \rangle$ is the one of the Kagomé sites.

A shortcoming of the mean-field analysis is that it leads to a second-order phase transition instead of the observed first-order one. This is almost certainly due to strong correlations which suppress charge fluctuations between different vanadium sites. In order to incorporate them at least approximately, one must allow for unrestricted mean-field solutions by breaking the spin symmetry. But the constraint of zero total magnetic moment has to remain. This is done by ascribing to the sites μ of a tetrahedron an occupational number n_μ and a magnetization $m_\mu = \sum_{v\sigma} \langle n_{l\mu v\sigma} \rangle \sigma$. The spins are assumed to be directed towards the center of the tetrahedron in the undistorted phase. For them the [111] direction is a convenient quantization axis. In the distorted phase the spins of the $V(1)$ sites are slightly tilted with respect to this axis so that the net magnetization remains zero. The free energy is a function of the different n_μ and m_μ and must be minimized with respect to both. The energy bands do now depend on spin σ. The hopping matrix elements between nearest neighbor sites must be transformed accordingly, so that the different spin directions are accounted for. For details see Ref. [314]. As a result the charge disproportionation is found as function of the various parameters. We show in Figure VII.8 the results for $T = 0$ as function of the ratio V/t. As usual the calculated disproportionation is larger, here by a factor of 2.5 than

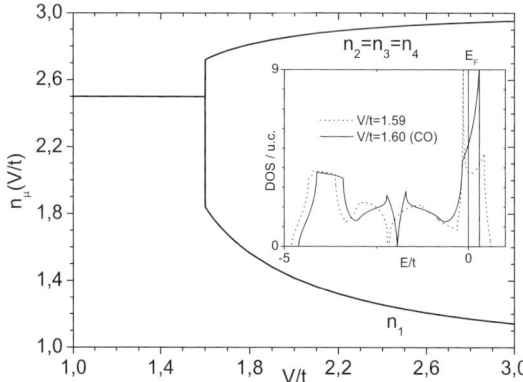

FIG. VII.8. Charge disproportionation based on Eq. (7.10) as function of V/t. The n_μ denote the occupation numbers of the four sites of a tetrahedron. In the inset the changes in the density of states are shown when V/t is just below and above the critical value at which charge ordering sets in. For $V/t = 1.67$ one has $n_1 = 2.5 - 3\delta$ and $n_i = 2.5 + \delta$ ($i = 2$–4) with a charge disproportionation $\delta \simeq 0.25$. Here $U = 3.0$ eV, $J = 1.0$ eV, $\lambda t = 1.0$ eV and $8t = 2.7$ eV. (After Ref. [314].)

---

[314] Y. Z. Zhang, P. Fulde, P. Thalmeier, and A. Yaresko, *J. Phys. Soc. Jpn.* **74**, 2153 (2005).

the one obtained from a valence band analysis of the measured distorted structure. This is due to the simplified model Hamiltonian which does not allow for screening by non-d electrons. The same difficulty was found for Yb_4As_3.[31,315] Also shown in Figure VII.8 as an inset is the change in the density of states in the vicinity of the critical ratio $(V/t)_{crit} = 1.6$ at which the transition to a charge ordered state does occur. The strong change which one can notice might explain the observed small but steep increase in the resistivity when charge ordering sets in. The phase transition is found to be of first order when the free energy is evaluated. For $V/t = 1.67$ the calculated transition temperature is $T_c = 660$ K which is reasonably close to the observed one of $T_{exp} = 700$ K. Of course, there is some arbitrariness in the particular choice of V/t. It should be pointed out that the LSDA calculations yield a value of $\delta = 0.17$ for the disproportionation and a small hybridization gap.

In the undistorted phase the point symmetry of the vanadium ions is D_{3d} and the t_{2g} degeneracy is reduced to an a_{1g} singlet and an e'_g doublet. In the distorted, i.e., charge ordered phase the symmetry of the triangular sites remains D_{3d} while the one of the Kagomé sites is lowered to C_{2h}. This lifts also the e'_g degeneracy. It is due to an elongation of the crystal in [111] direction and a contraction perpendicular to it. While in the distorted phase of AlV_2O_4 the energies of the three orbitals are nearly the same because of a very small distortion of the oxygen octahedra one expects that in LiV_2O_4 the energy of the a_{1g} orbital is highest. There the distortion of the octahedra is fairly large. Since the changes in orbital energies are small in AlV_2O_4 one expects that the system remains gapless even in the charge ordered state in agreement with LDA calculations. The LDA band structure can be reproduced by a proper choice of the hopping parameters. By determining self-consistently the occupational numbers of the three $V(1)$ sites one finds that $n_2^{xy} = n_2^{zx} > n_2^{yz}$, $n_3^{xy} = n_3^{yz} > n_3^{zx}$ and $n_4^{yz} = n_4^{zx} > n_4^{xy}$ in the t_{2g} basis. This shows that the ordered orbitals on sublattices 2, 3 and 4 are perpendicular to each other. There is no orbital ordering on the $V(2)$ sites in the charge ordered state because of the small energy difference between the a_{1g} and e'_g orbitals as compared with the bandwidth. The sharp decrease of the density of states near E_F as well as the order of the charge ordering phase transition are not influenced by orbital order. The same holds true for the absence of an energy gap at E_F. Therefore orbital ordering is of little importance for AlV_2O_4.

These findings should be compared with charge ordering observed in LiV_2O_4 under pressure. A crucial difference is the average d-electron number per V site which is 1.5 for LiV_2O_4 as compared with 2.5 in the case of AlV_2O_4 and the distinct role of the a_{1g} orbital. Consequently in the limit $t \to 0$ the a_{1g} orbital is

---

[315] U. Staub, B. D. Patterson, C. Schulze-Briese, F. Fauth, M. Shi, L. Soderholm, G. B. M. Vaughan, E. Dooryhee, J. O. Cross, and A. Ochiai, *Physica B* **318**, 284 (2002).

empty at the $V(1)$ sites of LiV$_2$O$_4$ while the split e'_g orbitals are singly occupied with $S = 1$. On the $V(2)$ sites with d^0 also the split e'_g orbitals are unoccupied. Therefore a gap at E_F is expected in the charge ordered state. This is what is found when the Hamiltonian (7.5) is treated in mean-field approximation with a filling factor of $1/4$ instead of $5/12$ as in the case of AlV$_2$O$_4$. The opening of a gap can also be inferred from measurements of $\rho(T)$.[297] Orbital ordering is obviously not relevant here. For further details we refer to Ref. [314]. The structure in the distorted phase on which the above theory has been based was recently called into question in Ref. [316]. Instead of a low-temperature structure consisting of triangular and Kagomé planes those authors interpret their data in terms of V$_7$ molecular clusters. The future has to show which structure is the correct one.

18. Fractional Charges Due to Strong Correlations

Consider the pyrochlore lattice with half as many electrons than there are number of sites. We take the limit $U \to \infty$ and neglect for simplicity the electron spin so that we are dealing with spinless fermions. In that case half of the lattice sites are singly occupied and half of them are empty. Restricting the inter-site Coulomb repulsion to nearest neighbors the Hamiltonian is

$$H = -t \sum_{\langle ij \rangle} (c_i^\dagger c_j + h.c.) + V \sum_{\langle ij \rangle} n_i n_j. \tag{7.12}$$

The $c_i^\dagger$ create spinless fermions at sites i and as usual $n_i = c_i^\dagger c_j$. As discussed before, the repulsion V is minimized when in each tetrahedron two sites are occupied and two sites are empty (tetrahedron rule). Let us now add one particle to the system. Since each site belongs to two tetrahedra, the above rule is violated for two neighboring tetrahedra which now contain three electrons each. The interaction energy is increased by $4V$. But when one of the four nearest neighbors of the added particle hops onto an empty site the two tetrahedra violating the rule are separated. This is shown in Figure VII.9. The important point is that the interaction energy remains unchanged by this separation, i.e., it is still $4V$. When the charge of the added particle is e, each fragment must be assigned a charge $e/2$.

Taking into account the kinetic energy term of the Hamiltonian lifts the exponentially large degeneracy of the ground state which is present when the kinetic energy term is absent. Deconfinement of the charges $e/2$ remains intact provided

---

[316] Y. Horibe, M. Shingu, K. Kurushima, H. Ishibashi, N. Ikeda, K. Kato, Y. Motome, N. Furukawa, S. Mori, and T. Katsufuji, *Phys. Rev. Lett.* **96**, 086406 (2006).

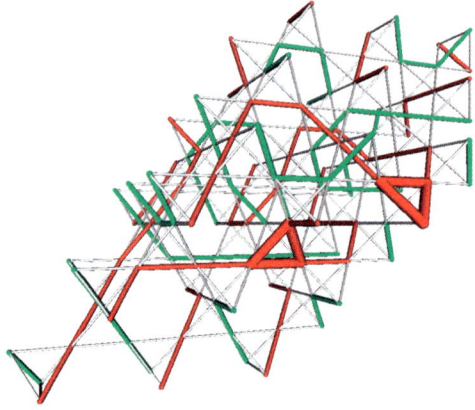

FIG. VII.9. Separated tetrahedra on a pyrochlore lattice after an electron has been added to the system. Occupied sites are connected by solid red lines and empty by green lines (courtesy of F. Pollmann).

t/V is sufficiently small.[317–319] Since a numerical study of a pyrochlore lattice is difficult we shall investigate instead the simpler checkerboard lattice. The latter can be considered a projection of the pyrochlore lattice onto a plane. But one must keep in mind that the lower dimension of the checkerboard lattice may result in different behaviors as regards confinement or deconfinement of the fractional charges $e/2$. This is known from lattice gauge theories which are closely related to the present problem[317,320] Also the statistics of fractionally charged excitations may differ in two and three dimensions. In two dimensions the wavefunction of the particles belongs to a representation of the braid group while in three dimensions it is one of the permutation group. Figure VII.10 shows one of the ground state configurations. They have the form of string nets (see also Ref. [321]). In the absence of hopping the degeneracy is $N_{\text{deg}} = (4/3)^{3N/4}$ where $N = N_x N_y$ and N_x, N_y are the number of lattice sites in x and y direction. In order to study numerically the ground state of the system for large ratios of V/t it is advantageous to introduce an effective Hamiltonian[42] which acts on the different configurations obeying the tetrahedron rule. They form what will be called the *allowed subspace*

---

[317] A. M. Polyakov, *Gauge Fields and Strings*, vol. 3 of *Contemporary Concepts in Physics*, Harwood Academic Publ. (1993), 5th print.
[318] P. Fulde, K. Penc, and N. Shannon, *Ann. Phys. (Leipzig)* **11**, 892 (2002).
[319] J. Betouras, F. Pollmann, and P. Fulde, to be published (2006).
[320] E. Fradkin, *Field Theories of Condensed Matter Systems*, Addison-Wesley Publ., Redwood City (1991).
[321] M. Levin and X.-G. Wen, *Phys. Rev. B* **73**, 035122 (2006).

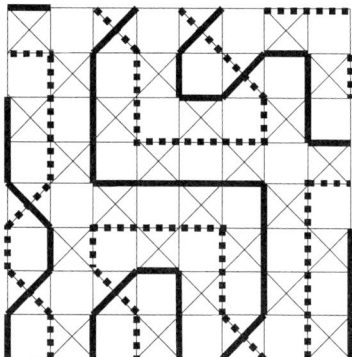

FIG. VII.10. Ground-state configuration of a checkerboard lattice with half-filling of spinless fermions. Occupied sites are connected by solid lines and empty sites by dashed lines. Hopping takes place along thin lines with matrix element $-t$.

of all configurations. An effective Hamiltonian allows for the diagonalization of much larger clusters than does the full H. To leading order in t/V one finds that the energy is lowered for all allowed configurations by an amount of

$$\Delta E = -\frac{4t^2}{V} \sum_i n_i. \tag{7.13}$$

Therefore it does not lift the degeneracy and one has to go to the next higher order term. Different allowed configurations are connected through ring hopping processes. The smallest non-vanishing process is

$$H_{\text{eff}} = \frac{-}{(+)} \frac{6t^3}{V^2} \left(\vcenter{\hbox{□}} + \vcenter{\hbox{□}} + \vcenter{\hbox{□}} + \vcenter{\hbox{□}} \right) + (\circlearrowleft \leftrightarrow \circlearrowright) \tag{7.14}$$

where filled dots denote occupied and empty dots empty sites, respectively. The site in between can be empty or occupied. The arrows stand for particle hopping. With

$$t_{\text{ring}} = \frac{12t^3}{V^2} \tag{7.15}$$

this Hamiltonian can be written as

$$H_{\text{eff}} = t_{\text{ring}} \sum_{\bigcirc} c^\dagger_{j_6} c^\dagger_{j_4} c^\dagger_{j_2} c_{j_5} c_{j_3} c_{j_1}. \tag{7.16}$$

Hopping on larger rings implies higher orders in t/V. The matrix elements of H_{eff} with respect to different allowed configurations $|i\rangle$ and $|j\rangle$ are

$$\langle j | H_{\text{eff}} | i \rangle = (-1)^{n_0} t_{\text{ring}} \tag{7.17}$$

where n_0 is the occupation number of the site inside the ring. It is worth realizing that the sign dependence of $\langle j|H_{\text{eff}}|i\rangle$ is absent when instead of a checkerboard lattice the pyrochlore lattice is considered. In the latter structure there is no lattice site inside a 6-ring loop. The partial lifting of the degeneracy of the ground-state manifold by H_{eff} can be understood with the help of the height representation. For that purpose one divides the crisscrossed squares of the checkerboard lattice into sublattices A and B and assigns a clockwise and counter-clockwise orientation to them. To each occupied site is attached a unit vector the direction of which is in accordance with the orientation of the corresponding crisscrossed squares. At each empty site the vector is pointing into opposite direction. This defines a vector field $\mathbf{f}(\mathbf{r})$ where $\mathbf{r}$ is defined with respect to the uncrossed plaquettes. Because of the tetrahedron rule, i.e., two sites are occupied and two are unoccupied on a crisscrossed square, the vector sum over a closed loop vanishes, implying curl $\mathbf{f} = 0$. Therefore $\mathbf{f}$ can be written as the gradient of a scalar field $h(\mathbf{r})$, i.e., a potential which is called height field. It allows for the introduction of two topological quantum numbers κ_x and κ_y. They quantify the difference in the potential at the upper and lower boundary and at the right and left one. Their values cover the range $-N_x/2 \leqslant \kappa_x \leqslant N_x/2$ and $-N_y/2 \leqslant \kappa_y \leqslant N_y/2$.

In Figure VII.11 an example is given of the change of the height field $h(\mathbf{r})$ defined on the uncrossed squares. On a torus κ_x and κ_y are winding numbers. An important point is that the application of H_{eff} does not change the topological quantum numbers of a configuration. Therefore the degenerate ground-state configurations are divided into different classes according to (κ_x, κ_y). The matrix $\langle i|H_{\text{eff}}|j\rangle$ reduces into block matrices characterized by (κ_x, κ_y). It turns out that H_{eff} does not necessarily connect all configurations within a class. Therefore the matrix $(H_{\text{eff}})_{ij}$ for the class (κ_x, κ_y) may reduce further to irreducible blocks $(\kappa_x^{(\alpha)}, \kappa_y^{(\alpha)})$. This holds true in particular for the class $(0, 0)$ which has the largest

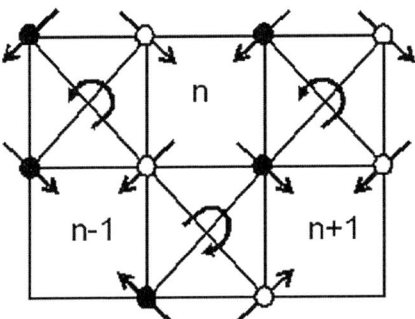

FIG. VII.11. Values of the height field h for a part of a given configuration: solid and empty dots mark occupied and empty sites, respectively.

number of elements. It is found that for a system of size 8×8 with $N_f = 32$ spinless fermions the ground state has topological quantum numbers (0, 0). The subclass to which it belongs has more than 10^5 elements. The ground state is two-fold degenerate and the two states are related by particle–hole symmetry. Note that H_{eff} has this symmetry while H does not have it. The two states are charge ordered along one of the diagonals and can be transformed into each other by a rotation of 90 degrees. This invalidates a supposition in Ref. [42]. The participation ratio (PR) of a wavefunction is a measure of how extended that function is. With

$$|\psi\rangle = \sum_\nu \alpha_\nu |\nu\rangle \qquad (7.18)$$

where $|\nu\rangle$ denotes different configurations the participation ratio is

$$PR[\psi] = \frac{1}{\sum_\nu |\alpha_\nu|^4}. \qquad (7.19)$$

For the ground state $|\psi_0\rangle$ of the 8×8 system this ratio is very large, i.e.,

$$PR[\psi_0] = 46.9 \cdot 10^3 \qquad (7.20)$$

indicating that many configurations contribute to it. The largest possible value of $PR[\psi_0]$ in a Hilbert space of dimension N is $PR[\psi_0] = N$. The total density of state is

$$\rho_{\text{tot}}(E) = \sum_l \delta(E - E_l) \qquad (7.21)$$

where E_l are the eigenvalues of H. From it we can compute the specific heat $C(T)$ according to

$$C(T) = \frac{\partial}{\partial T} \frac{\int dE\, E \rho_{\text{tot}}(E) e^{-\beta E}}{\int dE\, \rho_{\text{tot}}(E) e^{-\beta E}}. \qquad (7.22)$$

Numerical results for different system sizes are shown in Figure VII.12. One notices an almost linear in T behavior. This suggests that a large quasiparticle mass in geometrically frustrated lattices may possibly be due to charge instead of spin degrees of freedom, the usual source of heavy quasiparticles. The large number of low lying excitations causing the steep linear increase of the specific heat with temperature is here due to a release of entropy $(3/4) \ln(4/3) \approx 0.22$ per site over a temperature range of $k_B T \leqslant 2 t_{\text{ring}}$.

In the following we want to go beyond the space of allowed configurations which obey the tetrahedron rule. When a particle is hopping to a neighboring empty site (vacuum fluctuation) the tetrahedron rule is broken for two tetrahedra: now one tetrahedron contains three particles, while another contains one only (see

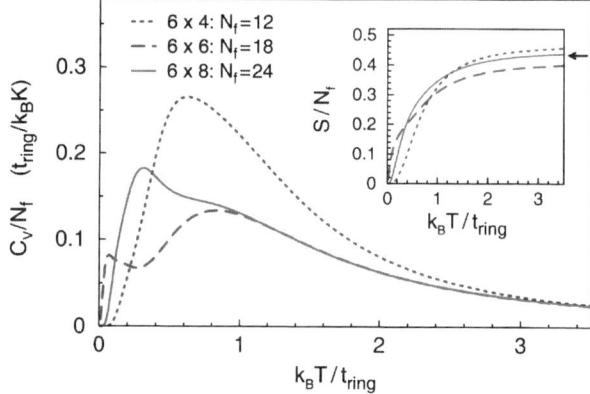

FIG. VII.12. Specific heat per particle for a checkerboard lattice of spinless fermions at half filling for various system sizes. Inset: corresponding entropies, the arrow marks the value of $2(3/4)\ln(4/3)$. (After Ref. [42].)

Figure VII.13a). For the corresponding uncrossed squares curl $\mathbf{f} \neq 0$. By subsequent hopping these two objects can separate carrying a charge of $\pm e/2$ each. As a result of the vacuum fluctuation the vector field $\mathbf{f}(n)$ contains a vortex–antivortex pair. An insulator to metal transition may be viewed as a proliferation of such pairs and resembles a Kosterlitz–Thouless transition. It is difficult to compute numerically the ratio V/t at which this transition takes place, but an estimate yields a critical ratio of $(V/t)_{\mathrm{cr}} \simeq 7$. This is considerably larger than the value obtained by conventional equations of motion methods in Hubbard I-type approximation.[322]

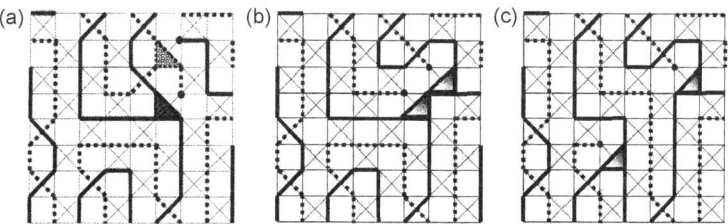

FIG. VII.13. (a) Vacuum fluctuation. One criss-crossed square (tetrahedron) contains three particles while another one contains one particle only; (b) particle added to the otherwise half-filled checkerboard lattice; (c) after a hop of one of the electrons the two squares with three particles each have separated. (After Ref. [318].)

---

[322] Y. Z. Zhang, M.-T. Tran, V. Yushankhai, and P. Thalmeier, *Eur. Phys. J.* **44**, 265 (2005).

When an extra particle is added to the system two neighboring criss-crossed squares contain three occupied sites each (see Figure VII.13b). These two special squares separate from each other when one of the electrons is hopping to a neighboring site (see Figure VII.13c). There is no change in the particle repulsions associated with this separation of the two. Since the charge of the added particle is e, each of the separated squares must have charge $e/2$. The reason for that is easily seen. When a square with three particles moves in the checkerboard lattice plane, the squares left behind must again satisfy the tetrahedron rule. Inspection shows that this requires the backflow of a charge $e/2$, thus reducing the charge flow due to hopping of an electron from e to $e/2$. For the checkerboard lattice it turns out that ring hopping processes cause a (weak) restoring force on the two particles with fractional charge $e/2$.[323] We have not discussed the issue of the spin here but instead have considered spinless or fully polarized fermions. In passing we point out that not only do we have spin-charge separation when the spin is included but also is the spin distributed over parts of the sample. For more details we refer to Ref. [318].

The spectral density is a quantity which is expected to show signatures of fractional charges. For that reason we determine the integrated spectral density

$$S(\omega) = \sum_{\mathbf{k}\nu} |\langle \psi_{\mathbf{k}\nu}^{N+1} | c_{\mathbf{k}\nu}^{\dagger} | \psi_0^N \rangle|^2 \delta(\omega - E_{\mathbf{k}\nu}^{N+1} + E_0^N). \tag{7.23}$$

Here $|\psi_{\mathbf{k}\nu}^{N+1}\rangle$ are the different excited states characterized by momentum $\mathbf{k}$ and band index ν of the $N+1$ particle system and $E_{\mathbf{k}\nu}^{N+1}$ are the corresponding energies. The energy of the N-particle ground state is E_0^N. The $c_{\mathbf{k}\nu}^{\dagger}$ create particles in quantum states $\mathbf{k}, \nu$.

For a 32 sites checkerboard cluster with $N = 16$ spinless fermions and $V/t = 25$ the integrated spectral density is shown in Figure VII.14a. When one includes only configurations of the $N+1$ particle system in which the added particle is not disintegrated like in Figure VII.14a one obtains the spectral density shown in Figure VII.14b. A comparison with Figure VII.14a shows a considerable reduction of the width of the excitation spectrum and an absence of the low frequency part. Therefore we may conclude that hopping processes leading to a separation of the charge e into two parts have an important effect on the spectral density and constitute further evidence for the appearance of fractional charges.

The above considerations suggest that in 2D there are also other lattices than the checkerboard one allowing for charge fractionalization. That holds particularly true for the Kagomé lattice which at 1/3 filling can support excitations with charge $e/3$ and $2e/3$ when a spinless fermion is added.[323] The requirement hereby is a short-range repulsion $V n_i n_j$ when i and j are on the same hexagon.

---

[323] F. Pollmann and P. Fulde, *Europhys. Lett.*, in print.

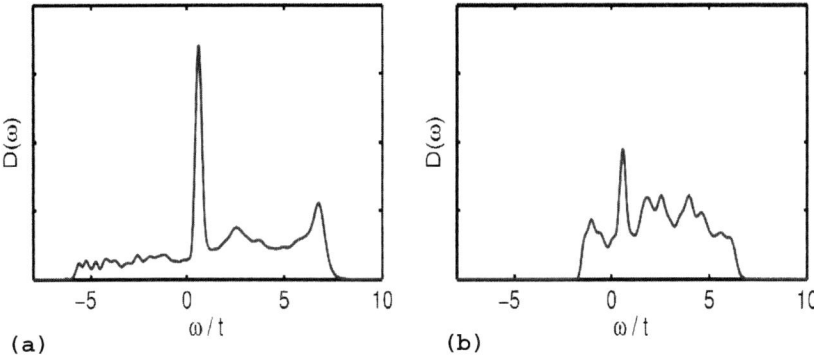

FIG. VII.14. Integrated spectral density as function of energy for a 36 sites checkerboard system with 18 particles and $V/t = 25$. (a) When the excited states are fully accounted for and (b) when only configurations are used in which the added particle remains an entity. (After Ref. [324].)

The fermionic character of the particles discussed here makes the present model different from related spin models and models for hard core bosons. Nevertheless, there exist also similarities between the problem of fractional charges and resonating valence bonds (RVB). For example, Kalmeyer and Laughlin[325] have shown that the ground-state wavefunction of an antiferromagnetic Heisenberg Hamiltonian with nearest-neighbor interactions on a triangular lattice is practically the same as the fractional quantum Hall effect (FQHE) wavefunction for bosons. As the ground-state of a Heisenberg Hamiltonian on a frustrated lattice is of the RVB type, the two phenomena, i.e., FQHE effect and RVBs are closely related. RVB systems support spinons, i.e., a spin flip with $\Delta s_z = 1$ breaks up into two spinons with spin 1/2 each.[326–329] The situation resembles the one in Figure VII.13 in particular when an Ising Hamiltonian us used. Spins up and down on a checkerboard lattice correspond to occupied and unoccupied sites and a state with total spin $S_{\text{tot}}^z = 0$ has a half-filled lattice of hard-core bosons as analogue. While a large body of work exists connecting RVB models with confined and deconfined phases of compact gauge theories (see, e.g., Refs. [320,326,330]) corresponding work for fermionic systems is still missing.

---

[324] F. Pollmann, E. Runge, and P. Fulde, *Phys. Rev. B* **73**, 125121 (2006).
[325] V. Kalmeyer and R. B. Laughlin, *Phys. Rev. B* **39**, 11879 (1989).
[326] R. Moessner and S. L. Sondhi, *Phys. Rev. B* **63**, 224401 (2001).
[327] L. Balents, M. P. A. Fisher, and S. M. Girvin, *Phys. Rev. B* **65**, 224412 (2002).
[328] O. I. Motrunich and T. Senthil, *Phys. Rev. Lett.* **89**, 277004 (2002).
[329] G. Misguich and C. Lhuillier, *Frustrated Spin Systems*, World Scientific, Singapore (2004).
[330] M. Hermele, M. P. A. Fisher, and L. Balents, *Phys. Rev. B* **69**, 064404 (2004).

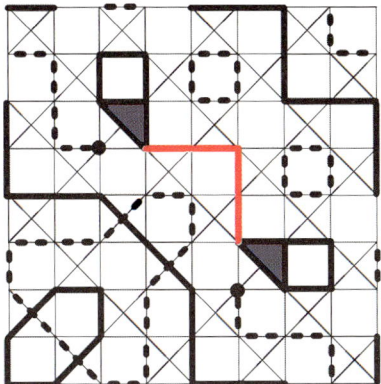

FIG. VII.15. String (marked in red) of an odd number of sites (here five) connecting two particles with charge $e/2$.

Finally we want to comment on some modifications which occur when the spin of the fermions is included. The question may be asked where the spin of an added electron goes when the excitation falls into two parts with charge $e/2$ each. The answer is found by looking at Figure VII.15 where periodic boundary conditions are used. The two charges $e/2$ are connected by a spin chain containing an odd number of sites. The ground state of that chain is two-fold degenerate and represents the spin degrees of freedom. Thus one may state that the spin is smeared over parts of the system as is the connected spin chain. This is a rather novel feature which we have not met before in condensed matter physics to the best of our knowledge. It should be also noticed that in the presence of spins the lowest order ring exchange process on a checkerboard lattice involves four sites instead of six. This is different for the pyrochlore lattice where the smallest possible ring involves six sites (see Figure VII.4). So the ground-state degeneracy is lifted in order t^2/V.

VIII. High-Energy Excitations

In Section II we demonstrated that strong electron correlations generate characteristic low energy scales. They are much smaller than the Fermi energy, the typical energy scale in a metal. But it is well known that strong correlations influence high-energy excitations as well. Hereby the expression *high energy* is used quite flexible. For example, it may include satellite structures or peaks as they are found in the photoelectron spectra of Ni metal[331,332] or of the high temperature superconducting cuprates. But it also includes energies which are larger than the one

---

[331] S. Hüfner and G. K. Wertheim, *Phys. Lett. A* **47**, 349 (1974).
[332] W. Eberhardt and E. W. Plummer, *Phys. Rev. B* **21**, 3245 (1980).

at which Fermi liquid behavior breaks down. In practice this may be a rather low energy or temperature. These effects are best studied by investigating the one-particle Green's function. In Dyson's representation it reads

$$G_\nu(\mathbf{k}, z) = \frac{1}{z - \epsilon_\nu(\mathbf{k}) - \Sigma_\nu(\mathbf{k}, z)}. \tag{8.1}$$

The energy $\epsilon_\nu(\mathbf{k})$ describes the dispersion of electrons in band ν within an effective single-electron approximation while the self-energy $\Sigma_\nu(\mathbf{k}, \omega)$ contains all effects beyond that approximation. In case that $\epsilon_\nu(\mathbf{k})$ describes the Hartree–Fock bands, $\Sigma_\nu(\mathbf{k}, \omega)$ contains the electronic correlations. When Landau's Fermi liquid theory does apply, the low-energy excitations near the Fermi energy are quasiparticles. They are obtained from

$$G(\mathbf{k}, \omega) = \frac{Z}{\omega - \epsilon_{qp}(\mathbf{k}) - i\gamma_\mathbf{k}\mathrm{sgn}\,\omega} + G_{\mathrm{inc}}(\mathbf{k}, \omega). \tag{8.2}$$

For convenience we have omitted the band index ν. The first term contains the quasiparticle pole at

$$\omega = \epsilon_{qp}(\mathbf{k}) + i\gamma_\mathbf{k} \tag{8.3}$$

with the quasiparticle dispersion $\epsilon_{qp}(\mathbf{k})$ and lifetime $\gamma_\mathbf{k}$. The second term is the incoherent part which is of special interest here. Usually it is not further discussed but we want to draw attention to its importance and to significant features when electron correlations are strong. We may think of it as being due to the internal degrees of freedom of the correlation hole which is surrounding an electron. The bare electron has weight $Z < 1$ in the quasiparticle. The reduction is due to the interaction of the electron (or hole) with the surroundings. The internal degrees of freedom of the modified surroundings, i.e., the correlation hole shows up in the integrated spectral density in form of quasiparticle damping and satellite structures. Those structures have generally a $\mathbf{k}$ dependence which is much weaker than the one of the quasiparticles. An example is the shadow band in the Hubbard model when the Hubbard I approximation[7] is made.

The internal degrees of freedom of the correlation hole are best described by the projection operator method. The idea is to select those particular operators which describe the most important microscopic processes in setting up the correlation hole. Green's function is determined within that restricted operator space. The Hubbard I approximation to the Hubbard model is the simplest example. Here only one operator, i.e., $c_{i\sigma}^+ n_{i-\sigma}$ where i is the site index is used in order to describe the correlation hole. More interesting cases are discussed in the following subsections.

We want to draw attention to the zero-point fluctuations of internal excitation modes of the correlation hole. A well known example is Gaskell's ansatz[333] for the ground-state $|\psi_0\rangle$ of a homogeneous electron gas. It is of the form

$$|\psi_0\rangle = \exp\left(\sum_{\mathbf{q}} \tau(\mathbf{q})\rho_{\mathbf{q}}^{\dagger}\rho_{\mathbf{q}}\right)|\Phi_0^{SCF}\rangle \quad (8.4)$$

where $|\Phi_0^{SCF}\rangle$ is the Hartree–Fock ground state and $\rho_{\mathbf{q}}$ are the density fluctuations of wavenumber q. Furthermore $\tau(q) \sim q^{-2}$ when $q \to 0$. Here the zero-point fluctuations of plasmons are taken into account which are the internal excitation modes of the correlation hole when the latter is described within the random phase approximation (RPA). It is well known that the RPA models very well the long-range part of the correlation hole but fails to describe properly the short-range part. Zero-point fluctuations of other modes, in particular of those associated with the short-range part of the correlation hole enter in a similar way the (Jastrow) prefactor in Eq. (8.4). We start out by describing the projection operator formalism before we discuss a number of applications.

19. Projection Operators

In the following we shall use the retarded Green function in order to determine the excitations of the system. It is given by

$$G_{\nu\sigma}^{R}(\mathbf{k}, t) = -i\Theta(t)\langle\psi_0|[c_{\nu\sigma}(\mathbf{k}, t), c_{\nu\sigma}^{\dagger}(\mathbf{k})]_{+}|\psi_0\rangle \quad (8.5)$$

where $|\psi_0\rangle$ is the exact ground state and $\Theta(t)$ is the step function, i.e., $\Theta(t) = 1$ for $t \geq 0$ and zero otherwise. The superscript R will be left out in the following discussion for simplicity. We assume that the Fourier transforms of the most important microscopic processes for the generation of the correlation hole are represented by a set of operators $\{A_{\mu}(\mathbf{k})\}$, the dynamical variables. But we want to include in this set also the original operators $c_{\nu\sigma}^{\dagger}(\mathbf{k})$. Explicit examples for proper choices of the $A_{\mu}(\mathbf{k})$ will be given when specific applications of this method are discussed. Within that reduced operator space $\mathfrak{R}_0$ spanned by the $\{A_{\nu}(\mathbf{k})\}$ we define the Green function matrix

$$G_{\mu\nu}(\mathbf{k}, t) = -i\Theta(t)\langle\psi_0|[A_{\mu}^{\dagger}(\mathbf{k}, t), A_{\nu}(\mathbf{k}, 0)]_{+}|\psi_0\rangle. \quad (8.6)$$

Note that the Green function (8.5) is just a diagonal element of that matrix. It is convenient to introduce the following notation

$$(A|B)_{+} = \langle\psi_0|[A^{\dagger}, B]_{+}|\psi_0\rangle. \quad (8.7)$$

---

[333] T. Gaskell, *Proc. Phys. Soc.* **72**, 685 (1958).

This enables us to rewrite (8.6) after a Fourier transformation in the condensed form

$$G_{\mu\nu}(\mathbf{k}, z) = \left(A_\mu \left| \frac{1}{z - L} \right| A_\nu \right)_+ \tag{8.8}$$

where $z = \omega + i\eta$ and η is a positive infinitesimal number. The Liouvillian L corresponds to H and is defined by its action on an arbitrary operator A through

$$LA = i\frac{dA}{dt} = [H, A]_- \tag{8.9}$$

so that

$$A(t) = e^{iLt} A(0). \tag{8.10}$$

By making use of the identity

$$\frac{1}{a+b} = \frac{1}{a} - \frac{1}{a} b \frac{1}{a+b}$$

we can rewrite (8.8) in the form of the following matrix equation

$$\left[z\mathbb{1} - (\mathbb{L} + \mathbb{M}(z))\chi^{-1} \right] \mathbb{G}(z) = \chi \tag{8.11}$$

with matrix elements

$$L_{\mu\nu} = (A_\mu | L A_\nu)_+, \qquad \chi_{\mu\nu} = (A_\mu | A_\nu)_+$$

$$M_{\mu\nu}(z) = \left(A_\mu \left| LQ \frac{1}{z - QLQ} QL A_\nu \right. \right)_+. \tag{8.12}$$

The matrix $\mathbb{M}(z)$ is called memory function.[334,335] It couples the relevant operators $\{A_\nu\}$ to the remaining degrees of freedom. The operator Q

$$Q = 1 - \sum_{ij} |A_i)_+ \chi_{ij}^{-1} (A_j| \tag{8.13}$$

projects onto those remaining degrees, i.e., onto an operator space perpendicular to the $\{A_\nu\}$, i.e., $Q|A_\nu)_+ = 0$. Setting $\mathbb{M}(z) = 0$ implies that the dynamics of the system is approximated by the $\{A_\nu\}$ and takes place within $\mathfrak{R}_0$. One may either choose a large basis $\{A_\nu\}$ and set $\mathbb{M}(z) = 0$ or work with a small basis and keep $\mathbb{M}(z)$. Examples are given below. If not stated otherwise we shall set $\mathbb{M}(z) = 0$.

The dimension of the matrix Eq. (8.11) equals the number of dynamical variables $A_\nu(\mathbf{k})$. The energy resolution of the excitation spectrum depends on the size of the set $\{A_\nu(\mathbf{k})\}$. For high-energy excitations a relatively small number of dynamical variables is sufficient. By increasing their number one can increase the energy resolution of the spectral density calculated from Eq. (8.8).

---

[334] H. Mori, *Progr. Theor. Phys.* **33**, 423 (1965).
[335] R. Zwanzig, *Lectures in Theoretical Physics*, Interscience, New York (1961), vol. 3.

20. THE HUBBARD MODEL: APPEARANCE OF SHADOW BANDS

The Hubbard model shows particularly well a number of generic features caused by strong electron correlations. Among them are the appearance of satellite structures, shadow bands and a marginal Fermi liquid like behavior close to half filling. The Hamiltonian is of the well known form

$$H = -t \sum_{\langle ij \rangle} \left(c_{i\sigma}^\dagger c_{j\sigma} + h.c. \right) + U \sum_i n_{i\uparrow} n_{i\downarrow}$$
$$= H_0 + H_I \qquad (8.14)$$

in standard notation. In the following we discuss solely for pedagogical and illustrative reasons the spin-density wave (SDW) and Hubbard I approximation by applying projection operators. The simplest case in which a shadow band does appear, is a square lattice at half filling when a spin-density wave (SDW) approximation is made. This simple approximation leads to an antiferromagnetic ground state $|\Phi_{AF}\rangle$. Charge fluctuations are suppressed here on a mean-field level by symmetry breaking. Breaking a symmetry can reduce intersite charge fluctuations similarly as strong correlations do without symmetry breaking. Therefore features of strong correlations show up already in this simple mean-field scheme.

From the kinetic energy term in (8.14) one obtains the dispersion $\epsilon(\mathbf{k}) = -2t(\cos k_x + \cos k_y)$. The lattice constant has been set equal to unity. Therefore, at half filling the Fermi surface is nested and the ground state in mean-field approximation is a spin density wave. One finds

$$|\Phi_{AF}\rangle = \prod_{\mathbf{k}\sigma} \left[u_\mathbf{k} c_{\mathbf{k}\sigma}^\dagger + \sigma v_\mathbf{k} c_{\mathbf{k}+\mathbf{Q}\sigma}^\dagger \right] |0\rangle \qquad (8.15)$$

where $\mathbf{Q}$ is a reciprocal lattice vector and $c_{\mathbf{k}\sigma}^\dagger$ is the Fourier transform of $c_{i\sigma}^\dagger$. Furthermore $u_\mathbf{k}^2 + v_\mathbf{k}^2 = 1$ with

$$u_\mathbf{k}^2 = \frac{1}{2}\left(1 - \frac{\epsilon(\mathbf{k})}{E(\mathbf{k})}\right), \qquad v_\mathbf{k}^2 = \frac{1}{2}\left(1 + \frac{\epsilon(\mathbf{k})}{E(\mathbf{k})}\right)$$

$$E(\mathbf{k}) = \left(\epsilon(\mathbf{k})^2 + \frac{m_0^2 U^2}{4} \right)^{\frac{1}{2}}. \qquad (8.16)$$

Here m_0 is the staggered magnetization. The latter has to be calculated self-consistently. In order to calculate the excitations we choose for the relevant dynamical variables $\{A_\nu\}$

$$A_1(\mathbf{k}) = c_{\mathbf{k}\sigma}^\dagger, \qquad A_2(\mathbf{k}) = c_{\mathbf{k}+\mathbf{Q}\sigma}^\dagger. \qquad (8.17)$$

When the 2×2 matrix (8.11) with $\mathbb{M}(z) = 0$ is evaluated one finds

$$G_{11}(\mathbf{k}, z) = \frac{u_\mathbf{k}^2}{z - (\frac{U}{2} - E(\mathbf{k}))} + \frac{v_\mathbf{k}^2}{z - (\frac{U}{2} + E(\mathbf{k}))} \qquad (8.18)$$

and for the spectral density

$$D(\mathbf{k}, \omega) = u_\mathbf{k}^2 \delta\left(\omega - \frac{U}{2} + E(\mathbf{k})\right) + v_\mathbf{k}^2 \delta\left(\omega - \frac{U}{2} - E(\mathbf{k})\right). \qquad (8.19)$$

For each **k** point there are two contributions to $D(\mathbf{k}, \omega)$, i.e., one δ-function peak with a large weight and one with a small one. The peaks with the smaller weight form a shadow band $E_{sb}(\mathbf{k}) = -E(\mathbf{k})$ which complements the band resulting from the peaks with large weight.[336,337] Because of the mean-field level the shadow band disappears for temperatures higher than the Néel temperature. This is in reality not the case and therefore one would like to reproduce a shadow band also for arbitrary filling factors and for the paramagnetic state. This can be done, of course, only by accounting for the strong correlations. They are described in the simplest way by the Hubbard I[7] approximation. In that case the following choice is made for the dynamical variables $\{A_\nu\}$

$$A_1(\mathbf{k}) = c_{\mathbf{k}\sigma}^\dagger, \qquad A_2(\mathbf{k}) = \frac{1}{\sqrt{N_0}} \sum_i e^{i\mathbf{k}\mathbf{R}_i} c_{i\sigma}^\dagger \delta n_{i-\sigma} \qquad (8.20)$$

with $\delta n_{i-\sigma} = n_{i-\sigma} - \langle n_{i-\sigma} \rangle$. The $\mathbf{R}_i$ denote the positions of the N_0 lattice sites. Again, we want to determine $G_{11}(\mathbf{k}, \omega)$ from (8.8) by setting $\mathbb{M}(z) = 0$. The 2×2 matrix equation can be easily solved and one finds

$$G_{11}(\mathbf{k}, z) = \left[z - \epsilon(\mathbf{k}) - \frac{U}{2} n \left(1 + \frac{\hat{U}}{z - \hat{U}}\right)\right]^{-1} \qquad (8.21)$$

where n is the number of electrons per site and $\hat{U} = U(1 - n/2)$. $G_{11}(\mathbf{k}, \omega)$ has two poles centered at $z = 0$ and $z = U$. Neglecting terms of order U^{-1} we can rewrite (8.21) as

$$G_{11}(\mathbf{k}, z) = \frac{1 - \frac{n}{2}}{z - \epsilon(\mathbf{k})(1 - \frac{n}{2})} + \frac{\frac{n}{2}}{z - U - \epsilon(\mathbf{k})\frac{n}{2}}. \qquad (8.22)$$

The poles give raise to two bands, i.e., the upper and the lower Hubbard band. Their widths differ except for $n = 1$. For $n = 0.25$ the upper Hubbard band is reduced to a satellite structure with a **k**-dependent dispersion. This is illustrated in Figure VIII.1 and is well known. The reason for repeating these facts here is

---

[336] A. P. Kampf and J. R. Schrieffer, *Phys. Rev. B* **42**, 7967 (1990).
[337] A. P. Kampf, *Phys. Rev.* **249**, 219 (1994).

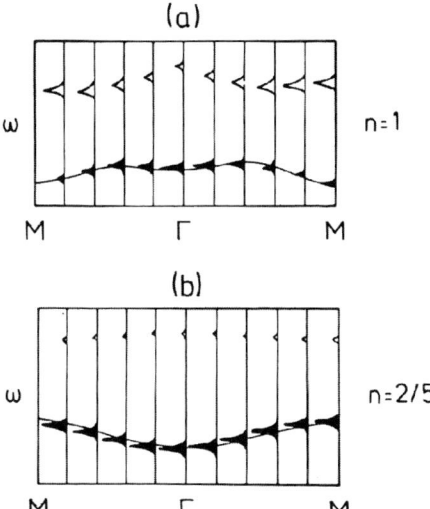

FIG. VIII.1. Schematic representation of the two bands resulting from the two variables of Eq. (8.20), i.e., the upper and the lower Hubbard band. The electron number per site is n, the broadening is artificial.

that we want to proceed similarly when we discuss the satellite structure in Ni or the spectral density of electrons in Cu–O planes. We want to emphasize the point of view that the incoherent part of a Green function is the superposition of satellite peaks which have a small **k**-dependence each. But before, we demonstrate that for special band fillings a Hubbard model can exhibit marginal Fermi liquid behavior. This is shown explicitly for a square lattice.

21. Marginal Fermi Liquid Behavior and Kink Structure

The Coherent Potential Approximation (CPA) has been widely used in electronic structure calculations of disordered systems.[338–343] Here we want to combine it with the projection operator technique and apply it to the Hubbard model. We remind the reader that a CPA was introduced first by Hubbard when treating his

---

[338] P. Soven, *Phys. Rev.* **156**, 809 (1967).
[339] D. W. Taylor, *Phys. Rev.* **156**, 1017 (1967).
[340] B. Velický, S. Kirkpatrick, and H. Ehrenreich, *Phys. Rev.* **175**, 747 (1968).
[341] F. Yonezawa, *Progr. Theor. Phys.* **40**, 734 (1968).
[342] R. J. Elliott, J. A. Krumhansl, and P. L. Leath, *Rev. Mod. Phys.* **46**, 465 (1974).
[343] H. Ehrenreich and L. M. Schwartz, in *Solid State Physics*, vol. 31, Academic Press, New York (1976).

Hamiltonian.[20] In the following we want to treat the many-electron problem as accurately as possible for a number of sites which need not be connected. This cluster is embedded in a medium with a coherent potential $\tilde{\Sigma}(\omega)$. The potential is determined self-consistently so that it agrees with the momentum integrated self-energy $\Sigma(\mathbf{k}, \omega)$ of the cluster. It is known[344,345] that the local projection operator method combined with the CPA[346] is equivalent to the Many-Body CPA,[347] the Dynamical CPA[348,349] as well as to the Dynamical Mean Field Theory (DMFT) which is based on many-body physics in infinite dimensions.[350,351] For reviews of the latter, see Refs. [186,352]. Here we want to discuss the more general non-local version of the projection operator method[353] combined with the CPA. It is based on a decomposition of the scattering matrix of the system into one-site, two-site etc. scattering matrices. This way the scattering matrix can be calculated successively in terms of increments, a method successfully applied in the theory of wavefunction based electronic structure calculations of solids. For a review of that method, see Refs. [354,41]. We apply the theory to the Hubbard model on a square lattice near half filling.[7,20] The results are interesting. A marginal Fermi liquid behavior is found for large U values and small hole doping and a corresponding phase diagram is worked out.[355] It is also found that for similar parameters the excitation spectrum (or real part of the self-energy) has a kink near E_F.[356] Such a structure was observed in a number of underdoped cuprates like $Bi_2Sr_2CaCuO_{8+\delta}$ or $La_{2-\delta}Sr_\delta CuO_4$.[357,358] It has been attributed to electron–phonon interactions[358–361] and also to the interaction with a magnetic resonance

---

[344] Y. Kakehashi, *Adv. Physics* **53**, 497 (2004).

[345] Y. Kakehashi, *Phys. Rev. B* **66**, 104428 (2002).

[346] Y. Kakehashi and P. Fulde, *Phys. Rev. B* **69**, 045101 (2004).

[347] S. Hirooka and M. Shimizu, *J. Phys. Soc. Jpn.* **43**, 70 (1977).

[348] Y. Kakehashi, *J. Magn. Magn. Mater.* **104–107**, 677 (1992).

[349] Y. Kakehashi, *Phys. Rev. B* **45**, 7196 (1992).

[350] W. Metzner and D. Vollhardt, *Phys. Rev. Lett.* **62**, 324 (1989).

[351] E. Müller-Hartmann, *Z. Phys. B: Cond. Matt.* **74**, 507 (1989).

[352] T. Pruschke, M. Jarrell, and J. Freericks, *Adv. Phys.* **44**, 187 (1995).

[353] Y. Kakehashi and P. Fulde, *Phys. Rev. B* **70**, 195102 (2004).

[354] H. Stoll, B. Paulus, and P. Fulde, *J. Chem. Phys.* **123**, 144108 (2005).

[355] Y. Kakehashi and P. Fulde, *Phys. Rev. Lett.* **94**, 156401 (2005).

[356] Y. Kakehashi and P. Fulde, *J. Phys. Soc. Jpn.* **74**, 2153 (2005).

[357] P. V. Bogdanov, A. Lanzara, S. A. Keller, X. J. Zhou, E. D. Lu, W. J. Zheng, G. Gu, J.-I. Simoyama, K. Kishio, H. Ikeda, et al., *Phys. Rev. Lett.* **85**, 2581 (2000).

[358] T. Chuk, D. H. Lu, X. J. Zhou, Z.-X. Shen, T. P. Devereaux, and N. Nagaosa, *Phys. Stat. Sol. (b)* **242**, 11 (2005).

[359] Z.-X. Shen, A. Lanzara, S. Ishihara, and N. Nagaosa, *Philos. Mag. B* **82**, 1349 (2002).

[360] S. Verga, A. Knigavko, and F. Marsiglio, *Phys. Rev. B* **67**, 054503 (2003).

[361] S. Ishihara and N. Nagaosa, *Phys. Rev. B* **69**, 144520 (2004).

mode observed by inelastic neutron scattering.[362–365] But as it turns out, also strong electron correlations in a 2D Hubbard model can produce it.

To reach our goal we start out by choosing for the $\{A_\nu\}$ simply the operators $c_{i\sigma}^\dagger$ but we keep this time the memory matrix, i.e., $\mathbb{M}(z) \neq 0$. In that case we write for a paramagnetic system

$$G(\mathbf{k}, z) = \frac{1}{z - \epsilon(\mathbf{k}) - U^2 M(\mathbf{k}, \omega)}. \tag{8.23}$$

The matrix $\mathbb{L}$ in (8.11) gives only a constant energy shift and is neglected here. For convenience a factor U^2 has been extracted from $M(\mathbf{k}, z)$. We decompose

$$M(\mathbf{k}, z) = \sum_j M_{j0}(z) e^{i\mathbf{k}\mathbf{R}_j} \tag{8.24}$$

and write for the reduced memory matrix according to (8.12)

$$M_{ij}(z) = \left(c_{i\sigma}^\dagger \delta n_{i-\sigma} \left| \frac{1}{z - \bar{L}} c_{j\sigma}^\dagger \delta n_{j-\sigma} \right. \right)_+, \tag{8.25}$$

where $\bar{L} = QLQ$ with the projector $Q = 1 - \sum_{i\sigma} |c_{i\sigma}^\dagger)_+ (c_{i\sigma}^\dagger|$. Stopping at this stage, i.e., neglecting the new memory function in the denominator of (8.25) would bring us back to the Hubbard I approximation. In order to establish a connection to the CPA we define the Hamiltonian of an effective medium through

$$\tilde{H}(z) = H_0 + \tilde{\Sigma}(z) \sum_i n_i \tag{8.26}$$

with the corresponding Liouvillian $\tilde{L}$, i.e., $\tilde{L}B = [\tilde{H}(z), B]_-$ for arbitrary operators B. The aim is to account efficiently for local as well as nonlocal correlations by means of an effective medium characterized by $\tilde{\Sigma}(z)$.

The coherent potential $\tilde{\Sigma}(z)$ is determined self-consistently from

$$\tilde{\Sigma}(z) = \frac{U^2}{N} \sum_\mathbf{k} M(\mathbf{k}, z) \tag{8.27}$$

---

[362] M. Eschrig and M. R. Norman, *Phys. Rev. Lett.* **85**, 3261 (2000).
[363] P. D. Johnson, T. Valla, A. V. Fedorov, Z. Yusof, B. O. Wells, Q. Li, A. R. Moodenbaugh, G. D. Gu, N. Koshizuka, C. Kendziora, et al., *Phys. Rev. Lett.* **87**, 177007 (2001).
[364] M. Eschrig and M. R. Norman, *Phys. Rev. B* **67**, 144503 (2003).
[365] E. Schachinger, J. J. Tu, and J. P. Carbotte, *Phys. Rev. B* **67**, 214508 (2003).

where N is the number of sites. With this in mind we decompose

$$\bar{L} = Q\tilde{L}Q + \sum_i Q\big(U\delta n_{i\uparrow}\delta n_{i\downarrow} - \tilde{\Sigma}(z)n_i\big)Q$$

$$= L_0(z) + \sum_i L_I^{(i)}(z)$$

$$= L_0(z) + L_I(z). \qquad (8.28)$$

Next we express $(z - \bar{L})^{-1}$ in terms of a scattering (super)operator T as

$$\frac{1}{z - \bar{L}} = \frac{1}{z - L_0} + \frac{1}{z - L_0} T \frac{1}{z - L_0}$$

$$= g_0(z) + g_0(z) T g_0(z) \qquad (8.29)$$

with

$$T = L_I + L_I g_0 L_I + \cdots.$$

The T-operator can be decomposed into a sequence of many-sites increments

$$T = \sum_i T_i + \sum_{\langle ij\rangle} \delta T_{ij} + \sum_{\langle ijk\rangle} \delta T_{ijk} + \cdots. \qquad (8.30)$$

Those increments are closely related to the T operators $T_i, T_{ij}, T_{ijk}, \ldots$ of single-site, 2-sites, 3-sites, etc. clusters. It is

$$\delta T_{ij} = T_{ij} - T_i - T_j$$
$$\delta T_{ijk} = T_{ijk} - \delta T_{ij} - \delta T_{ik} - \delta T_{jk} - T_i - T_j - T_k$$
$$\vdots \qquad (8.31)$$

This enables us to introduce retarded memory functions for clusters

$$M_{ij}^{(c)}(z) = \left(c_{i\sigma}^\dagger \delta n_{i-\sigma} \left| \frac{1}{z - \bar{L}^{(c)}} c_{j\sigma}^\dagger \delta n_{j-\sigma} \right.\right)_+. \qquad (8.32)$$

These matrices have dimensions 1×1 when $c = i$ (one-site cluster) and 2×2 when $c = (i, j)$, i.e., in the 2-site cluster approximation. The Liouvillian $\bar{L}^{(c)}(z)$ is given by

$$\bar{L}^{(c)} = L_0(z) + \sum_{n \in c}^{N_c} L_I^{(n)}(z). \qquad (8.33)$$

The sum over n involves all N_c sites belonging to a given cluster c. Within zeroth-order renormalized perturbation theory (RPT-0) only that part of $L_I^{(n)}(z)$

is used which projects onto the operators $c_{i\sigma}^{\dagger} \delta n_{i-\sigma}$, i.e., $\bar{P} L_I^{(n)}(z) \bar{P}$ with the projector.

$$\bar{P} = \sum_{i\sigma} |c_{i\sigma}^{\dagger} \delta n_{i-\sigma})_+ \chi_i^{-1} (c_{i\sigma}^{\dagger} \delta n_{i-\sigma}| \qquad (8.34)$$

and $\chi_i = \langle n_{i-\sigma} \rangle (1 - \langle n_{i-\sigma} \rangle)$. With this simplification the memory matrix can be expressed in terms of a "screened" one as

$$M_{ij}^{(c)}(z) = \left[\mathfrak{g} \cdot \left(1 - \mathbb{L}_I^{(c)} \mathfrak{g} \right)^{-1} \right]_{ij}. \qquad (8.35)$$

The screened memory matrix is given by

$$g_{ij}(z) = \left(c_{i\sigma}^{\dagger} \delta n_{i-\sigma} \left| \frac{1}{z - L_0(z)} c_{j\sigma}^{\dagger} \delta n_{j-\sigma} \right)_+ . \qquad (8.36)$$

The cluster memory matrix $M_{ij}^{(c)}(z)$ describes a Hubbard cluster embedded in a uniform medium with a Hamiltonian $\tilde{H}(z)$ (see the left side in Figure VIII.2). The interactions in the embedded cluster are given by $\sum_{i \in c}(U \delta n_{i\uparrow} \delta n_{i\downarrow} - \tilde{\Sigma}(z) n_i)$. We start from a uniform medium described by $\tilde{H}(z)$ (see Eq. (8.28) and the right side of Figure VIII.2). An alternative would have been to start from the same medium but with cavities at the sites of the cluster (see the middle of Figure VIII.2), i.e., from a Hamiltonian $\tilde{H}^{(c)}$ with

$$\tilde{H}^{(c)} = \tilde{H}(z) - \tilde{\Sigma}(z) \sum_{n \in c}^{N_c} n_i. \qquad (8.37)$$

In this case the interaction in the cluster would be $\sum_{n \in c} U \delta n_{i\uparrow} \delta n_{i\downarrow}$.

The diagonal matrix $\mathbb{L}_I^{(c)}$ in (8.35) describes the atomic excitations. The matrix elements are given by $[L_I^{(i)}, L_I^{(j)}, \ldots]$. It has one element $L_I^{(i)} = U(1 - 2\langle n_{i-\sigma} \rangle)/[\langle n_{i-\sigma} \rangle (1 - \langle n_{i-\sigma} \rangle)]$ when $c = i$ and a second one $L_I^{(j)}$ when $c = (i, j)$. The incremental cluster expansion is depicted in Figure VIII.3. With the above approximations the theory reproduces two limiting cases exactly, i.e., the limit of

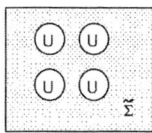
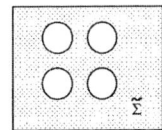
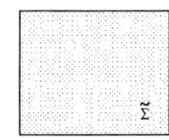

FIG. VIII.2. Left side: Cluster with on-site Coulomb repulsion U embedded in a medium. Middle: Cavities replacing the sites of the cluster. Right side: Uniform medium with self-energy $\tilde{\Sigma}(\omega)$. (After Ref. [353].)

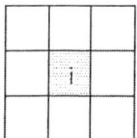 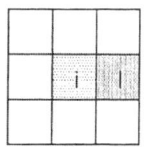 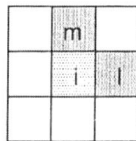

FIG. VIII.3. Schematic drawing of the multisite cluster expansion of Green's function. From left to right: single-site, two-sites and three-sites contributions. Note that the sites of a cluster need not be nearest neighbors.

small U when perturbation theory is applicable as well as the atomic limit. This is an important feature of the present theory.

In the limit of large clusters the memory functions are independent of the medium. But since the cluster expansion must be truncated in practical applications there is a dependence of the memory functions on the medium and we must make an optimal choice for it. When correlation effects on the *static* matrix elements are neglected one can write down an explicit expression for the screened memory function of the form

$$g_{ij}^{(c)}(z) = A_{ij} \int \frac{d\epsilon\, d\epsilon'\, d\epsilon''\, \rho_{ij}^{(c)}(\epsilon) \rho_{ij}^{(c)}(\epsilon') \rho_{ji}^{(c)}(\epsilon'') \chi(\epsilon, \epsilon', \epsilon'')}{z - \epsilon - \epsilon' + \epsilon''} \tag{8.38}$$

with $A_{ii} = [\langle n_{i-\sigma}\rangle(1 - \langle n_{i-\sigma}\rangle)]/[\langle n_{i-\sigma}\rangle_c(1 - \langle n_{i-\sigma}\rangle_c)]$ and $A_{i\neq j} = 1$. Here $\rho_{ij}^{(c)}(\epsilon)$ is the density of states of a system with one or two empty sites (depending on the cluster c) embedded in a medium with a coherent potential $\tilde{\Sigma}(z)$. More specifically

$$\rho_{ij}^{(c)}(z) = -\frac{1}{\pi} \operatorname{Im}\bigl[(z - \tilde{H}^{(c)})^{-1}\bigr]_{ij}. \tag{8.39}$$

Furthermore

$$\langle n_{i\sigma}\rangle_c = \int d\epsilon\, \rho_{ii}^{(c)}(\epsilon) f(\epsilon)$$

where $f(\epsilon)$ is the Fermi function. Moreover

$$\chi(\epsilon, \epsilon', \epsilon'') = f(-\epsilon) f(-\epsilon') f(-\epsilon'') + f(\epsilon) f(\epsilon') f(-\epsilon'').$$

These approximations are used in order so solve Eq. (8.27) self-consistently whereby $M(\mathbf{k}, z)$ is replaced by Eq. (8.35).

We reemphasize that in the two-site cluster approximation for fixed site i, all sites j are taken into account until convergence is achieved. For the Hubbard model on a square lattice this includes sites which are more than ten lattice vectors apart! In this respect the present theory resembles incremental schemes which have been applied in the treatment of solids by quantum chemistry methods.[366] For a review see Ref. [41]. The self-consistent projection method (SCPM) provides an interesting link between treatments of model Hamiltonians in solid state theory and true ab initio calculations with controlled approximations based on methods used in quantum chemistry. Our theory differs from extensions of the DMFT such as the Dynamical Cluster Approximations[367] or the Cellular Dynamical Mean-Field Theory[368] where clusters of *connected* sites are treated. They require a truly impressive amount of numerical work and have been carefully reviewed in Ref. [369].

Determining $M(\mathbf{k}, \omega)$ might look like a very demanding computational problem too, but that is not really the case. In fact, we can calculate $M(\mathbf{k}, \omega)$ directly without a numerical analytic continuation from the imaginary axis as is the case in other schemes.[369] Therefore a high numerical resolution as regards energy and momentum can be obtained.

The following results are obtained for $U = 8|t|$ and $T = 0$ in the underdoped regime, i.e., for $n \leqslant 1$. At half-filling, the system will always be an antiferromagnetic insulator due to nesting. But in the effective medium approach described here one finds in addition also a paramagnetic metallic solution. We use the latter here, being aware of the fact that an arbitrarily small temperature or doping concentration will destroy antiferromagnetic order in two dimensions (Mermin–Wagner theorem). A metallic solution does exist only for $U < U_{\text{crit}}$ where in single-site approximation $U_{\text{crit}} \simeq 14|t|$ is obtained. Therefore with a value of $U = 8|t|$ we are still in the metallic regime. We find that a flat quasiparticle band is crossing the Fermi energy E_F. There is also an empty upper Hubbard band found which is centered around the M point and there is incoherent spectral density near the Γ point resulting from the lower Hubbard band. The results compare well with those of Quantum Monte Carlo (QMC) calculations for finite temperatures.[370–372] This is seen in Figure VIII.4. The Fermi surface for $n = 0.95$ (underdoped regime) is found to be hole like. This is due to a collapse of the lower Hubbard band which causes a portion of the flat quasiparticle band near the X point to sink below E_F.

These findings are contrary to expectations from the Hubbard I approximation. From (8.22) it is seen that within that approximation the spectral weight of the lower Hubbard band increases while that of the upper Hubbard band decreases in case of hole doping $n < 1$. In the present, improved approximation this is not the

---

[366] H. Stoll, *Phys. Rev. B* **46**, 6700 (1992).
[367] M. H. Hettler, M. Mukherjee, M. Jarrell, and H. R. Krishnamurthy, *Phys. Rev. B* **61**, 12739 (2000).
[368] G. Kotliar, S. Savrasov, G. Pallson, and G. Biroli, *Phys. Rev. Lett.* **87**, 186401 (2001).
[369] T. Maier, M. Jarrell, T. Pruschke, and M. H. Hettler, *Rev. Mod. Phys.* **77**, 1027 (2005).
[370] N. Bulut, D. J. Scalapino, and S. R. White, *Phys. Rev. Lett., Phys. Rev.* **73, 50**, 748, 7215 (1994).
[371] R. Preuss, W. Hanke, and W. von der Linden, *Phys. Rev. Lett.* **75**, 1344 (1994).
[372] C. Gröber, R. Eder, and W. Hanke, *Phys. Rev. B* **66**, 4336 (2000).

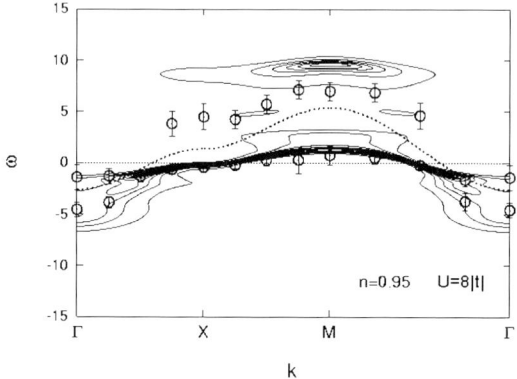

FIG. VIII.4. Single-particle excitations at zero temperatures for the Hubbard model on a square lattice in units of the nearest-neighbor transfer integral t. Dashed line: Hartree–Fock result. Open circles with error bars: QMC results for $T = 0.33$.[372] (After Ref. [355].)

case for small hole doping and implies that Luttinger's Theorem[30] does not apply here. It would require a reduction of the volume enclosed by the Fermi surface with increasing hole concentration. For $n = 0.8$ (overdoped regime) an electron-like Fermi surface is found. These results are in agreement with the Dynamical Cluster Approximation[373] and with QMC calculations.[370–372]

An important finding is that for doping less than 2% the effective mass $m_\mathbf{k}$ changes strongly between the M point, i.e., $[\pi, \pi]$ where it is minimal and the X point, i.e., $[\pi, 0]$ where it has its maximum. It is computed numerically from $m_\mathbf{k} = 1 - U^2 \operatorname{Re} \partial M(\mathbf{k}, \omega)/\partial \omega|_{\omega=0_+}$. Near the X point one finds that the smaller the $\delta\omega$ steps are in computing the derivative, the larger becomes $m_{\mathbf{k}=X}$. One finds that approximately $m_\mathbf{k} \sim \ln \delta\omega$, indicating marginal Fermi liquid behavior. The marginal Fermi liquid theory[46,47,374,375] had been designed in order to explain a number of features, observed in the normal state of the high-T_c cuprates where electron correlations are strong. They deviate from normal Fermi liquid behavior. Most noticeable is a linear temperature dependence of the resistivity $\rho(T) \sim T$ which has been observed, e.g., in doped La_2CuO_4 materials. It is suggestive to associate this dependence with a quasiparticle scattering rate $1/\tau(T) \sim \rho(T)$, i.e., $\tau^{-1} \sim \operatorname{Im} \Sigma(T) \sim T$. Then it is plausible to assume that for $T < |\omega|$ the corresponding expression is $\operatorname{Im} \Sigma(\omega) \sim |\omega|$. Indeed, optical reflectivity measurements

---

[373] T. A. Maier, T. Pruschke, and M. Jarrell, *Phys. Rev. B* **66**, 075102 (2002).
[374] C. M. Varma, P. B. Littlewood, S. Schmitt-Rink, E. Abrahams, and A. Ruckenstein, *Phys. Rev. Lett.* **64**, 497 (1990).
[375] G. T. Zimanyi and K. S. Bedell, *Phys. Rev. B* **48**, 6575 (1993).

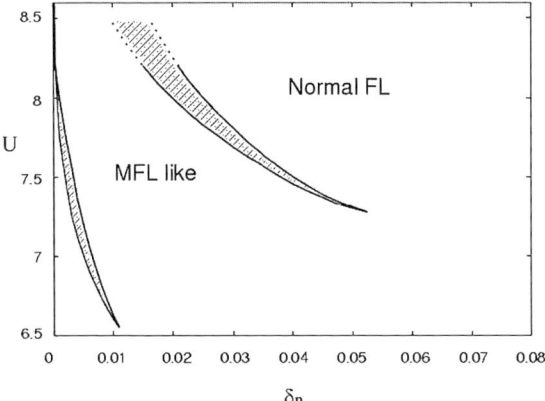

FIG. VIII.5. Phase diagram for the Hubbard model on a square lattice as function of hole doping δn. In the upper shaded region two self-consistent solutions are found one of which corresponds to a marginal and the other to a normal Fermi liquid. In the lower shaded region two slightly different marginal Fermi liquid solutions are found. (After Ref. [355].)

on $YBa_2Cu_3O_4$ are in accordance with this form.[376] But also the frequency-dependent spin susceptibility at low T shows marginal Fermi liquid behavior. Real and imaginary part of $\Sigma(\mathbf{k}, \omega)$ are related to each other via a Kramers–Kroning relation from which is follows that $\mathrm{Re}\,\Sigma(\omega) \sim \omega \ln(\omega/\omega_0)$ where ω_0 is a cut-off parameter. From this form of $\mathrm{Re}\,\Sigma(\omega)$ we conclude that the residuum of the Green's function pole (see Eq. (8.2)) is $Z(\omega) \sim 1/|\ln(\omega/\omega_0)|$ and that the quasiparticle mass $m_\mathbf{k}$ diverges at the Fermi energy like $m_\mathbf{k} \sim |\ln(\omega/\omega_0)|$.

The above treatment shows that a marginal Fermi liquid type of behavior can be obtained from a Hubbard Hamiltonian on a square lattice, provided one is in a certain U dependent doping regime. We expect that this region will be enlarged when long-range antiferromagnetic correlations are taken into account which have been neglected here. The phase diagram in the U vs. δn plane is shown in Figure VIII.5, where δn denotes hole doping. There are two small regions in which two self-consistent solutions are found. They separate the marginal Fermi liquid regime from the normal Fermi liquid one. For smaller U values the cross-over between the two regimes is continuous.

It is instructive to consider the origin of the marginal Fermi-liquid like behavior of the self-energy. For half-filling a van Hove singularity shows up in the density of states at E_F which leads, as is well known, to a self-energy of marginal Fermi liquid type. With hole doping this singularity moves rather fast away from E_F pro-

---

[376] R. T. Collins, Z. Schlesinger, F. Holtzberg, P. Chaudhari, and C. Feild, *Phys. Rev. B* **39**, 6571 (1989).

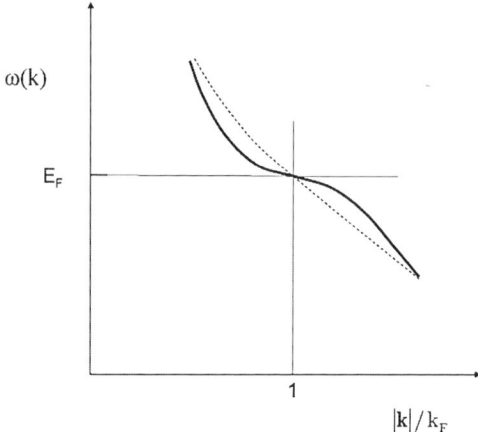

FIG. VIII.6. Schematic drawing of the dispersion $\omega(\mathbf{k})$ near k_F in the presence of a kink.

vided correlation effects are small.[377–379] But when U is large as assumed here, and for small hole doping spectral density moves to high energies and therefore E_F remains virtually pinned to the van Hove singularity. This changes at higher doping concentrations where E_F moves to lower energies and normal Fermi liquid behavior is recovered.

It is interesting that for small hole doping the calculations for $U = 8|t|$ show a kink in the excitation spectrum at approximately $\omega_{\text{kink}} = -0.8|t|$. A kink has been observed in high-resolution photoemission experiments in the quasiparticle band dispersion of high-T_c cuprates.[357,358] It was found that along the nodal direction $(0, 0) - (\pi, \pi)$ the effective Fermi velocity v_F as defined by the form $\omega(\mathbf{k}) = v_F(\mathbf{k})(|\mathbf{k}| - k_F)$ is neither sensitive to the type of cuprate, nor to doping concentration or isotope substitution[380] as long as $\omega < \omega_{\text{kink}}$ ($= 60$–70 meV). But when $\omega > \omega_{\text{kink}}$ a strong dependence of v_F on those quantities as well as on $\mathbf{k}$ is observed. The form of the dispersion is schematically shown in Figure VIII.6.

There have been two possible explanations advanced for this experimental observation. One is based on a coupling of electrons to the longitudinal optical phonon mode found in inelastic neutron scattering experiments.[358] The other is based on a coupling of the electronic quasiparticles to spin fluctuations, in particu-

---

[377] A. Virosztek and J. Ruvalds, *Phys. Rev. B* **42**, 4064 (1990).
[378] H. Schweitzer and G. Czycholl, *Z. Phys. B* **83**, 93 (1991).
[379] G. Kastrinakis, *Physica C* **340**, 119 (2000).
[380] G.-H. Gweon, T. Sasagawa, S. Y. Zhou, J. Graf, H. Takagi, D.-H. Lee, and A. Lanzara, *Nature* **430**, 187 (2004).

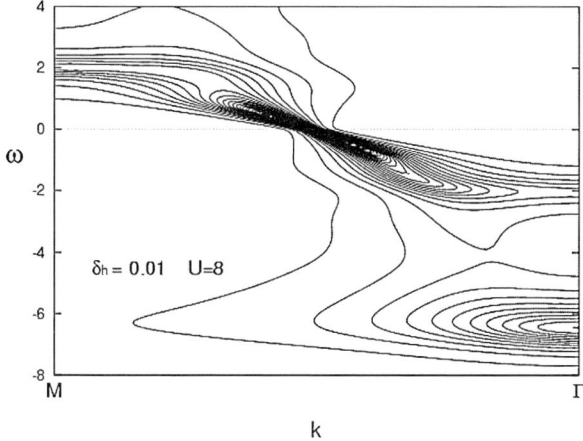

FIG. VIII.7. Hubbard model on a square lattice near half-filling. Contours of spectral function of excitations along M–Γ for hole doping $\delta n = 0.01$ showing a kink. (After Ref. [356].)

lar to an observed resonance mode.[362,363,381] No consensus has been reached yet. So it is interesting to note that a kink is also obtained from the Hubbard model on a square lattice for small doping concentrations. Here it is based on long-ranged electron correlations. In Figure VIII.7 we show the excitation spectrum near E_F for a small hole concentration of $\delta n = 0.01$ and $U = 8|t|$ as before[356] (compare with Figure VIII.4).

As before, the calculations assume a paramagnetic ground state. The kink structure becomes weaker as the doping concentration increases while the kink position changes little with δn. For $\delta n = 0.05$ the kink has disappeared together with the remnants of the lower Hubbard band. The ratio of the Fermi velocities v'_F/v_F above and below ω_{kink} is found to be $v'_F/v_F = 1.8$ for $\delta n = 0.01$ and 1.5 for $\delta n = 0.02$. The kink is caused by a hybridization of the quasiparticle excitation with short-range antiferromagnetic fluctuations. With decreasing antiferromagnetic correlations, e.g., by hole doping the kink structure does also decrease and eventually disappears. Whether or not it provides an explanation for the observed kink in the underdoped regime of $La_{2-x}Sr_xCuO_4$ remains an open question.

22. Nickel and its Satellite

Photoelectron spectroscopy has revealed a pronounced satellite structure in Ni metal which is approximately 6 eV below the Fermi energy E_F.[331] It is due to

---

[381] A. V. Chubukov and M. R. Norman, *Phys. Rev. B* **70**, 174505 (2004).

strong electron correlations and a number of different theoretical model calculations have been performed to explain it. The simplest way of including correlations is by calculating the self-energy in second-order perturbation theory.[382,383] This results in a d-band narrowing effect as well as in a satellite structure. However, for a quantitative comparison with experiments this is not sufficient. Another well known method is the t-matrix approach of Kanamori.[21] It accounts for multiple scattering of two d-holes, but it is strictly valid only in the cases of small or almost complete band filling. It has been used to explain the satellite structure[384,385] and was subsequently extended to include also multiple electron-hole scattering.[386–388] These calculations establish also a link to calculations based on couplings of a hole to magnons.[389–392] There have been also numerical diagonalizations based on a 4-sites tetrahedral cluster with five d-orbitals per site.[393] A correctly positioned multiplet structure was obtained for ferromagnetic Ni but more detailed information is beyond the scheme of that approach. Other calculations were based on a LDA+DMFT approach[394] and on a Gutzwiller projected wavefunction.[395] In the latter case band narrowing but no satellite structure was obtained.

Here we want to start from a five-band Hubbard model and to calculate the spectral density for paramagnetic Ni with the help of the projection operator method.[396] It gives good insight into the relevant microscopic processes which contribute to the satellite and are described by the set of operators $\{A_\nu\}$ (see Eq. (8.6)). The results are found to be similar to those obtained when the two-hole one-electron t matrix is solved by employing Faddeev's equations.[397]

---

[382] G. Treglia, F. Ducastelle, and D. Spanjaard, *J. Phys. (Paris)* **41**, 281 (1980).
[383] G. Treglia, F. Ducastelle, and D. Spanjaard, *J. Phys. (Paris)* **43**, 341 (1982).
[384] D. R. Penn, *Phys. Rev. Lett.* **42**, 921 (1979).
[385] A. Liebsch, *Phys. Rev. Lett.* **43**, 1431 (1979).
[386] A. Liebsch, *Phys. Rev. B* **23**, 5203 (1981).
[387] J. Igarashi, *J. Phys. Soc. Jpn.* **52**, 2827 (1983).
[388] J. Igarashi, *J. Phys. Soc. Jpn.* **54**, 260 (1985).
[389] L. Roth, *Phys. Rev. B* **186**, 428 (1969).
[390] J. A. Hertz and D. M. Edwards, *J. Phys. C* **3**, 2174 (1973).
[391] J. A. Hertz and D. M. Edwards, *J. Phys. C* **3**, 2191 (1973).
[392] H. Matsumoto, L. Umezawa, S. Seki, and M. Tachiki, *Phys. Rev. B* **17**, 2276 (1978).
[393] R. H. Victora and L. M. Falicov, *Phys. Rev. Lett.* **55**, 1140 (1985).
[394] A. I. Lichtenstein, M. I. Katsnelson, and G. Kotliar, *Phys. Rev. Lett.* **87**, 067205 (2001).
[395] J. Bünemann, F. Gebhard, T. Ohm, R. Umstätter, S. Weiser, W. Weber, R. Claessen, D. Ehm, A. Harasawa, A. Kakizaki, et al., *Europhys. Lett.* **61**, 667 (2003).
[396] P. Unger, J. Igarashi, and P. Fulde, *Phys. Rev. B* **50**, 10485 (1994).
[397] J. Igarashi, P. Unger, K. Hirai, and P. Fulde, *Phys. Rev. B* **49**, 16181 (1994).

We write the five-band Hubbard Hamiltonian for the d electrons of Ni in the following form

$$H = H_0 + \sum_l H_1(l)$$

$$H_0 = \sum_{\mathbf{k}\nu\sigma} \epsilon_\nu(\mathbf{k}) n_{\nu\sigma}(\mathbf{k})$$

$$H_1(l) = \frac{1}{2} \sum_{ijmn} \sum_{\sigma\sigma'} V_{ijmn} a_{i\sigma}^\dagger(l) a_{m\sigma'}^\dagger(l) a_{n\sigma'}(l) a_{j\sigma}(l). \quad (8.40)$$

Here l is a site index and i, j, m, n denote different d orbitals. The $\epsilon_\nu(\mathbf{k})$ with $\nu = 1, \ldots, 5$ are the energy dispersions of canonical d-bands obtained by solving a one-particle equation. We use for them the LDA bands being aware of the fact that they contain already some correlation effects. The $n_{\nu\sigma}(\mathbf{k}) = c_{\mathbf{k}\nu\sigma}^\dagger c_{\mathbf{k}\nu\sigma}$ are number operators for Bloch states with quantum numbers $\mathbf{k}, \nu$ and σ. Their creation operators are expressed in terms of the basis operators $a_{i\sigma}^\dagger(l)$ through

$$c_{\mathbf{k}\nu\sigma}^\dagger = \frac{1}{\sqrt{N_0}} \sum_{ln} \alpha_{\nu n}(\mathbf{k}) e^{i\mathbf{k}\mathbf{R}_l} a_{n\sigma}^\dagger(l) \quad (8.41)$$

where N_0 is the number of sites.

The interaction matrix elements in (8.40) are of the form

$$V_{ijmn} = U_{im} \delta_{ij} \delta_{mn} + J_{ij} (\delta_{in} \delta_{jm} + \delta_{im} \delta_{jn})$$

$$U_{im} = U + 2J - 2J_{im} \quad (8.42)$$

where U and J are average values of the Coulomb- and exchange interaction, respectively. The matrix J_{ij} can be expressed in terms of J and a single anisotropic parameter ΔJ provided we deal with a cubic lattice as is the case here. For more details we refer, e.g., to Refs. [398,399]. When electronic correlations are neglected the ground state is of the form

$$|\Phi_0\rangle = \prod_{\substack{|\mathbf{k}|<k_F \\ \nu\sigma}} c_{\mathbf{k}\nu\sigma}^\dagger |0\rangle. \quad (8.43)$$

The ground state with the inclusion of correlations $|\psi_0\rangle$ can be determined within the Hubbard model (8.14) by means of local correlation operators.[400,401]

---

[398] L. Kleinmann and K. Medrick, *Phys. Rev. B* **24**, 6880 (1981).
[399] A. Oleś and G. Stollhoff, *Phys. Rev. B* **29**, 314 (1984).
[400] G. Stollhoff and P. Thalmeier, *Z. Phys. B* **43**, 13 (1981).
[401] P. Fulde, Y. Kakehashi, and G. Stollhoff, *Metallic Magnetism*, vol. 42 of *Topics Curr. Phys.*, Springer-Verlag, Berlin, Heidelberg (1987).

They are applied on $|\Phi_0\rangle$ and generate a correlation hole around each d electron. We do not describe this process here in detail but refer to the original literature. Instead, we concentrate on identifying those operators $\{A_\nu\}$ which are needed in order to describe the satellite structure in Ni as well as possible. They can be obtained by simply calculating $[H_1(l), a_{i\uparrow}^\dagger(l)]_-$ and considering the operators which are generated by this commutator. They will be important to generate a local, i.e., on-site correlation hole for an electron in orbital i at site l. We find this way

$$A_{ij}^{(1)}(l) = \begin{cases} 2a_{i\uparrow}^\dagger(l)\delta n_{i\downarrow}(l), & i=j \\ a_{i\uparrow}^\dagger(l)\delta n_j(l), & i \neq j \end{cases}$$

$$A_{ij}^{(2)}(l) = \frac{1}{2}\left(a_{i\uparrow}^\dagger(l)s_j^z(l) + a_{i\downarrow}^\dagger(l)s_j^+(l)\right) \quad \text{and}$$

$$A_{ij}^{(3)}(l) = \frac{1}{2}a_{j\downarrow}^\dagger(l)a_{j\uparrow}^\dagger(l)a_{i\downarrow}(l). \tag{8.44}$$

Thereby the notation $\delta n_{i\sigma}(l) = n_{i\sigma}(l) - \langle n_{i\sigma}(l)\rangle$ with $n_{i\sigma}(l) = a_{i\sigma}^\dagger(l)a_{i\sigma}(l)$ and $\mathbf{s}_i(l) = \frac{1}{2}\sum_{\alpha\beta} a_{i\alpha}^\dagger(l)\boldsymbol{\sigma}_{\alpha\beta}a_{i\beta}(l)$ has been used. The operators $A_{ij}^{(1)}(l)$ describe density correlations while the $A_{ij}^{(2)}(l)$ and $A_{ij}^{(3)}(l)$ describe spin correlations of the added electron on site l. The relevant operators $\{A_\nu\}$ consist therefore of $A_\nu^{(0)}(\mathbf{k}) = c_{\mathbf{k}\nu\uparrow}^\dagger$ and

$$A_{ij}^{(\alpha)}(\mathbf{k}) = \frac{1}{\sqrt{N_0}}\sum_l A_{ij}^{(\alpha)} e^{i\mathbf{k}\mathbf{R}_l}, \quad \alpha = 1,2,3. \tag{8.45}$$

One checks easily that for each $\mathbf{k}$ point there are 66 relevant operators $\{A_m(\mathbf{k})\}$, i.e., $1 + 25 + 20 + 20$. Thus the (66×66) matrix

$$\mathbb{G}(z) = \chi(z\chi - \mathbb{L})^{-1}\chi \tag{8.46}$$

(see (8.8)) has to be diagonalized for each $\mathbf{k}$ point. The roots of the secular equation yield the d bands as well as the satellite structures.

The spectral density obtained this way for paramagnetic Ni is shown in Figure VIII.8. The parameters have been chosen as $U/W = 0.56$, $J/W = 0.22$ and $\Delta J/W = 0.031$ in units of the bandwidth W. Those values were obtained from a fit of experiments which measure the multiplet structure of Ni ions put into an Ag matrix.[397,402] A d electron number of $n_d = 9.4$ has been taken. A comparison with the calculations in SCF approximation, here identified with the LDA results show a narrowing of the bandwidth and a pronounced satellite structure.

---

[402] D. van der Marel and G. Sawatzky, *Phys. Rev. B* **37**, 10674 (1988).

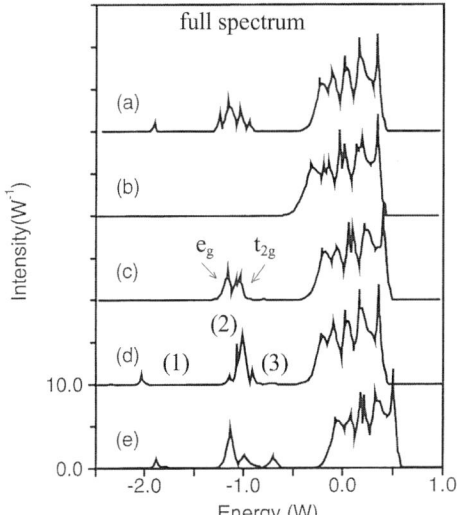

FIG. VIII.8. (a) Full spectrum of paramagnetic Ni for $U = 0.56$, $J = 0.22$ and $\Delta J = 0.031$ in units of W. (b) LDA results. (c) Spectrum when $J = \Delta J = 0$. (d) Spectrum when $\Delta J = 0$. The peaks (1)–(3) correspond to ^1S, ^1G and ^1D, ^3P and ^3F atomic d^8 configurations. (e) Spectrum when $|\Phi_0\rangle$ instead of $|\psi_0\rangle$ is used. (After Ref. [396].)

With a spin averaged bandwidth of $W = 4.3$ eV obtained from a local spin-density approximation this satellite is peaked at 6.8 eV below the top of the d bands (Figure VIII.8). This has to be compared with an experimental value of 6.3 eV. The shape of the satellite reflects the atomic d^8 multiplet. When the anisotropy parameter $\Delta J = 0$ is set to zero, it is split into three substructures representing a ^1S state, two degenerate singlets ^1G and ^1D and two degenerate triplets ^3P and ^3F. The energy difference between ^1G and ^3F is $2J$ and between ^1S and ^1G is $5J$. The three structures at $-1.9W$, $-1.1W$ and $-0.7W$ correspond to the three atomic levels. The main peak at $-1.1W$ is split by the anisotropy parameter into finer structures. But also the $e_g - t_{2g}$ splitting as well as the quasiparticle dispersion have an effect on the satellite structure. Also shown are the modifications in the satellite structure which arise when the ground state $|\psi_0\rangle$ is replaced by the one without correlations $|\Phi_0\rangle$. For completeness we show also the reductions in the widths of the different quasiparticle bands when correlations are taken into account (see Figure VIII.9). We want to mention that a satellite below the d bands is also found for fcc Co as well as for bcc Fe while for fcc Sc a satellite above the d bands is obtained. We repeat that our theoretical findings are based on a 5-band Hubbard model. It might not be sufficient for a satisfactory description in all cases because s electrons have been left out altogether and because intersite correlations are not taken into account in the choice for the set $\{A_\nu\}$.

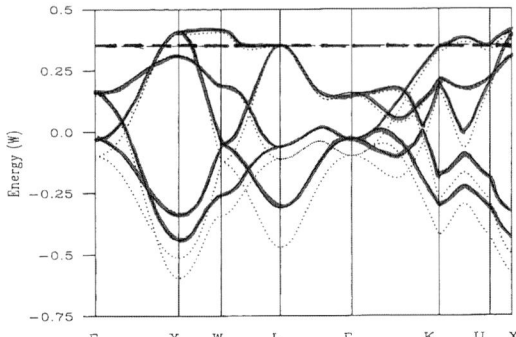

FIG. VIII.9. Narrowing of the quasiparticle bands of fcc paramagnetic Ni due to electron correlations (solid lines). The bands without correlations are identified here with the LDA bands (dotted lines). (After Ref. [396].)

23. MULTIPLET EFFECTS IN $5f$ SYSTEMS

The high-energy effects due to strong correlations discussed so far result from the overall suppression of charge fluctuations. They reflect the presence of atomic configurations with well-defined occupations of the partially filled inner valence shells. In a free atom or ion, the degeneracies of these configurations are (partially) lifted by the electron-electron interaction. This leads to the formation of atomic multiplets where the scale for the excitation energies is set by the exchange constant. The latter involves the anisotropic part of the Coulomb interaction which remains (almost) unscreened in a metal. As a consequence we expect pronounced multiplet effects in the single-particle spectra whenever the exchange constant exceeds the gain in kinetic energy as measured by the corresponding effective band width. This is the case for the high-energy satellites in Ni which were discussed in Section VIII.23. Multiplet effects are also strongly evident in the actinide compounds where the $5f$ exchange constant is of order 1 eV and exceeds the bare effective band width. Here we focus on U-based heavy-fermion compounds where strong intra-atomic correlations are responsible for the dual character of the $5f$ electrons. We restrict ourselves to qualitative features of the $5f$-spectral function

$$A_{j_z}(\mathbf{k}, \omega) = \begin{cases} \sum_n |\langle \Psi_n^{(N+1)} | c_{j_z}^\dagger(\mathbf{k}) | \Psi_0^{(N)} \rangle|^2 \delta(\omega - \omega_{n0}^{(+)}), & \omega > 0 \\ \sum_n |\langle \Psi_n^{(N-1)} | c_{j_z}(\mathbf{k}) | \Psi_0^{(N)} \rangle|^2 \delta(\omega + \omega_{n0}^{(-)}), & \omega < 0 \end{cases} \quad (8.47)$$

with

$$c_{j_z}(\mathbf{k}) = \sum_{\mathbf{a}} e^{i\mathbf{k}\mathbf{a}} c_{j_z}(a), \qquad \omega_{n0}^{(\pm)} = E_n^{(N\pm 1)} - E_0^{(N)}. \quad (8.48)$$

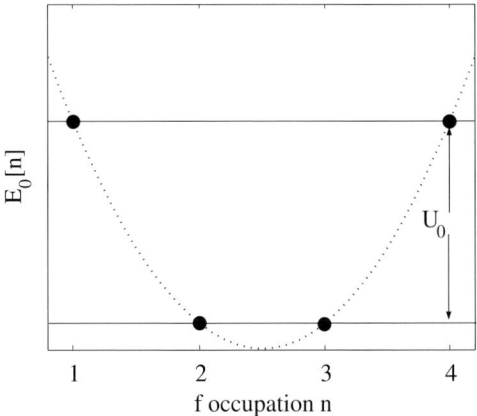

FIG. VIII.10. Configurational energies vs. $5f$-valence for the microscopic model adopted to describe the U sites. For a qualitative discussion, the value U_0 is treated as a parameter since it is screened in a metal and concomitantly will depend upon the crystallographic environment of the U site.

It yields the probability for adding ($\omega > 0$) or removing ($\omega < 0$) a $5f$ electron with energy ω in a state characterized by momentum $\mathbf{k}$ and angular momentum projection j_z to the N-particle ground state $|\Psi_0^{(N)}\rangle$ with energy $E_0^{(N)}$. The states with $N \pm 1$ and their energies are denoted by $|\Psi_n^{(N\pm1)}\rangle$ and $E_n^{(N\pm1)}$, respectively. The notation was introduced in Section V.

The structure of the spectra in the strong-coupling limit can be understood by considering the atomic limit where the system is modeled as a statistical ensemble of isolated $5f$ atoms. The absence of pronounced valence peaks in the photoemission and inverse photoemission spectra suggests that the energies of the f^2- and f^3-configurations, are (almost) degenerate i.e., $E(5f^2) = E(5f^3)$. The variation with $5f$-valence of the configurational energies $E(f^n)$ is shown schematically in Figure VIII.10.

The ground state will mainly involve $5f^2$ and $5f^3$ configurations. The corresponding spectral functions are obtained in close analogy to the classical work of Hubbard.[7] Let us first consider the zero configuration-width approximation which neglects intra-atomic correlations. The valence transitions $f^2 \to f^1$ and $f^3 \to f^4$ occur at high energies set by the isotropic average of the Coulomb interaction and, concomitantly, do not affect the low-temperature behavior. The latter is determined by the low-energy peak resulting from the transitions $f^2 \leftrightarrow f^3$ within the f^2- and f^3-configurations. This peak is a direct consequence of the intermediate-valent ground state. The distribution among the peaks can be estimated from combinatorial considerations. The weight $Z(f^2 \to f^1)$ of the transition $f^2 \to f^1$ equals the probability that a state with a given j_z is occupied in that

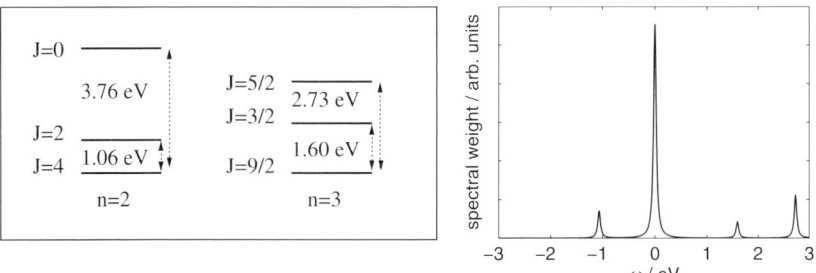

FIG. VIII.11. Left panel: Multiplets for f^2 and f^3 configurations of the U model in the atomic limit. The levels are obtained by diagonalizing the Coulomb matrix using j-j coupling and Coulomb parameters of UPt$_3$. Right panel: Spectral function for the model in the atomic limit.

f^2 contribution of the mixed-valent ground-state. Following these lines one finds

$$Z(f^2 \to f^1) = \frac{1}{6}, \qquad Z(f^2 \to f^3) = \frac{1}{3}$$
$$Z(f^3 \to f^2) = \frac{1}{4}, \qquad Z(f^3 \to f^4) = \frac{1}{4} \qquad (8.49)$$

The central focus is the evolution of the low-energy peak whose spectral weight sums up to $7/12 \, (= 1/3 + 1/4)$. Intra-atomic correlations which are usually described by Hund's rules yield the multiplets displayed schematically in Figure VIII.11. The values for the excitation energies are calculated from the Coulomb parameters of UPt$_3$ adopting j-j-coupling. In the atomic limit the model spectral function has a low-energy peak resulting from transitions between the Hund's rule ground-state manifolds $|f^2, J = 4, J_z\rangle$ and $|f^3, J = 9/2, J_z\rangle$ as well as peaks corresponding to transitions into excited multiplets. Due to the rotational invariance of the Coulomb interaction, the spectral functions do not depend upon the magnetic quantum number j_z. The spectral weights for the photoemission and inverse photoemission parts, $\frac{1}{10} \sum_{J_z} |\langle f^2, J', J_z - j_z | c_{j_z} | f^3, J = 9/2, J_z \rangle|^2$ and $\frac{1}{9} \sum_{J_z} |\langle f^3, J', J_z + j_z | c^\dagger_{j_z} | f^2, J = 4, J_z \rangle|^2$ are easily obtained by expressing the matrix elements in terms of reduced matrix elements and Clebsch–Gordan coefficients. The excitation energies as well as the corresponding spectral weights are listed in Table VIII.1. We should like to mention that there is no transition into the excited multiplet $|f^2, J = 0\rangle$.

The model therefore predicts $5f$ spectral weight at $\simeq 1$ eV binding energy in the photoemission spectra from U-based heavy fermion compounds. The position of these $5f$ structures should not depend sensitively on the actual crystalline environment since they result from intra-atomic excitations. The weights, however,

TABLE VIII.1. POSITIONS AND SPECTRAL WEIGHTS OF THE ATOMIC TRANSITIONS IN URANIUM. AT ZERO ENERGY POSITION THE PARTICLE AND HOLE PEAKS HAVE A TOTAL WEIGHT 0.414

	PES			BIS	
Position/[eV]	−1.06	0.00	0.00	1.60	2.73
Spectral weight	0.054	0.196	0.218	0.032	0.083

will be altered by the reconstruction of the ground state and the low-energy excitations in an extended solid and may vary with chemical composition and temperature. These predictions seem to be in agreement with recent experiments.[403] Recent photoemission studies by Fujimori et al.[404] seem to indicate the presence of features in the proper energy range. The fact that they cannot be attributed to LDA energy bands further supports our interpretation in terms of multiplet side bands. An unambiguous identification, however, will require resonant photoemission experiments.

24. EXCITATIONS IN COPPER-OXIDE PLANES

It is well known that the copper-oxide based perovskites which play a major role in high-temperature superconductivity are strongly correlated electron systems. This is immediately obvious by realizing that, e.g., La_2CuO_4 is an insulator and not a metal despite the fact that with one hole per unit cell one would expect a half-filled conduction band. Note that insulating behavior is found also above the antiferromagnetic transition temperature and therefore is unrelated to the doubling of the unit cell when the material becomes an antiferromagnet. Indeed, LDA calculations give an effective Coulomb repulsion of two d holes on a Cu site of $U_d = 10.5$ eV and a hopping matrix element of a hole from a Cu $3d_{x^2-y^2}$ orbital to an O $2p_{x(y)}$ orbital of $t_{pd} = 1.3$ eV. Therefore the bandwidth is small as compared with U_d.

The strong correlations lead to a single-particle spectral density which is quite distinct from the one of weakly correlated electrons. In fact, the excitations are dominated by the internal degrees of freedom of the correlation hole so that the one-particle or coherent part of Green's function plays a secondary role here. This will become clear towards the end of this section.

We shall use again a model Hamiltonian although an ab initio calculation would be highly desirable. We want to write it in hole representation in the form of a

---

[403] R. Eloirdi, T. Gouder, F. Wastin, J. Rebizant, and F. Huber, preprint (2005).
[404] S. Fujimori, private communication (2005).

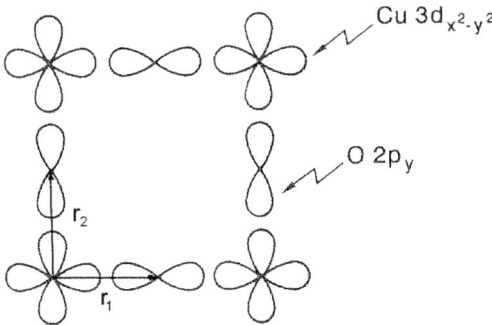

FIG. VIII.12. Cu and O orbitals accounted for in the 3-band Hubbard Hamiltonian (8.53).

three-band Hubbard or Emery model. The unit cell consists of one Cu $3d_{x^2-y^2}$ orbital and two O $2p_{x(y)}$ orbitals (see Figure VIII.12). The orbital energies are ϵ_d and ϵ_p. Two holes on a Cu site repel each other with U_d and on a O site with U_p. The hopping matrix element between a Cu orbital and an O orbital is t_{pd} and t_{pp} between the orbitals of the two O atoms. The parameter values can be derived from a constrained LDA calculation[405] with the following results: $U_d = 10.5$ eV, $U_p = 4.0$ eV, $t_{pd} = 1.3$ eV, $t_{pp} = 0.65$ eV, $\epsilon_p - \epsilon_d = 3.6$ eV. It is helpful to first introduce a basis of oxygen orbitals which is diagonal with respect to oxygen–oxygen hopping processes t_{pp}. Let us denote the corresponding creation operators by $c^\dagger_{m\mathbf{k}\sigma}$ where m is an oxygen band index. Those bands have a dispersion

$$\epsilon_m(\mathbf{k}) = \epsilon_p \pm 2t_{pp}\left[\cos[\mathbf{k}(\mathbf{r}_1 + \mathbf{r}_2)] - \cos[\mathbf{k}(\mathbf{r}_2 - \mathbf{r}_1)]\right] \quad (m = 1, 2) \quad (8.50)$$

where the vectors $\mathbf{r}_1$ and $\mathbf{r}_2$ are shown in Figure VIII.12. Due to the different orientations of the O $2p_{x(y)}$ orbitals, the sign of t_{pp} depends on the direction. It is positive in the direction $\mathbf{r}_1 + \mathbf{r}_2$, $-(\mathbf{r}_1 + \mathbf{r}_2)$ and negative in the direction $\mathbf{r}_1 - \mathbf{r}_2$, $\mathbf{r}_2 - \mathbf{r}_1$.

Next we introduce a linear combination of oxygen orbitals which possesses the same symmetry on a CuO$_4$ plaquette as the t_{pd}-matrix elements, i.e.,

$$p^\dagger_{\mathbf{k}\sigma} = \sum_m \phi_{m\mathbf{k}} c^\dagger_{m\mathbf{k}\sigma}, \quad (8.51)$$

with

$$\phi_{m\mathbf{k}} = -\frac{i}{\sqrt{2}}[\sin k\mathbf{r}_1 \pm \sin k\mathbf{r}_2]. \quad (8.52)$$

---

[405] M. S. Hybertsen, M. Schlüter, and N. E. Christensen, *Phys. Rev. B* **39**, 9028 (1989).

The model Hamiltonian we shall be using can thus be written as

$$H = \sum_{m\mathbf{k}\sigma} \epsilon_m(\mathbf{k}) c^\dagger_{m\mathbf{k}\sigma} c_{m\mathbf{k}\sigma} + U_p \sum_J n_{p\uparrow}(J) n_{p\downarrow}(J)$$
$$+ \epsilon_d \sum_{\mathbf{k}\sigma} d^\dagger_{\mathbf{k}\sigma} d_{\mathbf{k}\sigma} + U_d \sum_I n_{d\uparrow}(I) n_{d\downarrow}(I)$$
$$+ 2 t_{pd} \sum_{\mathbf{k}\sigma} (p^\dagger_{\mathbf{k}\sigma} d_{\mathbf{k}\sigma} + h.c.). \quad (8.53)$$

As usual $n_{p\sigma}(J)$ and $n_{d\sigma}(I)$ are the occupation number operator of the O $2p_{x(y)}$ orbital on site J and of the Cu $3d_{x^2-y^2}$ orbital on site I, respectively.

With the help of the above Hamiltonian we calculate the spectral density of the Cu–O planes, with and without hole doping. For that purpose we have to choose the right set of relevant operators $\{A_n(\mathbf{k})\}$. They must include the most important microscopic processes in the strongly correlated system which generate the correlation hole of an electron. First of all, the hole operators

$$A_p(m, \mathbf{k}) = p^\dagger_{m\mathbf{k}}, \quad A_d(\mathbf{k}) = d^\dagger_{\mathbf{k}\uparrow}, \quad m = 1, 2 \quad (8.54)$$

must be part of the set. They are supplemented by a number of *local* operators. In order to suppress double occupancies of the d and p orbitals due to the large values of U_d and U_p, we must include $\bar{d}^\dagger_\uparrow = d^\dagger_\uparrow(I) n_{d\downarrow}(I)$ and $\bar{p}^\dagger_\uparrow = p^\dagger_{J\uparrow} n_{p\downarrow}(J)$, i.e., their Fourier transforms

$$A_d(\mathbf{k}) = \frac{1}{\sqrt{N}} \sum_J e^{-i\mathbf{k}\mathbf{R}_I} \bar{d}^\dagger_{I\uparrow}$$
$$A_p(\mathbf{k}) = \frac{1}{\sqrt{2N}} \sum_J e^{-i\mathbf{k}\mathbf{R}_J} \bar{p}^\dagger_{J\uparrow} \quad (8.55)$$

where N is the number of Cu sites and there are two O sites per Cu site. In addition we want to account for spin flips of d holes at Cu sites in combination with spin flips at neighboring O sites. Those processes are important for the formation of a Zhang–Rice singlet.[406] This is a singlet state formed by a hole at a Cu site and another one at a nearest-neighbor O site.

The corresponding microscopic operator is

$$A_f(\mathbf{k}) = \frac{1}{\sqrt{N}} \sum_I e^{-i\mathbf{k}\mathbf{R}_I} \tilde{p}^\dagger_{I\uparrow} S^+_I \quad (8.56)$$

---

[406] F. C. Zhang and T. M. Rice, *Phys. Rev. B* **37**, 3759 (1988).

where

$$\tilde{p}_{I\sigma}^\dagger = \frac{1}{2}\left(p_{1\sigma}^\dagger + p_{2\sigma}^\dagger - p_{3\sigma}^\dagger - p_{4\sigma}^\dagger\right) \quad (8.57)$$

is a superposition of the four O $2p$ orbitals which surround the $3d$ orbital of Cu site I. The operator $S_I^+ = d_{I\uparrow}^\dagger d_{I\downarrow}$. For a possible formation of triplet states also the operator

$$A_a(\mathbf{k}) = \frac{1}{\sqrt{N}} \sum_I e^{-i\mathbf{k}\mathbf{R}_I} \tilde{p}_{I\uparrow}^\dagger n_{d\downarrow}(I) \quad (8.58)$$

is needed. In order to describe charge transfer in the vicinity of an added hole we also include

$$A_c(\mathbf{k}) = \frac{1}{\sqrt{N}} \sum_I e^{-i\mathbf{k}\mathbf{R}_I} p_{I\uparrow}^\dagger p_{I\downarrow}^\dagger d_{I\downarrow}. \quad (8.59)$$

This completes the choice of the set $\{A_\nu(\mathbf{k})\}$. With 9 operators $A_\nu(\mathbf{k})$ we have to diagonalize for each $\mathbf{k}$ point (9×9) matrices $L_{\mu\nu}$ and $\chi_{\mu\nu}$ (see (8.12)). The static expectation values which enter the matrix elements are determined from the spectral functions to which they are related via

$$\langle \psi_0 | A_m^\dagger(\mathbf{k}) A_n(\mathbf{k}) | \psi_0 \rangle = \int_{-\infty}^{+\infty} d\omega\, f(\omega) A_{mn}(\mathbf{k}, \omega) \quad (8.60)$$

where $f(\omega) = [e^{\beta\omega} + 1]^{-1}$ with $\beta = (k_B T)^{-1}$ is the Fermi distribution function. It can be replaced by a step function. By solving Eq. (8.60) self-consistently the static correlation functions are obtained. For more details we refer to the original literature.[407,408] The resulting density of states are shown in Figure VIII.13 for the half-filled case as well as for 25% of hole doping. The two cases are supposed to simulate La_2CuO_4 and $La_{2-x}Sr_xCuO_4$, respectively. The agreement with results of exact diagonalization of a small cluster containing four CuO_2 units[409] is very good. It is noticed that at half filling the system is insulating and a gap is present. Around $\omega = 0$ one notices the O $2p$ band and a $3d$ component due to hybridization. The upper Hubbard band is centered at $5t_{pd}$. The structure near $-7t_{pd}$ results mainly from $A_d(\mathbf{k})$ and is interpreted as the lower Hubbard band. The peak near $-4t_{pd}$ comes mainly from $A_c(\mathbf{k})$, i.e., from charge fluctuations. The structure close to $2.5t_{pd}$ marks the Zhang–Rice singlet state. It is important for the interpretation of spectroscopic experiments.[410] When the system is doped

---

[407] P. Unger and P. Fulde, *Phys. Rev. B* **48**, 16607 (1993).
[408] P. Unger and P. Fulde, *Phys. Rev. B* **47**, 8947 (1993).
[409] T. Tohyama and S. Maekawa, *Physica C* **191**, 193 (1992).
[410] H. Romberg, M. Alexander, N. Nücker, P. Adelmann, and J. Fink, *Phys. Rev. B* **42**, 8768 (1990).

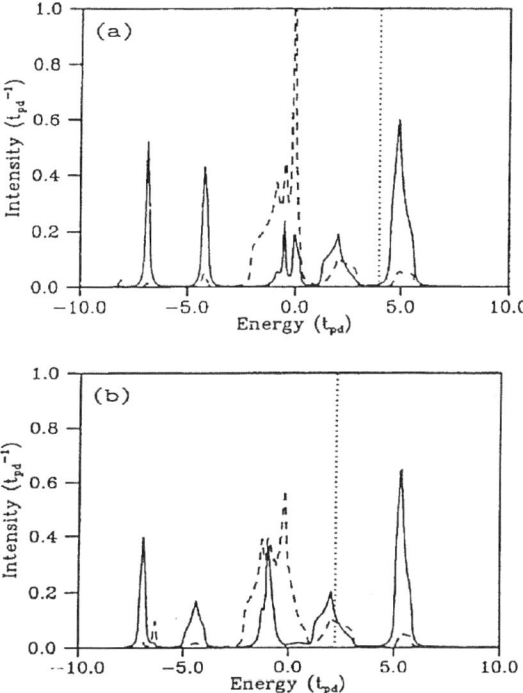

FIG. VIII.13. Spectral density of Cu–O planes described by the Hamiltonian (8.53); (a) at half-filling; (b) for 25% hole doping. Solid and dashed lines show the Cu and O contributions. In the hole doped case spectral density has been shifted from the upper Hubbard band to the region near E_F (dotted line). Parameters are $U_d = 8$, $U_p = 3$, $t_{pp} = 0.5$, $t_{pd} = 1$, $\epsilon_p - \epsilon_d = 4$. Compare also with Figure VIII.14. (After Ref. [407].)

with holes the Fermi energy moves into the Zhang–Rice singlet band. Simultaneously there is a transfer of spectral density taking place from the upper Hubbard band to energies below E_F.

As mentioned at the beginning of this section we are dealing here with a situation where the internal degrees of freedom of the correlation hole dominate the spectral density. This is most clearly seen by comparing the density of states for the half-filled case with the one obtained in the independent electron approximation. The latter is shown in Figure VIII.14a. It consists of a bonding, nonbonding and antibonding part. The splitting of the d electron contributions into a lower and upper Hubbard band is schematically shown in Figure VIII.14b and the singlet-triplet splitting in Figure VIII.14c. The difference between Figures VIII.13a and b demonstrates nicely the effect of the excitations of the correlation hole.

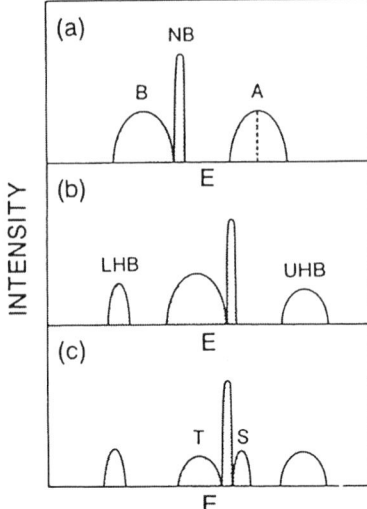

FIG. VIII.14. Schematic representation of the density of states of a Cu–O plane described within a 3-band Hubbard model. (a) Independent electron approximation. A, B, NB denote antibonding, bonding and nonbonding part. (b) Splitting of the d band into a lower (LHB) and upper (UHB) Hubbard band. (c) Singlet-triplet splitting (S, T) S = Zhang–Rice singlet.

IX. Summary and Outlook

The different topics we have discussed in this review concern mainly low energy effects of strongly correlated electrons, though not exclusively. In Section VIII we have also given examples of high energy features like shadow bands, satellites and kink effects which originate from strong electron correlations. Low energy effects are mainly due to heavy quasiparticles consisting of an electron with its (rigid) correlation hole. The latter is very pronounced when correlations are strong and therefore an electron moves with a reduced Fermi velocity through the system dragging the correlation hole with it. Low temperature properties are strongly influenced by the effective mass of the quasiparticles. Important examples are Ce-based heavy fermion metals where an interplay of strong on-site 4f Coulomb correlations and hybridization with conduction electrons leads to the heavy quasiparticle mass. These features may be described within Renormalized Band Theory by assuming periodic scattering with a resonant phase shift. This way one obtains the proper heavy quasiparticle bands and the Fermi surface. In recent years it has been realized that in U-based heavy fermion compounds like, e.g., UPd$_2$Al$_3$ heavy quasiparticles originate from a special mechanism: the 5f electrons form a two-component system with some itinerant and some localized orbitals and the renormalization of the former via intra-atomic excitations of the localized $5f$ electrons leads to strong mass enhancements. In fact, this mechanism has also been identified as the origin of superconductivity in UPd_2Al_3 which is not based on electron–phonon interactions.

We have also given examples where the low energy excitations cannot simply be described by a Fermi liquid of heavy quasiparticles. In general, even in weakly correlated metals, one expects non-Fermi liquid behavior close to quantum phase transitions. The earliest examples are the logarithmic corrections to the linear specific heat term coming from electrons dressed by paramagnon excitations close to a ferromagnetic quantum critical point. By now numerous examples of mostly Ce-based heavy fermion metals have been found where the vicinity to an AF quantum phase transition leads to pronounced non-Fermi liquid anomalies. Their microscopic understanding is still at an early stage.

A different type of deviation from the Fermi liquid picture emerges in strongly correlated electron compounds which exhibit Wigner-lattice type charge order as is the case, e.g., in Yb_4As_3. Here spin and charge degrees of freedom which are responsible for low energy excitations may belong to different types of electrons, i.e., they appear to be separated. In the example of Yb_4As_3 the spin degrees of freedom involve $4f$ holes of Yb while the charge degrees of freedom were found to be due to As $4p$ holes. The spins of $4f$ holes are aligned along chains and the 1D neutral spinon excitations cause the large linear specific heat term while light $4p$ holes carry the current.

Quite generally charge ordering in $3d$ or $4f$ mixed-valent systems is a promising route to obtain low-dimensional spin structures. An example is NaV_2O_5. There the 1D spin excitations that emerge via charge ordering interact with the lattice leading to a spin-Peierls type dimerization and spin-gap formation. An even stronger kind of lattice coupling occurs in the manganites which leads to polaron formation. The intersite Coulomb correlations then drive charge order of polaronic quasiparticles as observed in the bilayer-manganites. The nonstoichiometric compounds of this class are examples of strong Hund's rule correlations of itinerant (e_g) and localized (t_{2g}) electrons. This leads to almost ideal 2D ferromagnetic order driven by kinetic or double exchange mechanisms.

Another deviation from a simple Fermi liquid heavy-quasiparticle description was found for the Hubbard model on a square lattice at a particular range of doping and for U above the critical value of the metal–insulator transition at half filling. Here the excitations were of a form previously termed marginal Fermi liquid and suggested to explain some of the properties of layered cuprates.

High-energy features of strongly correlated electrons result on the other hand from excitations of internal degrees of freedom of the correlation hole. While the slowly moving quasiparticle has a low excitation energy the internal excitations of its cloud correspond to high energies and show up as satellite structures in photoelectron spectroscopy. Examples were given in Section VIII.

In which direction will research on the theory of strongly correlated electrons develop in the future? Here one can only speculate at the risk of being completely wrong. Firstly, there is the field of electronic structure calculations. After LDA based density functional calculations turned out to describe insufficiently strongly correlated electrons they were combined successfully with other approaches to improve on the shortcomings. As discussed in this review Renormalized Band Structure calculations showed considerable predictive power. But also LDA+U and LDA+GW schemes have been applied and yielded insight into strongly correlated electrons. However the more corrections are added to the original approximation to density functional theory like LDA, the less controlled the results are. Double counting of correlation contributions or screening is one of the uncertainties, accounting for the spatial extend of the correlation hole is another. The latter is a problem which also dynamical mean-field theory (DMFT) is facing in its present form. Extensions of that theory to clusters[369] are desirable but difficult to realize in practice. This suggests that over the long run wavefunction-based approaches using quantum chemical methods may be a possible solution to the problem. They are technically very demanding but they are well controlled and allow for a detailed understanding of different correlation processes. They may have a great future.[41]

Exact diagonalizations will certainly also play their role in the future. Usually they use model Hamiltonians with model parameters obtained from single particle LDA(+U) calculations. With increasing possibilities to treat larger clusters they will lead to much new insight into the nature of the correlated ground state as well as the excitations responsible for low temperature thermodynamic anomalies.

Analyses of the excitation spectrum of correlation holes seem to be a very promising field. By projecting the calculations onto a few selected variables which represent the most important degrees of freedom, insight may be gained about the relative importance of different correlations. We expect considerable extensions of our understandings of elementary excitations here. As was shown in Section VII in special lattices (here geometrically frustrated ones) it may happen that an electron added to the system may separate into two parts when correlations are strong. This may give raise to charge fractionalization, a phenomenon known from the fractional quantum Hall effect (FQHE).[40] Even if there is a weak restoring force between the two parts we would have "quasiparticles" of considerable spatial extent. They would have little in common with the usual picture of an electron surrounded by a correlation hole. Interestingly enough, certain analogies to field theories in elementary particle physics do appear here when dealing with confinement and deconfinement.[317,320] After all, why should nature restrict itself to realize certain basic properties only in one particular field of physics and not in others too?

Acknowledgment

We thank Prof. Y. Kakehashi for a critical reading of Section VIII and for helpful comments. Last not least we thank Mrs. Regine Schuppe for carefully preparing the manuscript.

LIST OF ACRONYMS

AF	antiferromagnet
ARPES	angle resolved photoemission spectroscopy
BEC	Bose–Einstein condensation
BZ	Brillouin zone
CDW	charge density wave
CEF	crystalline electric field
CMR	colossal magnetoresistance
CO	charge order
CPA	coherent potential approximation
DE	double exchange
dHvA	de Haas–van Alphen
DOS	density of states
DM	Dzyaloshinsky–Moriya
DMFT	dynamical mean-field theory
EHM	extended Hubbard model
e-p	electron–phonon
FL	Fermi liquid
FM	ferromagnet
FQHE	fractional quantum Hall effect
FS	Fermi surface
HF	heavy fermion or Hartree–Fock
ICM	incommensurate magnet
INS	inelastic neutron scattering
ISSP	Ising-spin-Peierls
ITF	Ising model in transverse field
JT	Jahn–Teller
JW	Jordan–Wigner
KS	Kondo singlet
LDA	local density approximation
LSDA	local spin density approximation
n.n.	nearest neighbor
n.n.n.	next nearest neighbor

NCA	non-crossing approximation
NFL	non-Fermi liquid
NMR	nuclear magnetic resonance
ODLRO	off-diagonal long range order
PM	paramagnet
QCP	quantum critical point
QMC	quantum Monte Carlo
QPT	quantum phase transition
RPA	random phase approximation
RKKY	Ruderman–Kittel–Kasuya–Yoshida
SCF	self-consistent field
SCPM	self-consistent projection method
SDW	spin density wave
TB	tight binding
WFM	weak ferromagnet

Defect-Induced Dynamic Pattern Formation in Metals and Alloys

Yves Bréchet

Laboratoire de Thermodynamique et Physico-Chimie Métallurgiques, Institut National Polytechnique de Grenoble, 38402, St. Martin D'Hères, France

Christopher Hutchinson

Department of Materials Engineering, Monash University, Clayton, 3168, Vic, Australia

I. Introduction	182
1. Chemical Patterning	185
2. Structural Patterning in Systems Prepared Far from Equilibrium	186
3. Structural Patterning in Systems Maintained Far from Equilibrium	187
4. Spatio-Temporal Patterning in Plasticity	187
II. Free Energy Changes, Driving Forces and Energy Input	188
5. Systems Prepared Far from Equilibrium	188
6. Systems Maintained Far from Equilibrium by the Continuous Input of Energy	191
III. Chemical Patterning: Interface-Mediated Transformations: Eutectoid Decomposition and Discontinuous Precipitation	192
7. Steady State Growth	197
8. The Spacing Selection Problem	213
9. Non-Steady State Growth	218
10. Coarsening and Related Morphological Evolution	225
11. Multilayer Stability	232
IV. Structural Defect Patterning: Grain Growth, Recovery and Recrystallization	233
12. Grain Growth	234
13. Recovery and Recrystallization of Dislocation Structures	240
V. Structural Defect Patterning: Irradiation and Plastic Deformation	248
14. Irradiation Induced Patterning	249
15. Plasticity Induced Patterning	255
VI. Spatio-Temporal Patterning in Plasticity: The Portevin–Le Chatelier Effect	273
16. Phenomenology of Portevin–Le Chatelier Effect	274
17. Statistical Analysis of Serrated Flow: An Example of Self Organized Criticality (SOC)	276
18. Portevin–Le Chatelier Effect: An Example of Deterministic Chaos	282
VII. Concluding Remarks	285

I. Introduction

There has recently developed within the non-linear physics community, a deep interest in the formation of 'patterns' in condensed matter systems driven far from equilibrium. Patterns are a concept somewhat vague and subjective. We will speak of spontaneous pattern formation when a spatial distribution of matter, or the temporal response of a solid to an external stimulus presents some sort of regularity which has a lower symmetry than its cause. The ideal regularity is periodicity, but one may also consider the possible order in apparently chaotic behavior, or the length scales appearing in an initially homogeneous body.

After the extensive work done in fluid mechanics where numerous examples of spontaneous pattern formation may be found (e.g., see Refs. [1,2]), and where the non-linear equations to be solved are well established, solidification seemed to be another of these areas to which the non-linear physicist could apply his methods to model the spontaneous emergence of patterns (dendritic forms and cellular and eutectic structures). Since the classic review paper by Langer,[3] a number of new mathematical techniques (e.g., solvability theory) or numerical methods (phase field, cellular automata)[4] have been developed to describe and predict the shapes and scales observed when going from the liquid to the solid. For both fluid mechanics and solidification, the dynamic equations to be solved (Navier–Stokes equation, and mass and heat transfer equation) are defined without ambiguity, and the problem is primarily to analyze and solve them.

Similarly, the spontaneous patterning (both in space and time) that can be observed in chemical reactions has also attracted significant recent attention, particularly from the 'Belgian school'. These systems have been labeled 'dissipative structures'[5–7] and in such situations, the transport of matter is often by a diffusion mechanism. The variety of possible 'intermediate reactions' leads to a much wider range of partial differential equations than in fluid mechanics or solidification, but the methods currently used to find the conditions for instabilities and their development are quite similar.

---

[1] S. Chandrasekhar, *Hydrodynamic and Hydromagnetic Stability*, Dover (1981).

[2] P. Berge, Y. Pomeau, and C. Vidal, *Order within Chaos: Towards a Deterministic Approach to Turbulence*, Wiley, New York (1987).

[3] J. S. Langer, *Rev. Mod. Phys.* **52**, 1 (1980).

[4] H. Muller-Krumbhaar, W. Kurz and E. A. Brener, in *Phase Transformations in Materials*, ed. G. Kostorz, Wiley–VCH, Weinheim (2001), p. 81.

[5] G. Nicolis and I. Prigogine, *Self-Organization in Nonequilibrium Systems: From Dissipative Structures to Order Through Fluctuations*, Wiley, New York (1977).

[6] D. Walgraef, *Spatio-Temporal Pattern Formation: With Examples from Physics, Chemistry, and Materials Science*, Springer, New York (1997).

[7] C. Vidal and H. Lemarchand, *La Reaction Creatrice*, Hermann, Paris (1988).

Comparatively less studied by the physics community is the spontaneous patterning in the solid state. This field is also very rich in examples, but there is no equivalent of a 'Navier–Stokes equation' which would apply to all the patterns observed. In addition, unlike liquids, crystalline solids also contain structural defects of various dimensions (point defects (0d), dislocations (1d), grain boundaries (2d)), and these defects bring both a new complexity and a new richness to the field. The approach we will take is to present an outline of the relevant phenomena and, where they exist, to summarize the possible equations used to describe the observations and the current models developed by materials scientists to capture the complexity of the problem.

The guide line of this chapter, which at the same time gives it its unity and limits its scope, is 'spontaneous pattern formation' in solids driven far from equilibrium, involving explicitly the role of structural defects.

It is well recognized that there is an energetic cost associated with pattern formation that involves a multiplication of interfaces. This is one fundamental reason for which being far from equilibrium seems to be a necessary condition for obtaining patterns. Solids can be either prepared out of equilibrium, or maintained out of equilibrium by a continuous injection of energy from outside the system. Both cases are possible for metallic alloys, and both lead to pattern formation. For example, preparing a solid far from equilibrium can be achieved by rapidly cooling below a phase boundary of the phase diagram: the high temperature structure is then out of equilibrium. Maintaining a system out of equilibrium can be accomplished either by plastic deformation or by irradiation: we refer to such systems as 'driven systems' and again, in these cases, patterns may spontaneously appear.

In metallic alloys, pattern formation in space can be manifested in two ways: it can be a patterning of phases of different chemical composition, such as in the decomposition of solid solutions, or it can be a patterning of the structural defects themselves as in the case of plastically deformed materials. Pattern formation in time is well know in chemistry, and 'chemical clocks' have been central to the development of the methods of non-linear physics.[5,8] To our knowledge there is no equivalent in the solid state (except perhaps in periodic oxidation, but it is translated in space under the form of layered oxides[9]). However, under special circumstances, we can observe an equivalent in the plastic behavior of materials: 'serrated yielding'.

---

[8] H. Haken, *Synergetics: An Introduction: Nonequilibrium Phase Transitions and Self-Organization in Physics, Chemistry, and Biology*, Springer, New York (1983).
[9] G. Bertrand, *Solid State Phenomena* **3–4**, 257 (1988).

TABLE I.1. CLASSIFICATION OF THE PATTERNING EXAMPLES PRESENTED IN THIS CHAPTER

	Structural defects patterning	**Chemical patterning**
Systems **prepared** far from equilibrium	*Grain growth, recovery and recrystallization* (Section IV)	*Interface-mediated phase transformations*: Pearlite and Discontinuous precipitation (Section III)
Systems **maintained** far from equilibrium	*Spatial patterning* (Section V) Dislocation and void self organization under irradiation. Pattern formation in plasticity (creep, fatigue...) *Temporal patterning* (Section VI) Serrated yielding (Portevin–Le Chatelier effect)	*Driven systems*[10] Irradiation induced precipitation and order/disorder transition, Ball milling amorphization[11] Fatigue induced dissolution[12]

We have classified the various phenomena to be presented in this chapter in Table I.1. We do not claim to be exhaustive (e.g., neither pattern formation in dynamic fracture nor pattern formation in martensitic transformations is treated here, and the case of chemical patterning under irradiation has recently been reviewed in the same series[10]), but we aim to present examples which are representative both of the phenomena occurring in the solid state involving structural defects as central actors, and of the current tools used for their understanding.

For systems prepared far from equilibrium that produce patterns whilst approaching the equilibrium state, thermodynamics is a natural guide: it indicates the end point of the transformation, and computing the 'driving force' for the evolution is meaningful. For systems maintained far from equilibrium, the situation is much more confusing, and the input of energy into the system is a key parameter, together with the ability of the system to store it or to dissipate it as heat. In these driven systems, the kinetics of the defects becomes a central issue, and therefore the modelling is bound to be much more 'system and problem specific'.

Depending on the nature of the departure from equilibrium (initially imposed or constantly maintained) the questions of driving forces and energy input must be addressed differently along with the consequences on relevant kinetic mechanisms. In a driven system, besides the natural atomic movements due to thermal activation, extra mobilities associated either with irradiation damage or with

---

[10] G. Martin and P. Bellon, in *Solid State Physics*, vol. 50, eds. H. Ehrenreich and F. Spaepen, Academic Press (1997), p. 189.
[11] P. Pochet, L. Chaffron, P. Bellon, and G. Martin, *Ann. Chimie* **22**, 363 (1997).
[12] Y. Brechet, F. Louchet, C. Marchionni, and J. L. Vergergaugry, *Phil. Mag. A* **56**, 353 (1987).

dislocation motion must also be taken into account. The evaluation of the contributions to the driving force and the relative magnitudes compared to the input energies will be discussed in Section II.

Within this scheme, the examples to be treated can be put into perspective with other systems more familiar to the physics community. We will consider successively: chemical patterning, structural defect patterning, and spatio-temporal patterning in plasticity.

1. CHEMICAL PATTERNING

There is often a thermodynamic requirement that alloys decompose, i.e., to produce, from an initially homogeneous mixture of different atomic species, phases with different compositions well separated in space. This process of decomposition, which reflects the competition between the entropic term which wins at high temperature and the enthalpic terms which dominate at low temperature, tends to bring the alloy closer to thermodynamic equilibrium. In these situations, there is a thermodynamic potential which 'shows the way', the free energy, and the possible kinetic paths for the alloy lead to a succession of microstructures that are spatially heterogeneous. One class of these phenomena, the diffusional decomposition of a solid solution by precipitation and/or spinodal decomposition, is familiar to the physics community. The process generates length scales and kinetic scaling laws which have been extensively studied, experimentally, theoretically, and with computer simulations (for a review, see Refs. [13,14]). At this level of generality, it does not really matter if the decomposition is in a solid or in a liquid. However, the specificity of the solid being able to sustain shear and usually having a crystalline structure introduces a new richness through the elastic effects and crystal anisotropy. Pattern formation under these conditions has been extensively studied,[15–17] and leads to specific behavior as far as coarsening rates and shapes are concerned. Again, the classical tools of statistical mechanics and computer simulations have led this field to a state of maturity, and it will not be reviewed here.

When decomposition is thermodynamically possible, there are various possible kinetic pathways, all compatible with the thermodynamics. One of them is to

---

[13] R. Wagner, R. Kampmann, and P. W. Voorhees, in *Phase Transformations in Materials*, ed. G. Kostorz, Wiley–VCH, Weinheim (2001), p. 309.

[14] K. Binder and P. Fratzl, in *Phase Transformations in Materials*, ed. G. Kostorz, Wiley–VCH, Weinheim (2001), p. 409.

[15] P. Fratzl, O. Penrose, and J. Lebowitz, *J. Stat. Phys.* **95**, 1429 (1999).

[16] F. Larche, in *Dislocations in Solids*, ed. F. R. N. Nabarro, North-Holland, Amsterdam, Netherlands (1979), p. 135.

[17] A. G. Khachaturian, *Theory of Structural Transformations in Solids*, Wiley, New York (1983).

take advantage of the defects in the material to find short circuits toward thermodynamic equilibrium. The decomposition, instead of taking place by a homogeneous nucleation and growth process, may take advantage of existing interfaces in the solid (most often grain boundaries) or it may proceed by creating an interface and propagating a front into the non-transformed region, to approach thermodynamic equilibrium more closely. These interface-mediated transformations lead to very nice regular patterns, which bear some similarities to those observed in solidification.[18] These patterns have attracted considerable interest in materials science and we will attempt to summarize this form of solid state patterning in Section III.

2. STRUCTURAL PATTERNING IN SYSTEMS PREPARED FAR FROM EQUILIBRIUM

Metals and alloys are rarely in their state of thermodynamic equilibrium. This would be given by the phase distributions predicted by the phase diagram, and given an infinite length scale, would, in principle, lead to an interfacial energy contribution equal to zero. Even the simplest case of a solidified pure metal, usually consists of a polycrystalline structure, i.e., it contains grain boundaries which are surface defects with an associated surface energy. This is the simplest possible out of equilibrium situation which may lead to defect patterning. The phenomenon of grain growth, in spite of its simplicity, presents some features of pattern forming systems which are worth further investigations.[19,20] The difference with the classical 'domain growth with non-conserved order parameter' familiar to the statistical physicist will be reviewed. Relying on recent simulations, some open questions on grain growth will be addressed in Section IV.12.

Another way of introducing extra energy in a simple system is through plastic deformation. The dislocations stored in the material after a deformation step tend to annihilate during further annealing, bringing the system closer to thermodynamic equilibrium. This procedure may be smooth and lengthy, or it may be discontinuous and rapid. The two phenomena, referred to as recovery and recrystallization,[21] are different pathways to thermodynamic equilibrium, and although they do not generate organized patterns, they lead to the evolution of patterns which in itself reveals some important differences between deformation

---

[18] G. R. Purdy and Y. J. M. Brechet, in *Phase Transformations in Materials*, ed. G. Kostorz, Wiley–VCH, Weinheim (2001), p. 481.

[19] D. Weaire and S. McMurry, *Solid State Physics* **50**, 1 (1997).

[20] C. V. Thompson, *Solid State Physics* **55**, 269 (2000).

[21] F. J. Humphreys, *Recrystallization and Related Annealing Phenomena*, Elsevier Science, New York (2003).

structures and equilibrium structures. These aspects will be discussed in Section IV.13 with a view to understanding the patterns formed during plastic deformation (Section V.15).

3. Structural Patterning in Systems Maintained Far from Equilibrium

A system may be maintained out of equilibrium by a continuous input of energy. Two cases will be investigated in this section. In Section V.14 we investigate the remarkable patterns arising from irradiation damage. The input of energy is transformed at least partially into point defects and the dynamical evolution of this population (its aggregation into dislocation loops or cavities) leads to patterns which are striking by their regularity (both in scale and in crystallographic orientation). The regular networks of dislocation loops or voids and bubbles is one of the most convincing cases of pattern formation in driven systems in metallurgy. The classical methods of 'chemical kinetics' have been remarkably successful in explaining them. This 'success story' is to be contrasted with the case described in Section V.15: pattern formation in plasticity. The patterns observed are also very nice although less regular, but our understanding of them is much less complete. The comparison with pattern formation under irradiation will help in identifying both the necessary ingredients to be included in a future theory for pattern formation during plastic deformation, and the limits of the 'chemical reaction' approach. Recent developments from intensive computer simulations as well as stochastic approaches will be presented.

4. Spatio-Temporal Patterning in Plasticity

Patterning at the microstructural scale is only one feature of plasticity. The distribution of plastic deformation in space may not be homogeneous, and in some situations, well defined 'plastic waves' may travel through the sample. The stress-strain curve is then no longer smooth and the observation of 'serrated yielding' presents some complex features strongly reminiscent of 'stick-slip' behavior. We will focus our attention on the Portevin–Le Chatelier (PLC) effect, and illustrate the existence of a regime of self-organized criticality (SOC),[22] and of deterministic chaos.[2]

---

[22] H. J. Jensen, *Self-Organized Criticality*, Cambridge University Press, Cambridge (1998).

II. Free Energy Changes, Driving Forces and Energy Input

We have classified the different flavors of defect-induced solid state patterning according to whether they occur in systems prepared far from equilibrium whilst approaching equilibrium or whether they occur in systems that are maintained far from equilibrium by some external input of energy (Table I.1). This classification not only provides a nice physical picture of the differences between the two types of patterning systems but also highlights the important fundamental difference that necessitates different theoretical approaches. As we shall see in the coming chapters, the relative success we have had in describing and explaining the patterning observed in the solid state, accurately reflects the relative maturity of the theoretical framework available for the two types of systems.

5. SYSTEMS PREPARED FAR FROM EQUILIBRIUM

The systems in which patterning is observed during an approach to equilibrium include interface-mediated chemical patterning and defect patterning in grain growth, recrystallization and recovery. Under isothermal conditions, these systems can be considered to be 'closed' in a thermodynamic sense and well defined initial and final states can be identified using classical thermodynamics. A thermodynamic potential, 'the free energy', can be defined and the change in this potential in going from the initial state to the final state can be quantified. The contributions to this free energy change that are interest to us are usually classified as chemical energy, strain energy and interfacial (surface) energy and we shall discuss these further in a moment. However, we are interested in dynamic pattern formation. This is the spontaneous patterning that occurs during the change from the initial to the final state and these changes are intrinsically irreversible. Even though the initial and final states may be well defined, classical thermodynamics tells us nothing about these irreversible changes. To describe the changes we resort to kinetic descriptions of the underlying physical processes (diffusion, dislocation glide, etc.) and then the theory of irreversible thermodynamics[23–25] provides us with a means of linking these irreversible microstructural changes to the equilibrium descriptions of the initial and final states. Our ability to describe the changes that occur depends on our ability to describe the initial and final states and on our understanding and ability to accu-

---

[23] W. Yourgrau, A. van der Merwe, and G. Raw, *Treatise on Irreversible and Statistical Physics*, MacMillan, New York (1966).
[24] S. R. DeGroot and P. Mazur, *Non-Equilibrium Thermodynamics*, North-Holland, Amsterdam, (1962).
[25] R. Haase *Thermodynamics of Irreversible Processes*, Addison-Wesley, Reading, MA (1969).

rately describe the fundamental physical processes underlying the changes that occur within the material. This approach has been very successful at describing, for example, the kinetics of phase transformations[13,14] where mass transport by diffusion is the primary process governing change. The kinetic descriptions of the underlying physical processes are usually based on the assumption that the rate of the process is linearly related to the gradient in a relevant thermodynamic potential. This is not a consequence of the theory of irreversible thermodynamics, but is a hypothesis that appears to be experimentally well supported. Fick's first law of diffusion (Eq. (2.1)) in an n component system is a good example:

$$J_i = \sum_{j=1}^{n} -L_{ij} \, \mathrm{grad}(\mu_j), \qquad (2.1)$$

L_{ij} are phenomenological coefficients and μ_j is the chemical potential of component j.

In this case the solute flux is proportional to the gradient in the chemical potential of the solute species. We usually refer to the gradient in the thermodynamic potential as the 'driving force' and the gradient in the chemical potential is the driving force for solute diffusion. These driving forces are not independent of the free energy change associated with the initial and final states of the material but neither are they the same. The driving forces are locally defined and should be associated with some particular physical process. Nevertheless, we usually categorize the driving forces in the same way as we categorize the free energy changes and it is worthwhile summarizing these different contributions.

a. Chemical Energy: The Free Energy of Strain-Free Regions of a Phase

When the compositions and/or the structures of the phases in a system deviate from the equilibrium values under the conditions examined a driving force for solute redistribution and/or crystal structure change arises. The classic example may be the cooling of pure Fe below the allotriomorphic phase transformation temperature of 1183 K. The change in free energy associated with the crystal structure change is of the order of 1 kJ/mol (140 MJ/m^3). Other microstructural changes driven by chemical free energy familiar to the physicist include the precipitation of a second phase from a metastable homogeneous solid solution.[13] In this case the chemical free energy change attending the reaction is typically somewhere in the range 0.1–3 kJ/mol ($\sim$10–300 MJ/m^3) and is used to drive both the crystal structure change and the necessary solute redistribution.

b. Strain Energy: The Increase in Free Energy Caused by Short and Long Range Elastic Strains

When a metal or alloy is deformed most of the work done is liberated as heat. However, a small amount (which is less than 10% and decreases with increasing strain) is stored within the material in the form of lattice strains and imperfections. The most important of these are dislocations which have a long range strain field. In most cases the entropic contribution to the free energy of the material from dislocations is negligible compared with the strain energy contribution and we can approximate the increase in free energy with the strain energy of the dislocation ensemble. The elastic stored energy of a single isolated edge dislocation is:

$$E_D = \frac{\mu b^2}{4\pi(1-\nu)} \text{Ln}\left(\frac{r}{r_0}\right), \tag{2.2}$$

μ is the shear modulus of the material, b is the burgers vector, r is the outer cut-off radius (which could be the specimen surface), r_0 is the inner cut-off radius and ν is Poisson's ratio.

In addition to the elastic stored energy of the long range strain field, we also associate a 'core' energy with the dislocation with a magnitude similar to the elastic contribution. Often, the total is approximated as $\frac{1}{2}\rho\mu b^2$. An ensemble of dislocations, ρ, can reduce its free energy contribution from ρE_D by reducing r through the rearrangement into stress-screening arrays. Clearly both the density and the distribution of dislocations must be known to calculate the free energy contribution from the dislocations. For a typical metal deformed at room temperature and containing a high dislocation density of $1 \cdot 10^{15}$ m^{-2} we may estimate the upper limit to the total strain energy (ρE_D) as of the order of 10–20 J/mol (1–3 MJ/m^3). It is this free energy that drives the recovery and recrystallization processes to be discussed in Section IV.13. Unlike solute diffusion, where the kinetic processes are relatively well described (Eq. (2.1)), dislocation motion is much more problematic. Dislocations motion is restricted to well defined crystallographic planes and dislocations are non-conserved quantities. They move by a dislocation glide process which occurs more readily than diffusion. These kinetic aspects as well as the difficulty in defining a local driving force for dislocation motion (which depends on the density and distribution of all other dislocations) will be discussed in later chapters and highlights the difficulties faced in describing dislocation patterning.

c. Interfacial (Surface) Energy: The Extra Free Energy of Atoms at Interfaces Between Phases and at Grain Boundaries

The extra energy associated with the atoms at the interfaces between phases and at grain boundaries is a free energy. This excess free energy is usually expressed as an energy per unit area of interface and gives rise to a driving force to reduce the interfacial area. This reduction occurs through interface migration and the local driving force is usually expressed in terms of a chemical potential difference on either side of an interface:

$$\Delta\mu = \frac{2\sigma}{r}, \tag{2.3}$$

σ is the interfacial free energy per unit area and r is the radius of curvature of the interface (which can be generalized to two principle radii of curvature for an arbitrary boundary). It is the interfacial contribution to the free energy of the system that drives the grain growth processes to be discussed in Section IV.12. A typical grain boundary energy in a metal may be expected to lie somewhere between 0.4–1 J/m^2 and for a grain size of 10 μm, this would give rise to a local driving force acting over the interface of the order of 1 J/mol (0.1 MJ/m^3). The kinetic description of interface motion is usually based on an assumed proportionality between this driving force and the interface velocity.

The above separation of the driving forces into chemical, strain and interfacial energies is useful for discussion purposes but it is of course artificial. Not only is there usually more than one type of contribution present, but the contributions often also represent retarding forces that need to be overcome for the physical process to proceed. For example, a concentration gradient in an otherwise perfect crystal will have contributions to the driving force from both chemical and strain energies. The latter arises from the concentration dependence of the molar volume of the lattice. In the case of the interface-mediated transformations to be discussed in Section III, the transformation (which is driven by the chemical free energy change) results in the generation of interfaces and strains (because of differences in molar volumes of the phases). The associated interfacial and strain energies must be 'paid for' by the chemical free energy driving the transformation.

When considering the contributions (driving or retarding) from chemical, strain and interfacial energies, it is useful to bear in mind the relative magnitudes. Although the driving forces usually evolve in time, in general we find that the typical chemical contribution is one or two orders of magnitude larger than a typical strain energy contribution which is usually at least an order of magnitude larger than the interfacial energy contribution.

6. Systems Maintained Far from Equilibrium by the Continuous Input of Energy

For those systems that are continuously maintained far from equilibrium by the input of energy, no general framework such as that offered by classical and irreversible thermodynamics discussed above exists. It is not meaningful to define an initial and final state of the material. The system is dynamic and the external input of energy input plays a vital role in the identification and the stability of any stationary states. The defect patterning that occurs during irradiation and during plastic deformation both belong to these types of systems.

The energetic contributions to the driving forces discussed above (chemical, strain and interfacial) all still apply to driven systems but an additional two factors must also now be considered. The first is the input of energy into the systems and the means of storing this energy. In the case of plastic deformation, the energy is stored in the form of dislocations and represents less than 10% of the energy expended in deformation. In the case of irradiation, the equivalent quantity is the displacement per atom induced by irradiation. The energy is stored in the form of point defects (vacancies and interstitials) and the only way to evaluate the quantities in an irradiation cascade is through molecular dynamics simulations. This continuous input of energy during plastic deformation or irradiation modifies the 'potential landscape' in which the system evolves. Any attempts at identifying stationary states must consider the effect of this input energy on the landscape being sampled. In addition, the defects introduced by plastic deformation and irradiation modify the kinetic processes governing the changes in the material. Therefore, not only is the 'potential landscape' modified but the kinetic pathways available for searching the landscape are also changed. Both of these aspects must be considered in dealing with microstructural evolution in driven systems.

As we shall see in Section V.14, some success has been achieved in dealing with patterning under irradiation. In these cases a 'chemical kinetics' approach was adopted which shows some good agreement with experiment. In Section V.15, we discuss patterning during plastic deformation and we shall see that our understanding is much less complete. The difficulties in identifying the local driving force for dislocation motion and in mimicking the dislocation transport (glide) process currently present some barriers to progress.

III. Chemical Patterning: Interface-Mediated Transformations: Eutectoid Decomposition and Discontinuous Precipitation

Probably the best known of all solid state patterns formed in the science of metals and alloys, and certainly the most thoroughly studied, is that arising from the pearlite phase transformation in steels. The reaction belongs to the class of eutectoid transformations (the solid state equivalent of the eutectic transformation) and involves the decomposition of the face-centered cubic austenite (γ) phase into a lamellar pattern of body-centered cubic carbon-poor ferrite (α) and carbon-rich orthorhombic cementite (Fe_3C) behind a migrating interphase boundary.

$$\gamma \to \alpha + Fe_3C \qquad (3.1)$$

In this case, the defect involved in the pattern formation is the interface which is created between the growing product phases and the parent phase. The transformation usually begins at the prior γ grain boundaries or other structural heterogeneities and an example taken from an Fe-0.8C (wt.%) alloy is shown in Figure III.1[26] together with the phase diagram illustrating the three phases involved in the transformation.

Although the pearlite transformation in steels is by far the best known and studied of the eutectoid transformations, the formation of lamellar patterns from eutectoid decomposition in systems other than those based on Fe-C are also well known (e.g., Cu-Al,[27–30] Zn-Al,[31] Cu-In,[32] Au-In[33]). In all cases, the parent phase transforms into two different product phases and growth involves the redistribution of solute from the solute-poor product phase to the solute-rich phase. An important input into the theoretical treatments of growth is an identification of the 'path' taken by the redistributing solute. In some cases, such as in the Cu-Al system,[29] it has been quite well established that the diffusion occurs in the parent phase ahead of the migrating interface. In other cases (e.g., Cu-In[32]), it has been concluded that the solute redistribution occurs by short-circuit diffusion within the migrating interphase boundary itself where the atomic arrangement is comparatively disordered.[34,35] In the overwhelming majority of cases of pearlite growth, the necessary solute redistribution from the solute poor phase to the solute rich is restricted to the vicinity of the transformation interface and no long range solute diffusion either to or from the parent phase is observed. The pearlite aggregate forms with the same overall composition as the parent phase and growth occurs at a constant rate with a constant interlamellar spacing. It is these two characteristics of the pattern formation in eutectoid decomposition that have attracted the interest of researchers in materials science.

In the solid state, a lamellar pattern resembling the pearlite transformation in appearance can also arise as the product of a simple precipitation reaction involving only two crystallographically distinct phases. The transformation is referred

---

[26] L. S. Darken and R. M. Fisher, in *Decomposition of Austenite by Diffusional Processes*, eds. V. F. Zackay and H. I. Aaronson, Interscience, New York (1962), p. 249.
[27] G. V. T. Ranzetta and D. F. R. West, *J. Inst. Metals* **92**, 12 (1964).
[28] M. K. Asundi and D. F. R. West, *J. Inst. Metals* **94**, 19 (1966).
[29] D. Cheetham and N. Ridley, *Metall. Mater. Trans. A* **4**, 2549 (1973).
[30] M. J. Whiting and P. Tsakiropoulos, *Acta mater.* **45**, 2027 (1997).
[31] D. Cheetham and N. Ridley, *J. Inst. Metals* **99**, 371 (1971).
[32] A. Das, S. K. Pabi, I. Manna, and W. Gust, *J. Mat. Sci.* **34**, 1815 (1999).
[33] A. Das, W. Gust, and E. J. Mittemeijer, *Mat. Sci. and Tech.* **16**, 593 (2000).
[34] Y. Mishin, C. Herzig, J. Bernardini, and W. Gust, *Int. Mat. Rev.* **42**, 155 (1997).
[35] I. Kaur and W. Gust, *Diff. and Def. Data A* **66–69**, 765 (1989).

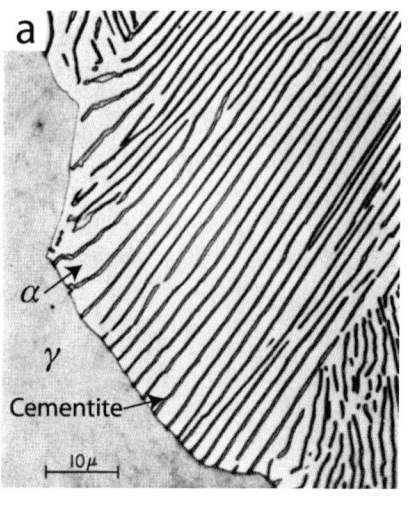

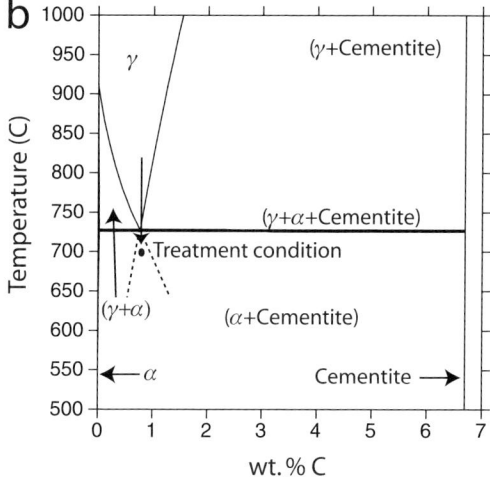

FIG. III.1. (a) Optical micrograph of a pearlite colony in an Fe-0.8C (wt.%) alloy partially transformed at 700 °C. The lamellar pattern consists of alternating plates of Fe_3C (dark) and α (light) separated from the parent γ phase by an interphase boundary. (b) The Fe-rich end of the binary Fe-C phase diagram illustrating the three phases involved in eutectoid decomposition. The C content of the α phase is so low that the single α phase field cannot be resolved for the domain of C contents chosen for illustration.

to as 'discontinuous' precipitation to distinguish it from the competitive 'continuous' mode of relieving the supersaturation of the parent phase, familiar to the physicist.[13,14] Discontinuous precipitation also usually begins at grain bound-

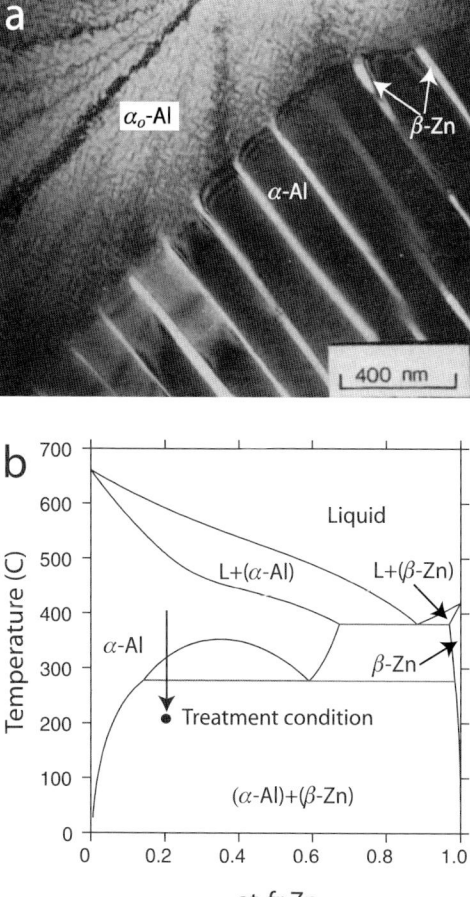

FIG. III.2. (a) Bright field transmission electron microscopy image of discontinuous precipitation in an Al-22Zn (at.%) alloy partially transformed at 478 K. The lamellar pattern consists of alternating plates of β-Zn and a solute depleted α-Al phase separated from the supersaturated α_0 phase (parent phase) by a grain boundary. (b) The binary Al-Zn phase diagram illustrating the two phases involved in discontinuous precipitation.

aries in the supersaturated parent phase (α_0) and forms as a regular aggregate of alternating compound phase (β) and a solute depleted matrix phase (α) growing behind a migrating grain boundary which separates the product from the parent phase (e.g., Figure III.2a).

$$\alpha_0 \to \alpha + \beta \quad (3.2)$$

The solute depleted matrix phase (α) is identical in structure to the parent phase (α_0) and differs only in solute content and crystal orientation across the transformation interface. The solute content and orientation appear to change 'discontinuously' across the grain boundary and it is from this change that the transformation derives its name. The solute redistribution from the solute depleted product phase to the solute rich compound phase occurs by short circuit diffusion within the transformation interface and volume diffusion in the parent phase is thought to play little role in the transformation. Like most cases of the pearlite transformation, under isothermal conditions the discontinuous precipitation product usually forms with the same overall composition as the parent phase and growth at a constant rate with a constant interlamellar spacing is observed.

Discontinuous precipitation has now been identified in more than 150 binary and ternary systems. A comprehensive review of theses systems was recently compiled by Manna et al.[36] and a thorough discussion of many aspects of the reaction can be found in the monograph by Pawlowski and Zieba[37] and in several recent reviews.[18,38] An example taken from the Al-Zn system[39] is shown in Figure III.2 along with a section of the corresponding binary phase diagram.

In the transmission electron micrograph shown, discontinuous precipitation of the β-Zn phase and the Zn poor α-Al matrix lamellar phase grow behind a migrating grain boundary into the parent phase. Unlike eutectoid decomposition, where pearlite formation is associated with a definite feature of the phase diagram (the eutectoid), it is not well understood under which conditions a supersaturated solution will decompose solely in a discontinuous manner, solely in a continuous manner or both together in a competitive manner. The result depends on the competition between the nucleation and growth of the continuous and discontinuous products. Although the growth aspects of both of these problems are now reasonably well understood, a reliable quantitative understanding of the rates of nucleation of the two products is lacking. This is especially true for the initiation of the discontinuous reaction. Several initiation mechanisms have been proposed over the years[36,37,39–42] but so far no single origin is consistent with all of the experimental observations.

---

[36] I. Manna, S. K. Pabi, and W. Gust, *Int. Mat. Rev.* **46**, 53 (2001).

[37] A. Pawlowski and P. Zieba, *Phase Transformations: Controlled by Diffusion at Moving Boundaries*, Polish Academy of Sciences, Warszawa (1991).

[38] D. B. Williams and E. P. Butler, *Int. Metals Rev.* **26**, 153 (1981).

[39] I. G. Solorzano, G. R. Purdy, and G. C. Weatherly, *Acta metall.* **32**, 1709 (1984).

[40] G. Meyrick, *Scripta metall.* **10**, 649 (1976).

[41] S. F. Baumann, J. Michael, and D. B. Williams, *Acta metall.* **29**, 1343 (1981).

[42] H. I. Aaronson and C. S. Pande, *Acta mater.* **47**, 175 (1998).

The pearlite and discontinuous reactions shown in Figures III.1 and III.2 give rise to lamellar patterns. The pattern takes the form of alternating plates of the product phases and this is indeed the most commonly observed morphology. Less frequently, products of both the discontinuous and eutectoid decomposition reactions have also been observed to adopt 'rod-like' morphology; examples include the discontinuous precipitation reaction in Cu-Co[43] and the pearlite ($\alpha + Mo_2C$) formed in Fe-C-Mo.[44] As is the case for eutectic solidification, the choice of morphology depends on the relative volume fraction of phases formed and the energy of the interface between the product phases.[4] In the following discussion, we shall restrict ourselves to treatments of the more common lamellar morphology.

The regularity of the patterns formed from the pearlite and discontinuous precipitation reactions has led many theoreticians to search for a quantitative description of the pattern repeat distance and the growth rate. In Section III.7, we review the most popular treatments and then in Section III.8 consider the attempts to resolve the central unanswered growth related question: 'the spacing selection problem'. A brief discussion of lamellar pattern formation under non-steady state conditions is presented in Section III.9 and some illustrations of the resulting patterns are provided. Since the products of both the pearlite and the discontinuous precipitation reactions are highly non-equilibrium structures, a significant driving force exists for coarsening processes. In Section III.10 we summarize the known mechanisms responsible for the changes in shape and scale of the lamellar products during coarsening. Finally, a short discussion on the possibility of using artificially created metallic multilayers to investigate some of the fundamental aspects of the selection and evolution of solid state patterns in interface-mediated transformations is presented.

7. Steady State Growth

a. Theoretical Treatments

The pearlite and discontinuous precipitation products shown in Figures III.1 and III.2 were formed isothermally under steady state conditions with an approximately constant (average) growth rate and interlamellar spacing. This problem of the steady state growth of a lamellar pattern has a long history and has been

---

[43] A. Perovic and G. R. Purdy, *Acta metall.* **29**, 53 (1981).
[44] R. E. Hackenberg and G. J. Shiflet, *Acta mater.* **51**, 2131 (2003).

considered by many authors.[45–66] In the following, we will present a selection from the sequence of models which have appeared in the literature, each model representing a response to the limitations of its predecessors. The reason for this approach is not only to provide a clearer insight into the necessary ingredients for a full solution of the problem, but also to give the reader access to the simplest model possible for interpreting their experimental observations. If only the growth kinetics and length scales are to be interpreted, then Zener's[47] approach is sufficient. If local chemical analysis between lamellae is to be analyzed, then the full solution of the diffusion equation provided by Cahn[50] is necessary. If the shapes of the moving boundaries are available to experimental investigation, then the treatment developed by Hillert[49,57,60,64] is necessary. If the full selection problem is to be investigated, then a complete solution of the diffusion equations, coupled with appropriate thermodynamic data and a local force balance at every position along the transformation interface is required. We start with the simplest approaches and progressively proceed toward the more sophisticated ones, but want to emphasize that for a given set of experimental observations the appropriate theory may not be the most sophisticated one.

At the outset it is worth emphasizing the difference between the pearlite and discontinuous precipitation reactions. Even though both processes give rise to a lamellar pattern and the diffusion geometry is similar in both cases, one reaction is

---

[45] W. H. Brandt, *J. Appl. Phys.* **16**, 139 (1945).
[46] E. Scheil, *Z. Metallkd.* **37**, 123 (1946).
[47] C. Zener, *Trans. AIME* **167**, 550 (1946).
[48] D. Turnbull, *Acta metall.* **3**, 55 (1955).
[49] M. Hillert, *Jernkont. Ann.* **141**, 757 (1957).
[50] J. W. Cahn, *Acta metall.* **7**, 18 (1959).
[51] K. A. Jackson and J. D. Hunt, *Trans. AIME* **236**, 1129 (1966).
[52] K.-N. Tu and D. Turnbull, *Scripta metall.* **1**, 173 (1967).
[53] H. I. Aaronson and Y. C. Liu, *Scripta metall.* **2**, 1 (1968).
[54] B. E. Sundquist, *Acta metall.* **16**, 1413 (1968).
[55] J. M. Shapiro and J. S. Kirkaldy, *Acta metall.* **16**, 579 (1968).
[56] J. Petermann and E. Hornbogen, *Z. Metallkunde* **59**, 814 (1968).
[57] M. Hillert, in *Mechanism of Phase Transformations in Solids*, vol. 33, The Institute of Metals (1968), p. 231.
[58] B. E. Sundquist, *Acta metall.* **17**, 967 (1969).
[59] M. Hillert, *Acta metall.* **19**, 769 (1971).
[60] M. Hillert, *Metall. Mater. Trans. A* **3**, 2729 (1972).
[61] M. P. Puls and J. S. Kirkaldy, *Metall. Mater. Trans. A* **3**, 2777 (1972).
[62] G. Bolze, M. P. Puls, and J. S. Kirkaldy, *Acta metall.* **20**, 73 (1972).
[63] B. E. Sundquist, *Metall. Trans. A* **4**, 1919 (1973).
[64] M. Hillert, *Acta metall.* **30**, 1689 (1982).
[65] A. Bogel and W. Gust, *Z. Metallkd.* **79**, 296 (1988).
[66] L. Klinger, Y. J. M. Brechet, and G. R. Purdy, *Acta mater.* **45**, 5005 (1997).

a precipitation reaction and involves only two different phases, whereas the eutectoid decomposition reaction involves three different phases. The thermodynamic situation in the two cases is very different and this difference must be considered in the development of models that attempt to explain the observed growth. Nevertheless, there are many common assumptions made in the derivations and it is also worth summarizing these commonalities.

In all cases, the model derivations are based on the isothermal growth of a perfectly regular lamellar aggregate. The isothermal assumption includes all areas of the transformation interface and it is assumed that the heat liberated by the reaction can be conducted away from the transformed volume quickly enough that any local rises in temperature can be ignored. It is usually assumed that all of the phases involved in the reaction have equal molar volumes and that any local stresses developed upon transformation have a negligible influence upon the growth. Finally, it is usually assumed that any interfacial energies that enter into the derivation are isotropic.

One of the earliest theoretical treatments was due to Zener[47] who was interested in describing the kinetics of the pearlite transformation in steels (Eq. (3.1), Figure III.1). Zener provided an approximate treatment that assumed the growth rate of the pattern was controlled by the redistribution of carbon from the ferrite (α) to the cementite (Fe_3C) and that this occurred entirely by volume diffusion in the parent austenite (γ) phase, Figure III.3. The diffusion was driven by the difference in carbon composition in the γ at the γ/α and γ/Fe_3C interfaces, ($X^{\gamma/\alpha} - X^{\gamma/Fe_3C}$).

If the γ/α and γ/Fe_3C interfaces are planar and assuming dilute solution thermodynamics, then the driving force for carbon redistribution is given by the difference in interfacial carbon contents which are usually assumed to be the equilibrium values (local equilibrium assumption) found from the metastable extensions of the $\gamma/(\gamma + \alpha)$ and $\gamma/(\gamma + Fe_3C)$ phase boundaries into the ($\alpha + Fe_3C$) two phase field of the phase diagram (dotted lines, Figure III.1b). Zener realized that not all of this driving force could be used to drive the C diffusion because during growth some must be spent on the creation of the new α/Fe_3C interfaces. The driving force, ΔG, spent on the creation of the interlamellar interface is

$$\Delta G_m^{\text{surface}} = \frac{2\sigma V_m}{\lambda}, \qquad (3.3)$$

λ is the interlamellar spacing (pattern repeat distance), σ is the surface energy of the α/Fe_3C interface and V_m is the molar volume assumed to be the same for all phases.

A critical interlamellar spacing, λ_c, can be identified for which all the free energy available would be spent on creating α/Fe_3C interfaces and the growth rate of the aggregate would fall to zero. For interlamellar spacings greater than λ_c,

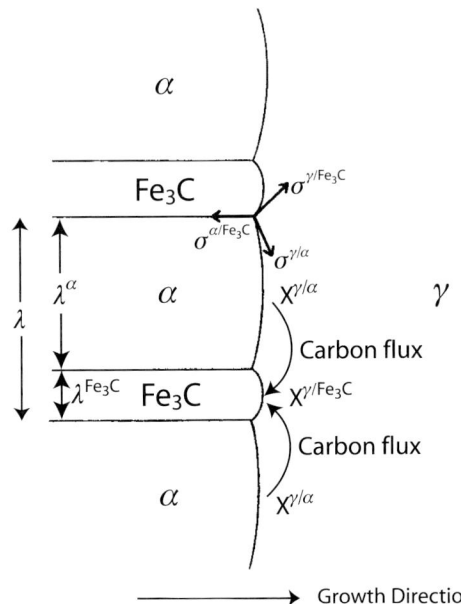

FIG. III.3. Schematic illustration of the diffusion geometry for lamellar pattern formation under conditions where the solute redistribution occurs by volume diffusion in the parent phase.

the net driving force for diffusion could then be approximated by multiplying the total driving force for the reaction by a scaling factor, $(1 - \lambda_c/\lambda)$. Combining this driving force for C diffusion with a characteristic diffusion distance (which was assumed proportional to the lamellar thickness), a carbon flux can be evaluated and when combined with a solute mass balance across the interface, an expression for the growth rate of the regular lamellar aggregate can be written:[47,60]

$$v = \frac{2D^\gamma}{f^\alpha f^{Fe_3C}} \left(\frac{X_e^{\gamma/\alpha} - X_e^{\gamma/Fe_3C}}{X^{Fe_3C} - X^\alpha} \right) \frac{1}{\lambda} \left(1 - \frac{\lambda_c}{\lambda} \right), \qquad (3.4)$$

v is the growth rate, D^γ is the C diffusivity in γ, λ^α and λ^{Fe_3C} are thicknesses of the α and Fe_3C lamellae respectively, the interlamellar spacing $\lambda = \lambda^\alpha + \lambda^{Fe_3C}$, f^α and f^{Fe_3C} are the mol fractions of α and Fe_3C in the aggregate, X^α and X^{Fe_3C} are the C compositions of the growing α and Fe_3C phases, both assumed to be constant, and $X_e^{\gamma/\alpha}$ and X_e^{γ/Fe_3C} are the C compositions in the γ at the γ/α and γ/Fe_3C interfaces, usually assumed to be the local equilibrium values found from the phase diagram.

Zener's treatment, like almost all subsequent treatments, provides a relationship between the growth velocity, the interlamellar spacing and the boundary conditions of the diffusion problem. After a choice of appropriate interfacial compositions, an additional criterion is still required to decouple the interface velocity from the interlamellar spacing to obtain a unique velocity/spacing pair. This is referred to as the 'spacing selection' problem. Zener invoked a maximum growth rate hypothesis to remove the degeneracy of the problem and it is clear from Eq. (3.4) that the maximum growth rate occurs at an interlamellar spacing, $\lambda = 2\lambda_C$, (i.e., when half of the free energy driving the process is stored as interfacial energy). According to Zener's reasoning this would be the real pearlite growth rate. A consequence of Zener's assumed proportionality between the observed lamellar spacing and the critical spacing (Eq. (3.3)) is that we should expect a plot of the reciprocal interlamellar spacing versus the driving force to give a straight line. The driving force usually shows good proportionality with the undercooling from the eutectoid temperature and indeed plots of the reciprocal spacing versus undercooling do fall on a straight line describing well the smaller interlamellar spacings that are observed in pearlite formed at lower and lower temperatures (e.g., Figures III.6b and III.8b). In principle, either the interlamellar interfacial energy (σ) or the proposed proportionality between the observed lamellar spacing and the critical spacing (and therefore the optimization principle) could be extracted from the slope of such plots but unfortunately neither is currently known with sufficient certainty for meaningful conclusions to be drawn. Certainly, reliable theoretical calculations of the interlamellar interfacial energy could shed some valuable light on the question of optimization principles.

The first treatment applicable to discontinuous precipitation appeared in 1955 by Turnbull[48] who was interested in interpreting his growth kinetic results for a Pb-Sn alloy.[67] Turnbull first used Zener's treatment for the pearlite transformation with boundary conditions for the diffusion problem appropriate for discontinuous precipitation and found that his kinetics was around eight orders of magnitude faster that those calculated assuming volume diffusion of Sn. Turnbull subsequently modified Zener's treatment by assuming that the diffusion path for the solute was the transformation interface itself (Figure III.4) instead of volume diffusion in the parent phase as assumed by Zener for the pearlite transformation.

He neglected the effect of surface energy and made some approximations in his evaluation of the net driving force for diffusion and obtained an expression for the pattern growth rate in discontinuous precipitation:

$$v = \frac{D^i \delta}{(\lambda^\alpha)^2} \left(\frac{X^{\text{bulk}} - X_e^{\alpha/\beta}}{X^{\text{bulk}}} \right), \qquad (3.5)$$

X^{bulk} is the initial composition of the supersaturated parent phase, D^i is the solute diffusivity within the reaction interface and δ is the width of the interface itself.

---

[67] D. Turnbull and H. N. Treaftis, *Acta metall.* **3**, 43 (1955).

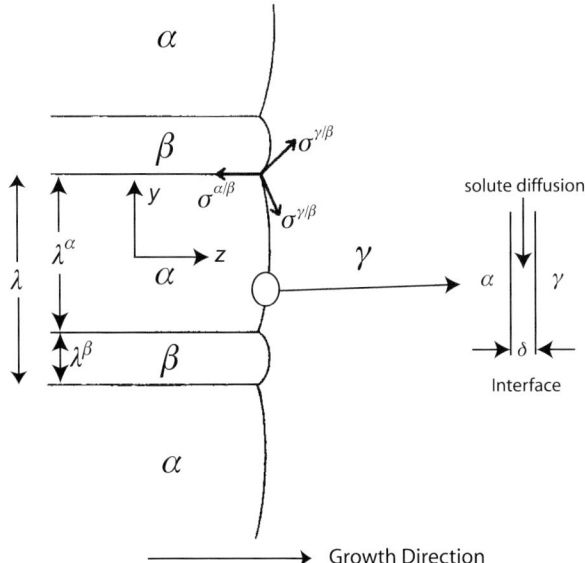

FIG. III.4. Schematic illustration of the diffusion geometry for lamellar pattern formation under conditions where the solute redistribution occurs by short-circuit diffusion within the migrating interface.

λ^α is the width of the solute depleted lamella and $X_e^{\alpha/\beta}$ is the local equilibrium solute concentration in the α at the α/β interface.

Turnbull found satisfactory agreement with his experimentally observed growth rates when using an activation energy for grain boundary diffusion that was a little less than half of that for bulk diffusion. This was consistent with the early findings for other metals and has since been found to be typically representative.[34,35,68] However, the experimentally observed lamellar spacing was much larger than that predicted using Zener's maximum growth rate hypothesis. Even though the diffusional treatment of the problem seemed adequate, Zener's solution to the spacing selection problem did not appear to be satisfactory. Furthermore there was an indication that the discontinuous precipitation reaction resulted in an incomplete relief of the supersaturation of the parent phase and therefore the local equilibrium assumption for the boundary conditions of the diffusion problem may not be appropriate.

The next steps in the theoretical treatments of lamellar growth came very soon after Turnbull's treatment and represented major theoretical advances that put the diffusion problem on a much sounder mathematical basis. Hillert[49] was the first

---

[68] P. Zieba and W. Gust, *Interface Sci.* **10**, 27 (2002).

to present his treatment which was developed with the pearlite transformation in mind, again assuming that the C redistribution occurred only through the γ. He realized that in solving the diffusion problem all the carbon concentrations in the austenite should be considered and not only the compositions at the γ/α and γ/Fe_3C interface (Figure III.3) as was done in Zener's approximate treatment. Hillert, building on the work of Brandt[45] and Scheil[46] obtained a Fourier series expression for the C diffusion field in the γ as a solution to the steady state two-dimensional carbon diffusion equation. He included the effect of capillarity from the curved γ/pearlite transformation interface but to apply the boundary condition at the interface and solve for the unknown Fourier coefficients in the C field expression required knowledge of the shape of the transformation interface which was, *a priori*, unknown. By assuming a relatively flat interface and mechanical equilibrium at the $\gamma/\alpha/Fe_3C$ triple junction (Figure III.3), Hillert was able to obtain an analytic growth rate expression which turned out to be very similar to the approximate expression obtained by Zener (Eq. (3.4)):

$$v = \frac{\pi^3 D^\gamma f^\alpha f^{Fe_3C}}{b} \left(\frac{X_e^{\gamma/\alpha} - X_e^{\gamma/Fe_3C}}{X^{Fe_3C} - X^\alpha} \right) \frac{1}{\lambda} \left(1 - \frac{\lambda_c}{\lambda} \right), \qquad (3.6)$$

b is a constant that depends on the ratio of the lamellar widths and typically has a value ~ 0.5.

Furthermore, Hillert's treatment allowed for a detailed calculation of the interphase boundary shape. This had previously only been calculated by Brandt.[45] To solve the spacing selection problem, Hillert, like Zener, assumed a maximum growth rate hypothesis.

Soon after Hillert's treatment, Cahn[50] provided a general treatment for lamellar growth applicable to both pearlite and discontinuous transformations under conditions where it was assumed that the solute redistribution occurs solely by short-circuit diffusion in the migrating interface (Figure III.4). Turnbull's experiments in the Pb-Sn system[67] had provided an indication that during the discontinuous precipitation reactions, the solute redistribution between the growing phases was incomplete and a non-negligible supersaturation remained in the matrix lamellar phase. With this in mind, Cahn wrote his differential equation for grain boundary interface solute transport but solved it under conditions where the composition of the product phases could vary across the lamellar (i.e., perpendicular to the growth direction). Cahn realized that equilibrium could not be maintained at the moving curved interface and that this resulted in non-equilibrium solute partitioning between the product phases. He solved his steady state diffusion problem and obtained an expression for the solute concentration profile within the solute depleted lamellae as a function of the growth rate, v, the interlamellar spacing, λ, and the solute transport properties of the interface (Figure III.4):

$$\frac{X^{\text{bulk}} - X^{\alpha}(y)}{X^{\text{bulk}} - X^{\alpha/\beta}} = \frac{\cosh \frac{y\sqrt{\alpha}}{\lambda^{\alpha}}}{\cosh \frac{\sqrt{\alpha}}{2}} \quad \text{where } \alpha = \frac{kv\lambda^2}{D^i \delta}, \quad (3.7)$$

$X(y)$ is the solute content in the solute depleted α lamellae behind the migrating interface as a function of position, $X^{\alpha/\beta}$ is the solute concentration in the α at the α/β interlamellar interface which is usually taken as the equilibrium value from the phase diagram, D^i is the solute diffusivity in the migrating boundary, δ is the boundary width and k is a proportionality constant that relates the concentration in the thin slab of interface, X^i, to the solute concentration inherited by the product phase, X^{α}.

Cahn assumed that the interface velocity was simply proportional to the net Helmholtz energy change across the interface, ΔF, which can be calculated from a knowledge of the system thermodynamics, the solute profiles retained within the lamella and the interfacial energy of the interlamellar interfaces:

$$v = M \Delta F, \quad (3.8)$$

M is a macroscopic kinetic parameter (friction) which is a measure of the response of the interface to the applied driving force. Again, Cahn adopted an optimization principle (optimal rate of free energy dissipation) to remove the degeneracy of the problem and select a velocity/spacing pair.

Cahn's derivation of the solute profile left behind the migrating interface is applicable to both pearlite and discontinuous precipitation products. However, by nature of the system thermodynamics, the composition gradients which can be sustained in pearlite, and still result in a net free energy decrease during growth, are very limited. In contrast, in the discontinuous reaction, the ranges of available compositions allowed for a supersaturated solution are much wider. The difference is shown in Figure III.5.

In the top figure the free energy curves for the precipitation reaction are shown. For a bulk alloy composition X_0, the chemical part of the free energy will decrease in a precipitating system as long as the precipitate, β, forms with a composition greater than X', regardless of how little precipitates and therefore how little the α phase changes composition. For the eutectoid composition shown below, the new α and β phases cannot take compositions anywhere near the original γ matrix composition. Segregation to at least X'_{α} and X'_{β} is necessary to give a decrease in the free energy of the system. The exact solution for the solute diffusion profile between two lamellae provided by Cahn's approach proved to be both theoreti-

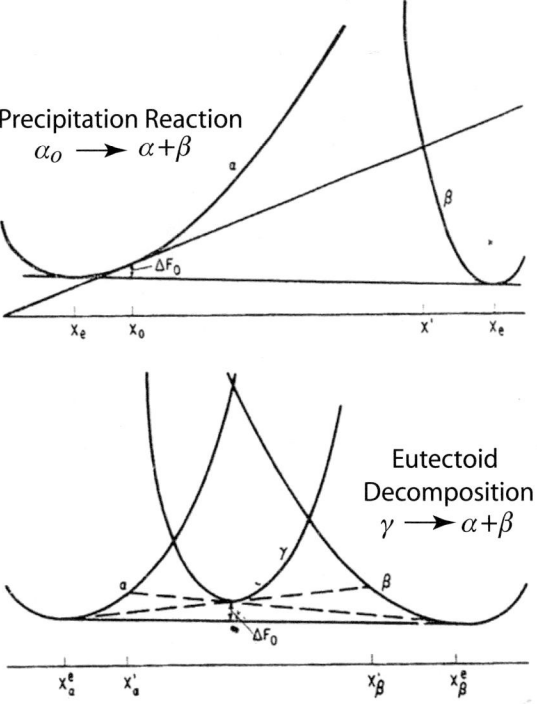

FIG. III.5. Schematic free energy curves for (a) precipitating and (b) eutectoid systems.[50]

cally important, and possible to test experimentally by local chemical analysis, in discontinuous reaction products, e.g., see Refs. [69–73].

Hillert criticized Cahn's treatment because it did not consider the force actually pulling the grain boundary in the first place and therefore did not really explain how discontinuous precipitation occurs. In addition, Cahn prescribed a flat interface, whereas, in principle, not only the spacing between lamellae, but also the shape of the moving grain boundary should be part of the solution of the problem. Hillert[57,60,64] then provided treatments for the pearlite and discontinuous transformations under conditions where the solute redistribution was restricted to the

---

[69] D. A. Porter and J. W. Edington, *Proc. Roy. Soc. London A* **358**, 335 (1977).
[70] P. Zieba and W. Gust, *Int. Mat. Rev.* **43**, 70 (1998).
[71] P. Zieba and W. Gust, *Mikrochimica Acta* **132**, 295 (2000).
[72] P. Zieba and W. Gust, *Z. Metallkd.* **92**, 645 (2001).
[73] P. Zieba *Mat. Chem. and Phys.* **62**, 183 (2000).

transformation interface, by considering the necessary diffusion and a detailed balance of surface tensions across the interface during growth.

The treatments of Cahn and Hillert can respectively be characterized as 'global' and 'detailed' in the sense that, in Cahn's treatment the interface is a high diffusivity reaction path and its mobility controls the rate at which the interface moves in response to the existing driving force; whilst in Hillert's treatments of lamellar growth, detailed knowledge of the interface and its properties is often necessary for application of the growth rate expressions. The theoretical treatments of Hillert and Cahn, made in the 1950s and 1960s, remain to this day the treatments used most frequently. Since this time, contributions to the theory of lamellar growth have concentrated on extending these original treatments to include such effects as: ternary solute additions,[58,59] non-uniformity in the compositions of the lamella phases in pearlite,[62] coherency stresses[60] and resolution of the spacing selection problem.[61,66,74]

b. Experimental Observations

Many aspects of the theoretical treatments discussed above can be compared with experimental data. These include metallographic measurements of lamellar growth rates and interlamellar spacings and more recently, analytical electron microscopy measurements of the solute concentrations inherited by the growing phases. For both the pearlite transformation and the discontinuous precipitation reaction, much of the experimental data has been collected and compared with the theoretical predictions in periodical reviews of the subject (pearlite: see Refs. [61, 75,76]; discontinuous precipitation: see Refs. [36–38]). Here we present a couple of selected examples to illustrate the types of comparisons that are usually made.

Pearlite. Traditionally, researchers have been interested in describing both the growth rate and the repeat distance of the evolving pattern. These are the two quantities that are experimentally measured. The growth rates are usually calculated from a representative width of the pearlite colonies in a series of isothermally transformed specimens and the results are then compared with the predictions of one of the models discussed above. The first example is taken from the Cu-Al system. The transformation rate in a Cu-12Al (wt.%) alloy has been measured by several authors[27–29] and the experimental growth rates and interlamellar spacings as a function of temperature are shown in Figure III.6.

---

[74] J. S. Kirkaldy, *Phys. Rev. B* **30**, 6889 (1984).
[75] R. F. Mehl and W. C. Hagel, *Prog. Met. Phys.* **6**, 74 (1956).
[76] J. W. Cahn, in *Decomposition of Austenite by Diffusional Processes*, eds. V. F. Zackay and H. I. Aaronson, Interscience, New York (1962), p. 131.

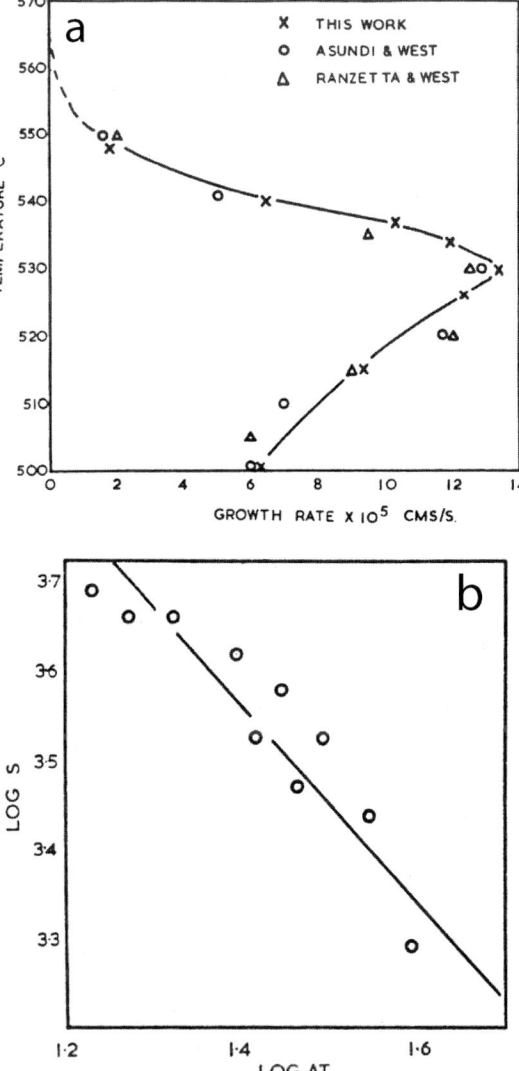

FIG. III.6. Experimentally measured (a) pearlite growth rates and (b) interlamellar spacings in a Cu-11.8Al (wt.%) alloy as a function of temperature.[29].

For the data shown, Cheetham and Ridley[29] then applied Hillert's volume diffusion growth rate equation (Eq. (3.4)) using the experimentally observed interlamellar spacings to examine the agreement between theory and observation. To

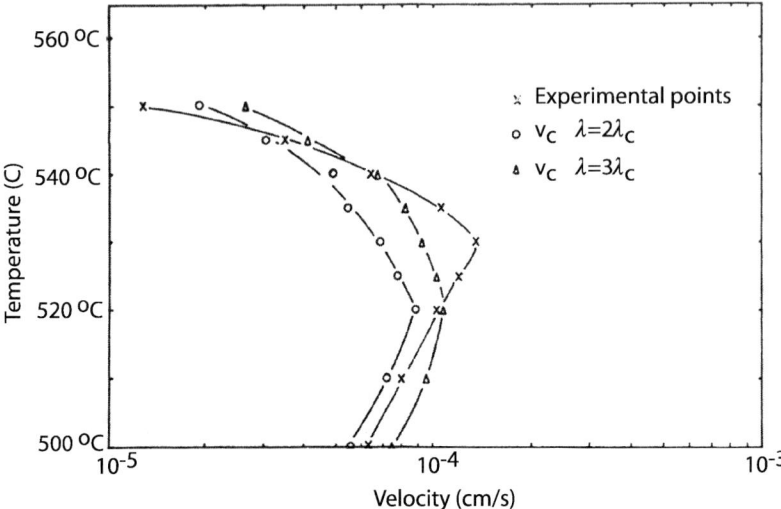

FIG. III.7. Comparison of the growth rate data in Figure III.6a with the theoretical predictions calculated assuming local equilibrium interfacial compositions, volume diffusion of the components and using the experimentally observed interlamellar spacings.[29] The results using both the maximum growth rate hypothesis of Zener[47] and the optimal rate of entropy production of Kirkaldy[61] are shown.

make the comparison, a choice must be made concerning the relationship between the observed lamellar spacing and the critical spacing. The calculated results using both Zener's maximum growth rate hypothesis ($\lambda = 2\lambda_c$) and Kirkaldy's maximum rate of entropy production ($\lambda = 3\lambda_c$) are compared with the experimental results in Figure III.7. The agreement in both cases is considered to be quite good.

On the basis of such comparisons it is now fairly well accepted that the solute redistribution in this reaction occurs predominantly by volume diffusion in the parent phase ahead of the interface but considering the uncertainty in experimental measurements and the thermodynamic and kinetic quantities entering into Eq. (3.4), solid conclusions cannot be drawn concerning the operative optimization principle.

The second example concerns the pearlite reaction in binary Fe-C steels. There has been much discussion in the literature concerning the diffusion path for C in this reaction.[47,49,50,54,57,76] During the 50s, 60s and 70s when much of the experimental work was carried out, the accuracy with which the relevant thermodynamic and kinetic data was known was insufficient to draw any solid conclusions about the C diffusion path through comparisons with theory. Nevertheless, many authors had tentatively suggested that the experimental growth rates appear to be at least an order of magnitude faster than those predicted by the volume diffusion con-

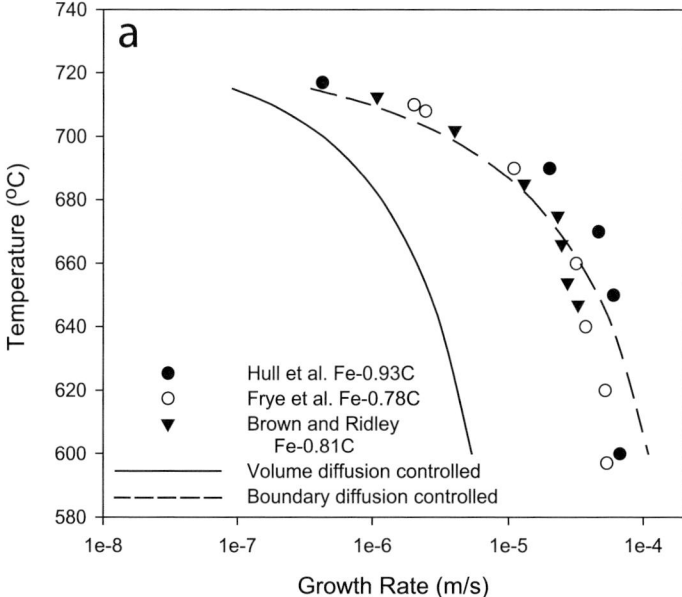

FIG. III.8. Experimentally observed pearlite (a) growth rates and (b) interlamellar spacings in a series of binary Fe-C alloys as a function of temperature. The theoretical predictions assuming local equilibrium interfacial conditions, a maximum growth rate hypothesis and both volume diffusion control and boundary diffusion control are shown in (a). The volume diffusion controlled kinetics is approximately an order of magnitude slower than the experimental observations. The boundary diffusion controlled kinetics was calculated using a $sD^i\delta$ triple product with an activation energy ~ 0.5 of that for bulk C diffusion. (*Continued on next page.*)

trolled models.[47,49,54,76] Today, our knowledge of the relevant thermodynamic and kinetic parameters is much better and we have carried out a new series of calculations using the existing data to shed some light on this issue. The experimentally measured growth rates[77–79] and interlamellar spacings[79] are shown in Figure III.8.

The theoretical prediction using the experimental lamellar spacings and assuming both volume diffusion of C and Zener's maximum growth rate hypothesis is also shown. It is clear that the kinetics are indeed at least an order of magnitude slower than the experimental values, suggesting a diffusion path for C other than the parent γ phase must be contributing to solute redistribution. Hillert[49]

---

[77] F. C. Hull, R. A. Colten, and R. F. Mehl, *Trans. AIME* **150**, 185 (1942).
[78] J. H. Frye, E. E. Stansbury, and D. L. McElroy, *Trans. AIME* **197**, 219 (1953).
[79] D. Brown and N. Ridley, *J. Iron and Steel Inst.* **204**, 811 (1966).

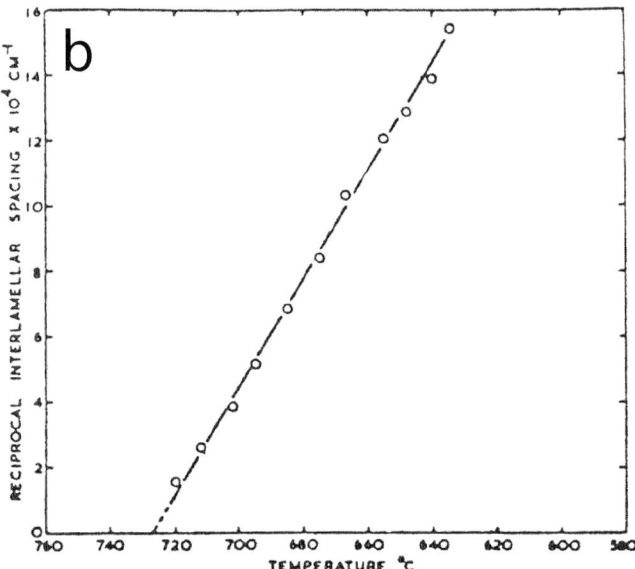

FIG. III.8. (*Continued.*)

and Cahn[76] have discussed the possibility of different diffusion paths for the C and it now seems the most likely candidate is the moving interphase boundary itself (as is the case for the discontinuous precipitation reaction). The theoretical treatments of growth according to this diffusion path have also been made but the problem lies in the lack of knowledge of the thermodynamic and kinetic quantities of the moving boundary region. Hillert's[57] expression for the growth rate of pearlite according to boundary diffusion control is:

$$v = \frac{12 s D^i \delta}{f^\alpha f^{Fe_3C}} \left(\frac{X_e^{\gamma/\alpha} - X_e^{\gamma/Fe_3C}}{X^{Fe_3C} - X^\alpha} \right) \frac{1}{\lambda^2} \left(1 - \frac{\lambda_c}{\lambda} \right). \tag{3.9}$$

Our problem is that neither D^i, δ or s are known with any certainty. Instead, what is usually done, is to extract an effective triple product, $sD^i\delta$, from the experimental results using one of the models for growth and then to make a judgment about whether this seems reasonable for boundary diffusion control. If so, it is then usually concluded that the diffusion path for the solute is probably the moving boundary. We have made such calculations and find that using a segregation coefficient (s) of 1 and a grain boundary width (δ) of 0.5 nm, the experimental data can be adequately described by assuming Zener's maximum growth rate hypothesis ($\lambda = 3\lambda_c/2$ in the case of boundary diffusion control) with an activation energy for boundary diffusion of C that is approximately 0.7 of that for C diffusion in the bulk (Figure III.8a). In the case of substitutional elements diffusing in migrating interfaces, experimental measurements have found that the activation energy for diffusion is indeed approximately 0.5–0.7 of the corresponding value in the bulk.[34,35,68] Whether the same should be true for C which diffuses interstitially in the bulk it is not clear. Nevertheless, this example illustrates the difficulty encountered in making any solid conclusions about the transformation when considerable uncertainty exists in the knowledge of either the optimization principle or the interlamellar interfacial energy and the relevant thermodynamic and kinetic data for the migrating boundary region. Kaur et al.[80] have compiled a large set of interphase boundary diffusion data for those systems that have been experimentally examined and as we shall see in the next section, much of this data is obtained through back calculation from the results obtained in discontinuous precipitation reactions.

During the last 20 years, it has become possible to use advanced electron microscopy techniques to directly measure solute concentrations in the vicinity of the transformation interface. Indeed, the state of the art instruments today are capable of making measurements in the sub-nm range. This would be an ideal, direct way of identifying the solute diffusion path in these reactions. This has been done for several cases and it has recently been shown that for the pearlite formed in Fe-C-Mn alloys,[81] the diffusion path for the redistributing substitutional Mn element is the reaction interface itself and volume diffusion of Mn plays little role in the reaction. However, because of the very low C contents that are encountered in eutectoid steels and the contamination of the surface of the thin-foil samples for electron microscopy during preparation, it is currently not possible to measure with sufficient accuracy the C concentrations in these specimens. In this respect, progress in our understanding of some of the fundamental aspects of the pearlite transformation in steels has not been able to take advantage of the technological advances that techniques like analytical electron microscopy have been able to bring to other transformations, such as the discontinuous precipitation reaction to be discussed below.

Discontinuous Precipitation. In making comparisons between theory and experiment in the discontinuous precipitation reaction we are not faced with the question of the solute diffusion path. From the earliest theoretical treatments it

---

[80] I. Kaur and W. Gust, *Fundamentals of Grain and Interphase Boundary Diffusion*, John Wiley & Sons, New York (1995).
[81] C. R. Hutchinson, R. E. Hackenberg, and G. J. Shiflet, *Acta mater.* **52**, 3565 (2004).

was assumed that the path was the reaction interface itself and this is now experimentally well established. However, the comparisons are complicated by the non-equilibrium segregation of solute at the transformation interface. The solute depleted matrix lamella does retain some level of residual supersaturation. For comparisons between experimentally measured growth rates and interlamellar spacings with theoretical predictions, many of the models proposed require a quantification of this non-equilibrium segregation. Cahn provided an expression describing the solute profile across the lamella behind the transformation interface as a function of the growth rate, the interlamellar spacing, the bulk solute content and the transport properties of the boundary. This is one area where progress in our understanding has benefited from recent advances in analytical electron microscopy. There are now many experimental studies that have directly measured the solute profiles behind the migrating boundary and have found that Cahn's equation is well obeyed. Many of these studies have recently been summarized in a series of reviews, e.g., Refs. [36,37,72]. An example taken from the work of Duly et al.[82] on the discontinuous precipitation reaction in an Mg-18.8Al (at.%) alloy is shown in Figure III.9. The data shows solute profiles measured at two distances behind the reaction interface and the solid curve is the best-fit (in α and $X^{\alpha/\beta}$, Eq. (3.7)) using Cahn's equation. From the value of α obtained, and the experimental measurements of the growth rate and interlamellar spacing of the lamellar pattern, values for the $sD^i\delta$, triple product can be extracted and a judgment can be made concerning the reasonableness of these values. Clearly, depending on which of the theoretical models for growth is chosen for comparison, different values of this triple product can be obtained.

Amir and Gupta[83] have nicely presented the possible differences calculated for each of the different models using data for an Mg-7Al (at.%) alloy. They metallographically measured the growth rate and interlamellar spacing as a function of temperature and used X-ray diffraction techniques to estimate the level of non-equilibrium solute segregation, all as a function of temperature. They then extracted the triple products necessary to reproduce the experimental data for each of the different models and compared these. They are shown in Figure III.10.

The three most sophisticated models (Cahn,[50] Hillert[60,64] and Sundquist[63]) all give quite similar activation energies for boundary diffusion, ~ 70 kJ/mol. The activation energy for chemical diffusion in dilute Mg-Al alloys is reported to be ~ 143 kJ/mol. The proposed boundary diffusion value is then ~ 0.5 of that for volume diffusion and is consistent with the findings in other alloys. The extraction of boundary diffusion parameters from the analysis of the discontinuous precipitation reaction has now become one of the prime sources of boundary diffusion data. However, as is clear from Figure III.10, the values obtained depend on the model used to describe the discontinuous reaction.

---

[82] D. Duly, M. C. Cheynet, and Y. Brechet, *Acta metall mater.* **42**, 3843 (1994).
[83] Q. M. Amir and S. P. Gupta, *Can. Metall. Qrty* **34**, 43 (1995).

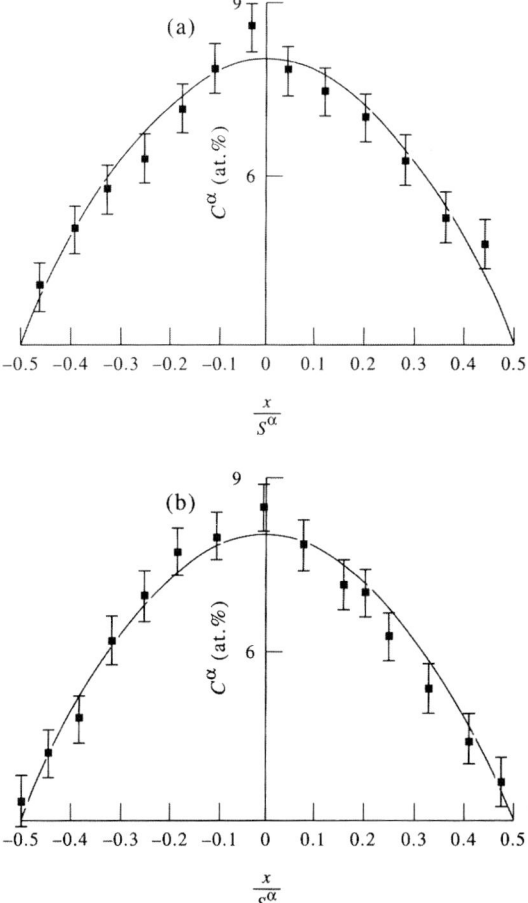

FIG. III.9. Plots of Al composition as a function of position across the solute depleted lamellae at two distances behind the reaction interface; (a) at 380 nm and (b) at 200 nm during discontinuous precipitation in an Mg-18.8Al (at.%) alloy treated at 220°C for 16 h. The profiles were measured using electron energy loss spectroscopy and a scanning transmission electron microscope.[82] In both cases, the profiles are well described by Cahn's equation (solid line) using the same $X^{\alpha/\beta}$ and α values.

8. The Spacing Selection Problem

In each of the theoretical treatments of steady state lamellar pattern growth discussed above, a relation between the growth velocity, v, the interlamellar spacing, λ, and the boundary conditions for the diffusion problem is obtained. After ap-

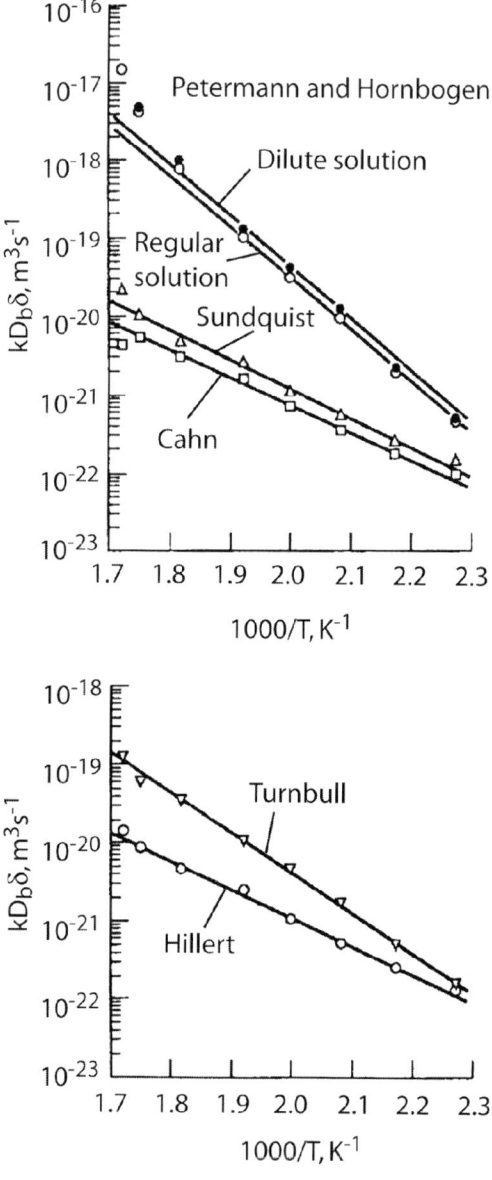

FIG. III.10. Extracted triple product ($sD^i\delta$) values as a function of $1/T$ (K^{-1}) according to each of the different discontinuous precipitation models for experimental growth rate and interlamellar spacing data from a Mg-7Al (at.%) alloy treated at a variety of temperatures.[83]

propriate boundary conditions have been chosen, an additional criterion must still be invoked to decouple the growth velocity, v, and the interlamellar spacing, λ. This problem of 'spacing selection' is the central unanswered question in the formation of lamellar patterns in the solid state. Traditionally a global optimization principle has been invoked. Zener[47] and Hillert[49] used the maximum growth rate hypothesis, Kirkaldy[61] has considered an optimal rate of entropy production and Cahn[50] has considered an optimal rate of free energy dissipation. If we consider, for example, the pearlite transformation, according to volume diffusion control, the optimal lamella spacing according to Zener's maximum growth rate hypothesis and Kirkaldy's maximum rate of entropy production are $\lambda = 2\lambda_c$ and $\lambda = 3\lambda_c$. For boundary diffusion controlled growth, the corresponding values are $\lambda = 1.5\lambda_c$ and $\lambda = 2\lambda_c$. The differences in spacing predicted by the two hypotheses differ only by a small factor. The accuracy with which diffusion coefficients are known does not allow us to differentiate between these two heuristic principles and, as is often the case with these pattern selection problems, the very question of the existence of a selected spacing is still open. Gust et al.[84] have carefully measured the interlamellar spacing in selected discontinuous precipitation reactions and found that under isothermal conditions, not a single spacing but a Gaussian distribution of spacings is experimentally observed (Figure III.11). Similar observations were made by Duly et al.[85] in their investigations of the discontinuous reaction in Mg-Al alloys.

Solorzano and Purdy[86] have examined Cahn's maximum rate of free energy dissipation hypothesis by comparing the theoretical predictions with experimental results from the Mg-Al system. These authors calculated a dissipation function, $\Delta G^{NET} \cdot v$ as a function of interface velocity and interlamellar spacing. They did indeed find a maximum in the dissipation function for a particular velocity and that this maximum was highest for a particular interlamellar spacing. Example calculations for the Mg-9Al (wt.%) alloy at two temperatures are shown in Figure III.12.

The numbers on the curves refer to the interlamellar spacings. Given the experimentally observed growth rates and interlamellar spacings and an estimate of the thermodynamic and kinetic properties of the boundary extracted from experimentally measured solute concentration profiles, Solorzano and Purdy estimated the remaining supersaturation in the lamella behind the moving interface and calculated ΔG^{NET}. The observed interlamellar spacings could then be compared with those predicted by a maximum rate of free energy dissipation by identifying the theoretical spacing which gave the highest value in the dissipation function for the

---

[84] W. Gust, B. Predel, and U. Roll, *Z. Metallkd.* **68**, 117 (1977).
[85] L. Klinger, Y. Brechet, and D. Duly, *Scripta mater.* **37**, 1237 (1997).
[86] I. G. Solorzano and G. R. Purdy, *Metall. Trans. A* **15A**, 1055 (1984).

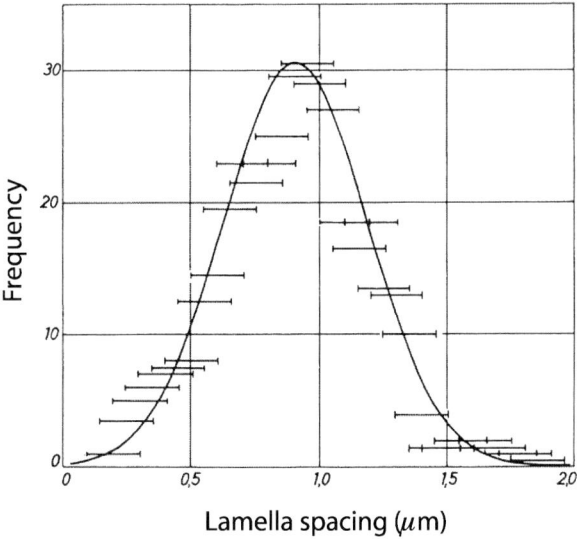

FIG. III.11. Experimentally measured distribution of lamellar spacings in a Ni-31.7Zn (at.%) alloy treated for 24 h at 600 °C.[84] The distribution is well described as Gaussian.

experimentally observed velocity. It was found that from the myriad of possible kinetic states, the velocity/spacing pair selected were within a factor of ~ 2 of that which would give a maximum in the rate of free energy dissipation. Considering the uncertainties in the thermodynamic quantities, in particular interfacial energies, the agreement was deemed encouraging.

a. Klinger's Alternative for Discontinuous Precipitation

Only recently, an alternative treatment of the diffusion problem for the discontinuous precipitation reaction has been supplied by Klinger et al.[66] which provides a unique velocity/spacing pair and for which an additional 'optimization' criteria is not necessary. Klinger et al. realized that the degeneracy of the problem could be removed by invoking an additional local condition in the derivation rather than imposing an optimization principle on a relation between the interface velocity and the lamella spacing. The conditions imposed were all local in nature and relied on thermodynamic assumptions of local equilibrium, which may be arguable, but are at least clearly stated. The starting point of the approach is that both the α and the β lamellae must grow at the same rate, and that both these rates are diffusion controlled. The thermodynamic difference between the α_0/α interface and the α_0/β interfaces, which parallels the previously stated difference between

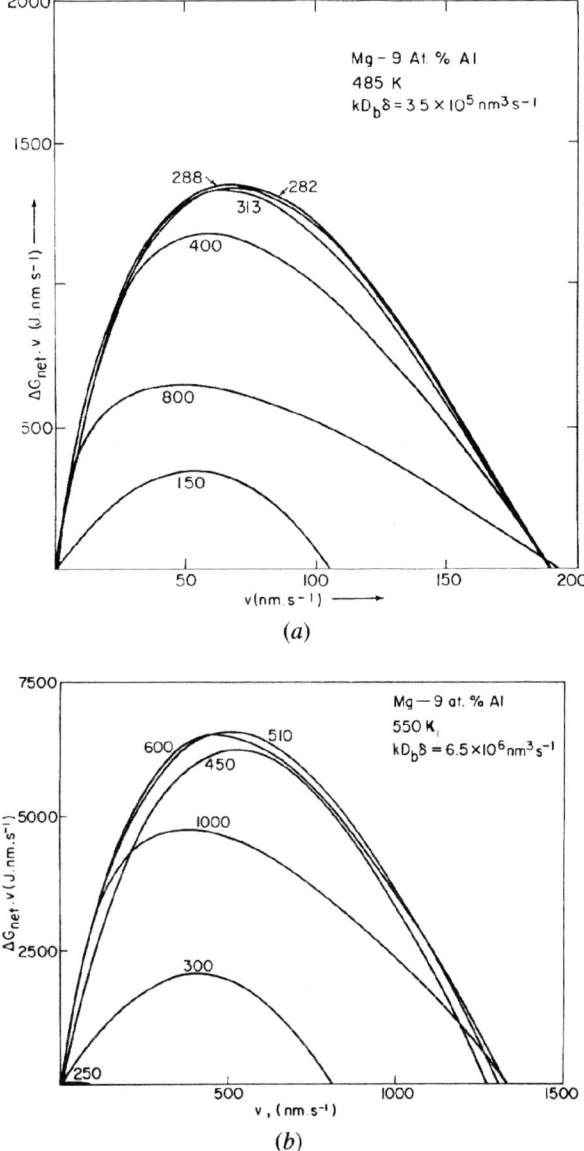

FIG. III.12. Computed variation of the dissipation function, $(\Delta G^{\text{NET}}.v)$, as a function of discontinuous precipitation cell velocity in an Mg-9Al (wt.%) alloy. The calculations are performed at (a) 485 K and (b) 550 K for a range of interlamellar spacings (labelled).[86]

pearlite and discontinuous precipitation, requires different treatments for the diffusion problem.

In previous treatments of the reaction, the solute diffusion from the solute poor lamella to the solute rich lamellae was assumed to occur within the α/α_0 grain boundary and was driven by concentration gradients. The nature of the β/α_0 interface and the driving force for solute diffusion within this interface was not usually considered. Klinger et al. treated the growth of the α and β phases simultaneously, subject to conditions of equal growth velocities, but assumed that the diffusion in the α/α_0 grain boundary was driven by concentration gradients and diffusion within the β/α_0 interface boundary was driven by gradients in curvature. Such an assumption for the β/α_0 interface boundary is physically based since the compound phase in discontinuous precipitation is very often a stoichiometric phase which offers very little range for concentration gradients. The treatment of the grain boundary followed the classical treatments of Cahn[50] and Hillert[57] and that of the interphase boundary followed the treatment of Mullins.[87] An assumption of local equilibrium and diffusional flux continuity at the $\alpha/\beta/\alpha_0$ triple junction was sufficient to remove the degeneracy of the problem and select a unique velocity/spacing pair. Klinger et al. compared their treatment with the available experimental data in the Al-Zn system and found good agreement using reasonable values of the unknown parameters.

Klinger's approach can be seen as an idealized, but tractable version of the problem. Each of the assumptions made can be relaxed through a more sophisticated treatment: local equilibrium assumptions can be modified to account for attachment kinetics, the thermodynamic model for the solid solution can be made more realistic, the shape of the moving interface can be left free: however, the solution is no longer analytic and numerical calculations are needed which show a rich variety of behavior. The important issue is that the intrinsic degeneracy of the spacing selection problem can be relieved by local assumptions and does not require the introduction of an additional optimization principle.

9. NON-STEADY STATE GROWTH

In each of the theoretical treatments of growth discussed above, it is assumed that the lamellar pattern formed under steady state conditions of constant interlamellar spacing and growth rate. There are at least two sets of circumstances where the steady state hypothesis has been called into question and each will be briefly discussed. The first concerns the recent suggestions that growth is not a continuous process with a constant rate but rather a 'stop and go' process and as a result, the average growth rate, determined for example from metallographic measurements,

---

[87] W. W. Mullins, *J. Appl. Phys.* **28**, 333 (1957).

must be differentiated from the instantaneous growth rate. The second case concerns the formation of so-called divergent lamellar aggregates. Observations of isothermal lamellar growth with a rate that decreases in time and an interlamellar spacing that increases in time have been observed in both the pearlite and discontinuous precipitation reactions and the theory is qualitatively well understood.

a. 'Stop and Go' Interface Motion

On the basis of preliminary *in-situ* high voltage electron microscopy studies of the discontinuous precipitation reaction in an Al-Zn alloy, Bogel and Gust[65] proposed that the reaction interface does not migrate in a continuous manner with a constant velocity but rather by a 'stop and go' process. Subsequent electron microscopy studies[88] showed that this was observed experimentally and the plot of interface displacement versus time in Figure III.13 clearly shows the start-stop nature of the interface.

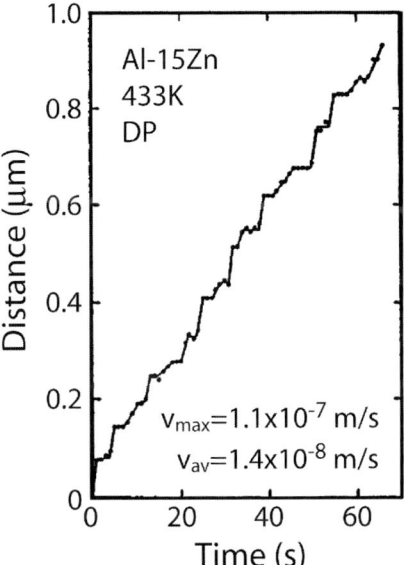

FIG. III.13. Interface migration distance as a function of time during discontinuous precipitation in an Al-15Zn (at.%) alloy transformed at 160 °C.[88] The measurements were made during the *in-situ* elevated temperature high-voltage electron microscopy observations. The stop-start nature of the interface migration is clearly shown in the plot.

---

[88] S. Abdou, G. Solorzano, M. ElBoragy, W. Gust, and B. Predel, *Scripta mater.* **34**, 1431 (1996).

Bogel and Gust claim that the difference between the instantaneous interface velocity and the average value can be up to three orders of magnitude.

Further indirect experimental evidence in support of the 'stop and go' motion was collected in Duly's studies of the discontinuous precipitation reaction in the Mg-Al system.[85] They found, as with previous investigators, that within a given lamellar colony, a range of lamella spacings is observed. When the solute profiles within the lamellae behind the reaction interface were measured, they could be fitted to Cahn's equation (Eq. (3.7)) using an approximately constant α parameter. If cooperative motion does occur in a continuous and constant manner, then everywhere along a reaction front where the lamellar spacing changes, so too should the α parameter necessary to describe the solute profile left behind. For lamellar spacings that range from $\langle S \rangle/2$ to $2\langle S \rangle$, we should expect to see a variation in the α parameter of $\langle a \rangle/4$ to $4\langle a \rangle$, a factor of 16. Klinger et al.[85] reinterpreted the observed chemical profiles in terms of a 'stop and go' motion of the interface and found that the approximately constant α parameter across a lamellar colony with varying lamellar spacing could be rationalized assuming the non-steady state type of growth proposed by Bogel and Gust.[65]

b. Divergent Lamellar Growth

The second case of non-steady state lamella growth worth considering is the formation of aggregates with diverging lamellae. The lamellar product of eutectoid decomposition or discontinuous precipitation usually inherits the same overall composition as the parent phase and therefore it is assumed that a constant (average) growth prevails. Indeed, this is experimentally confirmed in most cases. However, if for some reason the growing aggregate is constrained to inherit a composition different from that of the parent phase then long range solute diffusion is necessary between the growing lamellar aggregate and the parent phase and growth with a constant rate cannot prevail. Non-steady state growth of both pearlite and discontinuous precipitation is possible and both have been observed. These observations are restricted to multi-component systems and the examples considered below are each taken from Fe-based systems containing both a substitutional and interstitial solute. We first consider non-steady state growth of pearlite, using the Fe-C-Mn system as an example before examining a counterpart discontinuous precipitation reaction found in high N austenitic stainless steels.

Pearlite. Under conditions of constant temperature and pressure, the three phase equilibrium between the γ, α, and Fe_3C in the binary Fe-C system occurs at a single 'eutectoid' temperature (Figure III.1b). At temperatures below the eutectoid, the γ phase is unstable and the growing ($\alpha + Fe_3C$) pearlitic aggregate consumes 100% of the parent γ phase and therefore is obliged to inherit the bulk composition. However, in multi-component systems, the three phase equilibrium between the γ, α and M_3C ($M = Fe + X$), exists over a temperature range and a $(\gamma + \alpha + M_3C)$ three phase field is observed. It is within this $(\gamma + \alpha + M_3C)$ three phase field that non-steady state pearlite formation can occur. Two sections of the ternary Fe-C-Mn phase diagram are shown in Figure III.14.

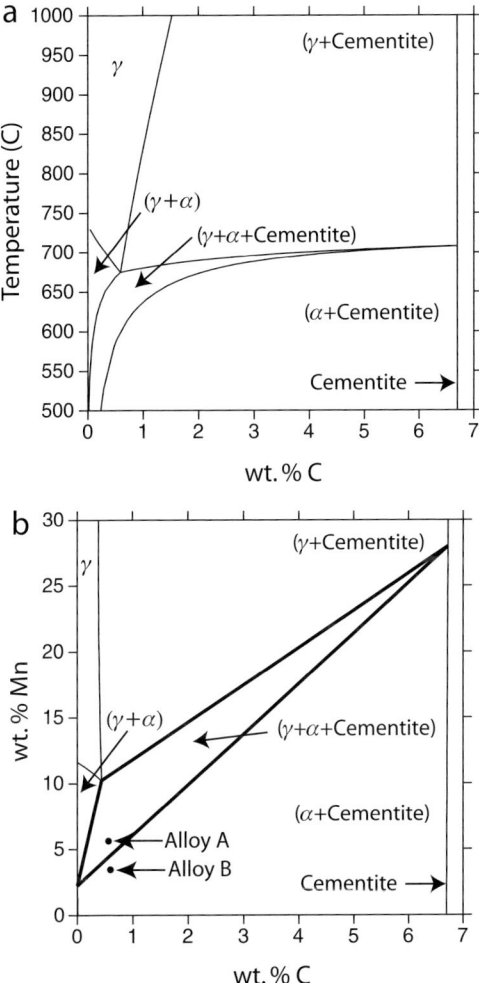

FIG. III.14. (a) Isoplethal section of the Fe-C-Mn system sectioned at 5.42 (wt.%) Mn. The $(\gamma + \alpha + M_3C)$ three phase field is clearly labelled. (b) Fe-rich corner of an isothermal section of the Fe-C-Mn phase diagram sectioned at 625 °C. Two alloy compositions are labelled. Alloy A lies within the $(\gamma + \alpha + M_3C)$ three phase field and Alloy B, within the $(\alpha + M_3C)$ two phase field.

The isopleth shown in (a) is taken at a constant Mn content of 5.4 (wt.%) and the isotherm shown in (b) corresponds to a temperature of 625 °C. In both diagrams the ($\gamma + \alpha + M_3C$) three phase field is clearly labelled. For an alloy with the composition labelled A in Figure III.14b, the equilibrium compositions of the three phases correspond to the three corners of the three phase tie-triangle (bold). It is clear that for equilibrium to be achieved, significant Mn redistribution must occur from the α to the M_3C, and to a lesser degree to the γ and substantial C redistribution must occur from the α to the M_3C and a smaller redistribution from the γ to the growing lamellar aggregate. The case of Fe-C-Mn is a little more complicated because C diffusion in γ occurs 10^6 times faster than Mn at these temperatures and, in practice, the growing pearlite colony does inherit the bulk Mn content of the austenite (there is no long range Mn diffusion to or from the γ). The Mn redistribution from the α to the M_3C occurs via short circuit diffusion in the reaction interface. Nevertheless, the short circuit Mn redistribution is still much slower than the significant long range C diffusion which must occur from the γ to the growing pearlite aggregate in addition to the short range C redistribution from the α to the M_3C, and the lamellar aggregate grows into what it sees to be a parent phase with a C content that decreases in time. As a result, the local equilibrium interfacial compositions (the boundary conditions of the diffusion problem) at the pearlite/γ interface change in time. The consequence is that the compositions inherited by both phases in the lamellar aggregate also continually changes in time and growth occurs with a rate that decreases in time and an interlamellar spacing that increases in time. The effect was first observed by Cahn and Hagel in 1962[76] and has since been observed in Fe-C-Si[89] and Fe-C-Mo[90] systems and has been discussed in detail by Hillert.[91] If instead, an alloy composition such as B in Figure III.14b is chosen, which lies within the ($\alpha + M_3C$) two phase field

---

FIG. III.15. (a) Optical micrograph of 'divergent pearlite' formed in an Fe-0.55C-5.42Mn (wt.%) alloy transformed for 384 h at 625 °C. The M_3C phase contrast is dark and the α phase is light. The alloy composition is marked, A, in the Fe-C-Mn isothermal section (Figure III.14b). Pearlite growth for compositions lying within the ($\gamma + \alpha + M_3C$) three phase field occurs with an interlamellar spacing that increases in time and a growth rate that decreases in time. (b) Scanning electron microscopy secondary electron image of a section of a pearlite colony in an Fe-0.54C-3.51Mn (wt.%) alloy treated at 625 °C for 5 h. In this case, the M_3C phase contrast is light and the α phase is dark. The alloy composition, B, lies within the ($\alpha + M_3C$) two phase field as marked in the isothermal section (Figure III.14b). For such alloy compositions, growth occurs with an approximately constant interlamellar spacing and rate.

---

[89] J. Fridberg and M. Hillert, *Acta metall.* **18**, 1253 (1970).
[90] J. Fridberg and M. Hillert, *Acta metall.* **25**, 19 (1977).
[91] M. Hillert, in *Solid Solid Phase Transformations*, eds. H. I. Aaronson, D. E. Laughlin, R. F. Sekerka, and C. M. Wayman, AIME (1981), p. 789.

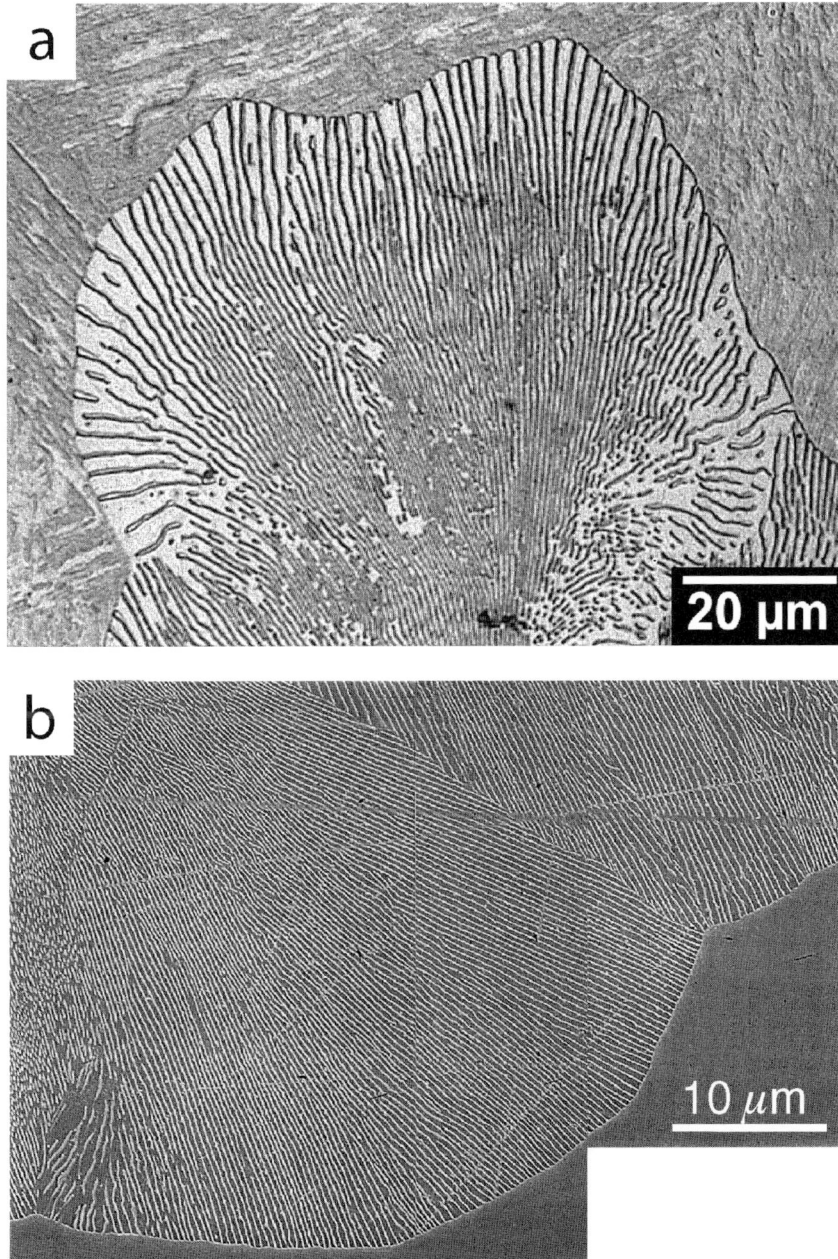

of the phase diagram, γ is not stable, and as was the case for the binary Fe-C system, the growing pearlite is obliged to inherit the bulk alloy composition and steady state growth can be established. The Fe-C-Mn system has recently been analyzed in detail[81] and micrographs of the pearlite formed in alloys A and B (Figure III.14b) and measurements of their growth rates at 625 °C are shown in Figures III.15 and III.16, respectively.

The divergency of the lamellar spacing of alloy A (Figure III.15a) and a growth rate which decreases in time (Figure III.16) are clearly contrasted with the steady state conditions which prevail for Alloy B (Figure III.15b).

Discontinuous Precipitation. As was the case for the non-steady state pearlite growth described above, the occurrence of non-steady state discontinuous precipitation arises from conditions that result in the growing aggregate inheriting an overall composition different from the bulk parent phase (i.e., long range diffusion to or from the parent phase is necessary in addition to the short range redistribution between the lamella phases). The most thoroughly studied case

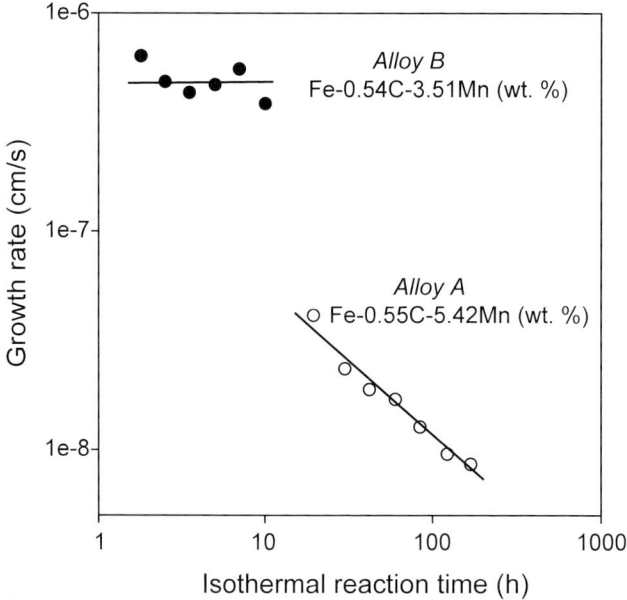

FIG. III.16. Experimentally measured pearlite growth rates as a function of time at 625 °C in the alloy compositions marked A and B in the Fe-C-Mn isothermal section in Figure III.14b. Alloy A, located within the ($\gamma + \alpha + M_3C$) three phase field forms with a growth rate that decreases in time. Alloy B, located within the ($\alpha + M_3C$) two phase field, forms with an approximately constant growth rate.

is the precipitation of Cr_2N in some Cr-Ni austenitic stainless steels. In a series of studies, Kikuchi[92–94] examined the discontinuous precipitation of Cr_2N in several Fe-20Ni-25Cr (wt.%) alloys with varying N contents and measured the growth rates and interlamellar spacings at several temperatures. These are shown for a series of alloys treated at 1073 K in Figure III.17. For a given alloy, the isothermal growth rate of the lamellar aggregate can change by 4 orders of magnitude and the interlamellar spacing can increase four fold during the reaction.

Using arguments based on Hillert's theoretical discussion of the formation of divergent pearlite,[91] Kikuchi et al.[94] argues that for the alloys examined, under conditions where no long range diffusion of Ni or Cr occurs between the parent γ and the growing lamellar aggregate, to maintain local equilibrium between the lamellar phases, long range diffusion of N from the parent γ phase to the aggregate is necessary. Kikuchi et al.[93,94] have shown experimentally that indeed no bulk diffusion of the substitutional solutes is observed for most of the reaction time and that the N content of the parent phase does decrease with time. As a result of the long range N diffusion, the lamellar aggregate grows into a parent phase whose composition is continually changing. The consequence is a continuous change in both the interfacial conditions at the moving boundary and the compositions inherited by the growing phases and results in lamellar aggregate growth with a velocity that decreases in time and an interlamellar spacing that increases in time.

10. COARSENING AND RELATED MORPHOLOGICAL EVOLUTION

The lamellar patterns arising from eutectoid decomposition and discontinuous precipitation are highly non-equilibrium structures. In both cases, energy is stored in the form of the interlamellar interfaces. In the case of the pearlite transformation, the compositions of the phases formed are usually relatively close to the equilibrium compositions found from the phase diagram and little chemical supersaturation generally exists (Figure III.5). However, the solute depleted lamellar phase in the discontinuous reaction necessarily retains some level of supersaturation and the non-equilibrium nature of this product is due to both the free energy stored in the interfaces and the residual chemical supersaturation. The relative contribution to the driving force from the chemical and surface terms will be system specific

---

[92] M. Kajihara, S. Choi, M. Kikuchi, R. Tanaka, Y. Seo, T. Okumura, and Y. Kondoh, *Z. Metallkd.* **77**, 515 (1986).
[93] M. Kikuchi, T. Urabe, G. Cliff, and G. W. Lorimer, *Acta metall.* **38**, 1115 (1990).
[94] M. Kikuchi, M. Kajihara, and S. Choi, *Mat. Sci. and Eng. A* **146**, 131 (1991).

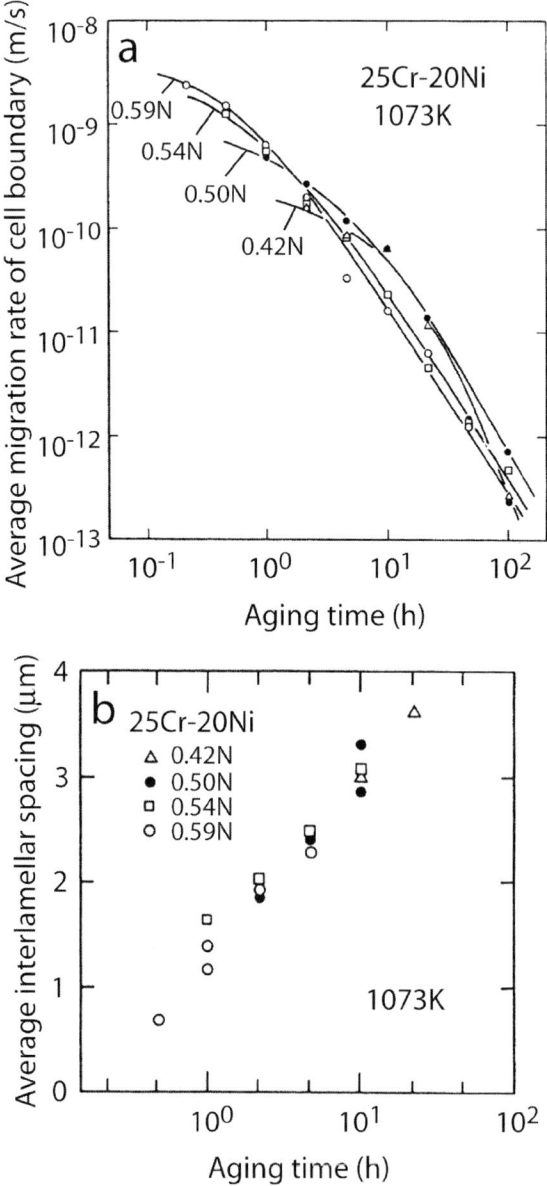

FIG. III.17. (a) Experimentally measured discontinuous precipitation growth rates in a series of N containing Fe-25Cr-20Ni (wt.%) alloys transformed at 800 °C.[94] The growth rate decreases during growth. (b) Corresponding interlamellar spacings showing an increase in spacing during growth.

and depend on the primary transformation conditions, but in most cases it is expected that the volumetric chemical term will dominate the surface term. Fournelle[95] has estimated the magnitudes of the two contributions to the driving force for coarsening in a Fe-30Ni-6Ti (wt.%) alloy and found that for his conditions, the chemical term was two orders of magnitude larger than the surface term.

For both types of lamellar products, the available stored energy provides a driving force for coarsening processes. The return towards equilibrium can take on two aspects: (a) the decrease in interfacial energy and (b) additionally for the discontinuous precipitation product, a decrease in the remaining supersaturation. The decrease in interface energy has obvious consequences in terms of the morphology and scale of the product phases. Just as is the case for precipitation, it can occur either continuously, or discontinuously through the migration of an interface.

a. Spheroidization

The reduction in total interface area can be obtained at almost constant scale through a shape change. Since Lord Rayleigh's[96] seminal work on the stability of fluid jets, it is known that a cylinder can reduce its surface area by a continuous growth of perturbations towards the production of a series of spheres: this process is spontaneous (i.e., there is no energy barrier to be overcome) and is called 'spheroidization'. The equivalent in the solid state has been analyzed by Nicols and Mullins.[97] By contrast, a perfect lamellar structure of infinite extent is stable with respect to infinitesimal fluctuations (although the sphere of equivalent volume has an obvious advantage in terms of reduced surface energy). However, real lamellar structures are not perfect: the lamellae contain both boundaries and defects. The morphological instabilities of lamellar structures are bound to initiate at boundaries and defects and several different types of instabilities leading to spheroidization have been observed. Werner[98] has recently reviewed these processes.

By nature of the difference in interface curvature, a driving force exists for solute transport from the edge of a lamella plate to the flat face of the plate. This solute transport in itself can lead to the direct cylinderization of a plate, and the cylinder is then unstable with respect to the type of perturbations discussed by Lord Raleigh, giving rise to the formation of an aligned row of spheres, Figure III.18a.[99]

---

[95] R. A. Fournelle, *Acta metall.* **27**, 1147 (1979).
[96] Lord Rayleigh, *Proc. Lon. Math. Soc.* **10**, 4 (1879).
[97] F. A. Nichols and W. W. Mullins, *Trans. AIME* **233**, 1840 (1965).
[98] E. Werner, *Z. Metallkd.* **81**, 790 (1990).
[99] T. H. Courtney and J. C. M. Kampe, *Acta metall.* **37**, 1747 (1989).

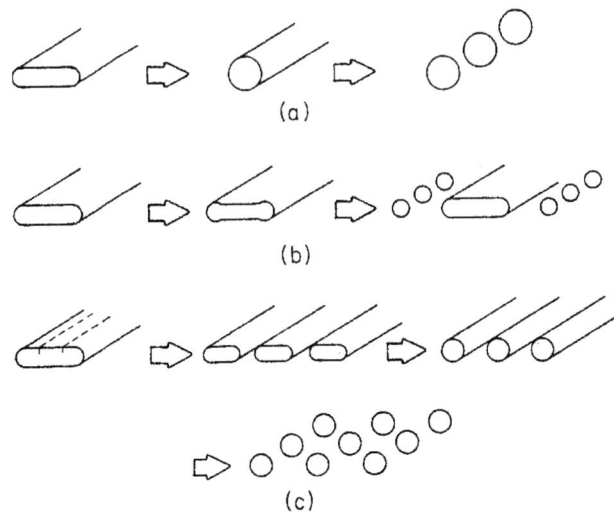

FIG. III.18. Schematic illustration of the primary instability modes for an isolated plate.[99] (a) A terminated plate may evolve into a cylinder as a result of capillarity driven mass transfer from the edges to the broad faces of the plate. Following 'cylinderization', a Raleigh instability causes the cylinder to decompose into a row of spheres. (b) The shape evolution gives rise to the pinching-off of a row of spheres along the edges of the plate (edge spheroidization). The spheres form as a result of perturbations developed along the ridge which forms at the edges during the first stages of mass transfer to the broad faces. (c) If boundaries are present in the plate, plate splitting may occur by sustained mass transport from boundary grooves to the broad face of the plate. Cylinderization and Raleigh instabilities can then follow.

Alternatively, part way through the cylinderization process of a plate, ridges may form along the edges and the partial cylinders at the edge of the plate may then decompose to rows of spheres directly on either side of the plate through the Raleigh instability. This process is shown schematically in Figure III.18b and is usually referred to as 'edge spheroidization'. Both of these processes lead to a change in shape at approximately the same scale and arise naturally from the solute transfer driven from the curved boundaries of plates to the flat faces. Clearly, the boundaries from which transport occurs could equally be holes in the plates themselves rather than external edges. Lamella plates also often contain boundaries. These can be the result of faulting during growth, or boundaries formed during recovery and recrystallization. In either case, these boundaries can also act as sites for the splitting of plates through a 'boundary grooving' mechanism which has been discussed in detail by Mullins.[87,100,101] The curvature of the

---

[100] W. W. Mullins, *Acta metall.* **6**, 414 (1958).
[101] W. W. Mullins, *Trans. AIME* **218**, 354 (1960).

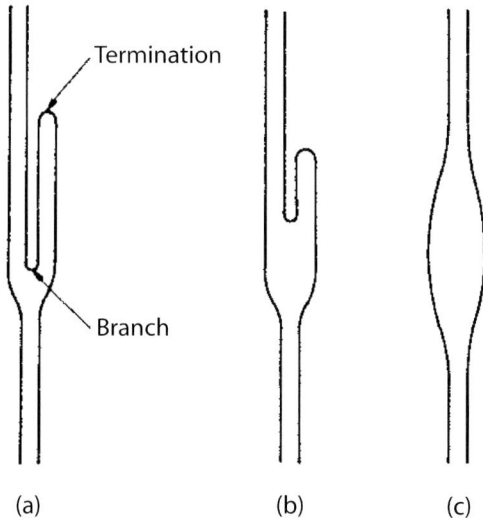

FIG. III.19. Schematic illustration of 'fault migration' during coarsening of lamellar structures.[102] (a) Initial geometry, (b) migration of the faults and (c) bulged plate. The driving force for mass transfer is the difference in curvature at the termination and the branch.

surface of the plate in the vicinity of the boundary can be sufficient to drive solute away from the boundary region. To maintain the balance of surface tensions at the triple point between the boundary and the interlamellar interface, in the presence of solute transfer, the groove continually deepens along the boundary until it is entirely removed and the plate is split in two. This process is shown schematically in Figure III.18c. Each of the instabilities shown in Figure III.18 involves the splitting of individual plates; it is a change is shape at approximately the same scale and we shall refer to them all under the umbrella of spheroidization. The reduction in surface energy can also be obtained by increasing the scale of the microstructure. This increase in scale requires diffusional mass transport and the standard means is by bulk diffusion. This is the process of Oswald ripening which has been extensively studied, both experimentally and theoretically, and has been excellently reviewed many times (e.g., see Ref. [103]).

An additional mechanism for accomplishing the change in scale involves the interactions between adjacent plates and is referred to as fault migration.[102] In this case, the difference in curvature at the end of a terminated plate and on the broad face of an adjacent plate drives the dissolution of the faulted plate and the

---

[102] H. E. Cline, *Acta metall.* **19**, 481 (1971).
[103] P. W. Voorhees, *Ann. Rev. Mat. Sci.* **22**, 197 (1992).

thickening of the adjacent plate by volume diffusion of the atomic species until the faulted region is entirely removed (Figure III.19).

As we shall see below, when volume diffusion of solute is too slow, an alternative mechanism can take over.

b. Discontinuous Coarsening

When bulk diffusion is too slow, the change in scale can be accomplished through discontinuous coarsening. This is a solution involving interface migration and interface diffusion and can take over from continuous coarsening (Ostwald ripening), just as discontinuous precipitation can take over from continuous precipitation in similar situations.

Discontinuous coarsening is a moving boundary reaction that converts a fine lamellar pattern into a structurally and morphologically similar, but distinctly coarser pattern. As is the case for the very similar discontinuous precipitation, the solute redistribution during the discontinuous coarsening reaction occurs via short circuit diffusion in the migrating boundary and a discontinuous change in crystal orientation is observed across the boundary. Coarsening by this mode has been observed for lamellar products of both discontinuous precipitation and eutectoid decomposition and Manna et al.[36] reports that it has been identified in more than 30 different alloy systems.

A classic example[104] of the discontinuous coarsening of a eutectoid lamellar aggregate in the Co-Si system is shown in Figure III.20.

The kinetics of the coarsening process were first treated theoretically by Livingston and Cahn[104] who were interested in the thermal stability of Co-Si, Cu-In and Ni-In aligned eutectoids. They were able to derive an expression for the velocity of the interface based on a solution to the diffusion equation and considering the change in interlamellar spacing, λ.

$$v_{DC} = \frac{8X^\beta}{X^\beta - X^e} \frac{s\delta D^b \sigma V_m}{f_\alpha^2 f_\beta^2 \lambda_{DC}^2 \lambda_{OR} RT} \left(1 - \frac{\lambda_{OR}}{\lambda_{DC}}\right), \qquad (3.10)$$

X^β is the equilibrium composition of the β phase, λ_{OR} is the lamellar spacing of the initial pattern and λ_{DC} is the spacing of the coarsened pattern.

Livingston and Cahn were considering the coarsening of eutectoid products. They assumed that the driving force in this case arises mostly from a reduction in the interfacial energy stored in the interlamellar interfaces. However, for discontinuous coarsening of discontinuously precipitated lamellar aggregates the driving force for the coarsening process must include the contribution from the chemical

---
[104] J. D. Livingston and J. W. Cahn, *Acta metall.* **22**, 495 (1974).

FIG. III.20. An optical micrograph of the discontinuous coarsening reaction in a Co-Si alloy treated for 4 days at 1000 °C.[104] The direction of grain boundary motion is indicated by the arrows.

supersaturation remaining in the aggregate. Fournelle,[95] modified a previous treatment for discontinuous precipitation,[56] and derived an expression for the coarsening rate that allows for the inclusion of both contributions to the driving force.

$$v_{\text{DC}} = \frac{8\Delta G_{\text{DC}}}{RT} \frac{s\delta D^b}{\lambda_{\text{DC}}^2}, \qquad (3.11)$$

ΔG_{DC} is the total driving force for the coarsening process and includes contributions from both the remaining chemical supersaturation and the interlamellar interfaces.

Under conditions where the chemical supersaturation remaining in the lamellar aggregate after the initial reaction is very close to zero, Eq. (3.11) reduces to an expression very similar to that derived by Livingston and Cahn (Eq. (3.10)).

It is clear from Eqs. (3.10) and (3.11), that a relation between the interface velocity and the interlamellar spacing is obtained. The two quantities are coupled and an identical spacing selection problem exists for the discontinuous coarsening reaction as for the discontinuous precipitation and pearlite transformations discussed in Section III.8.

11. MULTILAYER STABILITY

The pearlite and discontinuous precipitation reactions discussed in this chapter give rise to a solid state lamellar pattern with a characteristic spacing behind a migrating interphase or grain boundary. Some aspects of the problems are still not well understood (especially aspects concerning the initiation processes) and in an effort towards this goal, Klinger et al.[105–107] has suggested that several fundamental questions could be addressed through studies of the thermal stability of multilayers. Carefully controlled multilayer fabrication using techniques such as molecular beam epitaxy (MBE) provide a convenient means of examining the evolution of heterogeneous systems far from equilibrium. To date, much of the work has concentrated on the evolution of the systems controlled by bulk diffusion (an example is the precise determinations of low temperature solute diffusivities[108,109]), but multilayer growth on bi-crystal substrates that would result in a boundary that penetrates the entire structure could provide a means of examining interface-mediated pattern evolution under conditions where the boundary can act as a short circuit diffusion path for solute. Klinger et al.[105] have considered the homogenization of a fully miscible multilayer by boundary migration and have computed the boundary shape and velocity (pattern selection) as a function of the system dimensions and the relevant thermodynamic and kinetic parameters. They identify different homogenization 'modes' illustrated schematically in Figure III.21.

At one extreme (a) the homogenization may occur by the steady state motion of the boundary and at the other (c) through the instability of the boundary and a resulting 'fingering' solution. They were concerned with the conditions that would lead to the different patterns. More recently, they have considered cases were a third phase may form at the interfaces in the multilayer and have calculated the resulting shape and velocity of the interface between the three phases.[107]

Experiments of the type proposed by Klinger et al. may well provide a means of addressing such questions as velocity and pattern selection in well controlled systems as a function of the thermodynamics and kinetics of the system chosen and could certainly lead to a deeper understanding of the evolution of the lamellar patterns discussed in this chapter.

---

[105] L. Klinger, Y. Brechet, and G. Purdy, *Acta mater.* **45**, 4667 (1997).
[106] L. Klinger, Y. Brechet, and G. Purdy, *Acta mater.* **46**, 2617 (1998).
[107] L. Klinger, Y. Brechet, and G. Purdy, *Acta mater.* **47**, 325 (1999).
[108] A. L. Greer, *Curr. Opin. Sol. State. Mat. Sci.* **2**, 300 (1997).
[109] A. L. Greer, *Ann. Rev. Mat. Sci.* **17**, 219 (1987).

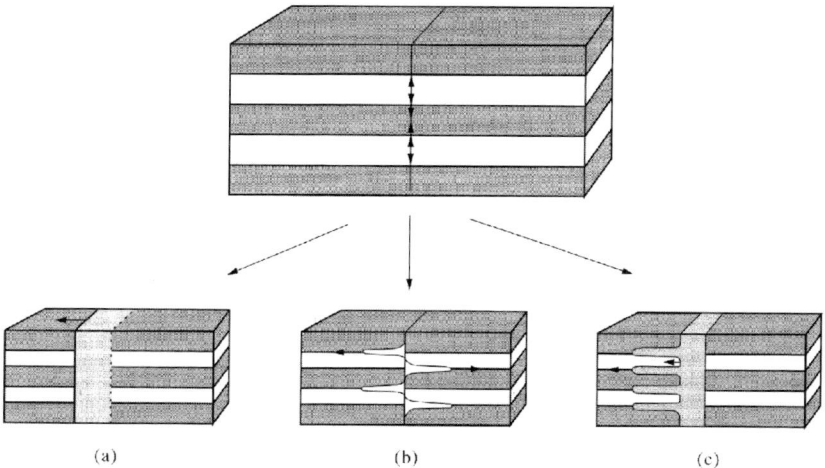

FIG. III.21. The different possibilities for discontinuous homogenization of a multilayer containing a moving grain boundary:[105] (a) steady state motion of the grain boundary, (b) sinusoidal instability of the grain boundary and (c) fingering instability of a moving grain boundary.

IV. Structural Defect Patterning: Grain Growth, Recovery and Recrystallization

In the previous section, the patterning examined was of a chemical nature. The structural defect involved was an interface which separated the transformed (patterned) region from the untransformed and in some cases provided a short-circuit diffusion path for solute. In the following sections we will consider the patterning of the structural defects themselves in systems that can be considered to be nominally chemically homogeneous.

The structural defects encountered in metals and alloys are of three types: point defects (vacancies and interstitials), line defects (dislocations), and planar defects (grain boundaries, interfaces and surfaces).

Point defects have a low energetic cost and a high configurational entropy: they are weakly interacting and only at short distances. Vacancies have a finite equilibrium concentration which can be reached when the material is cooled very slowly. Transport occurs by diffusion, in response either to a gradient in concentration, or to a gradient in stress (the latter being responsible for diffusional creep).

Line defects (dislocations) are much more disturbing for the theoretician: they have a high energetic cost and their configurational entropy is restricted by topological requirements (a dislocation line can't end within the crystal). Their equilibrium density is virtually zero at any temperature, but nevertheless, even if the material is cooled very slowly, dislocations are almost always present (e.g., Frank network). The reason for their persistence is the threshold force required to move them. The existence of this 'solid friction', together with the topological requirements and the crystallographic constraints to their motion (glide and cross slip on definite planes) prevents the simple definition of a relevant thermodynamic potential which drives their evolution. To make the things worse, dislocations are associated with a long range stress field (which decreases as $1/r$) and present short range reactions (junctions). Both of these aspects are crucial to the theory of plasticity: the long range stress field couples to the applied stress to give the Peach–Kohler force for dislocation motion under stress and the short range interaction and junction formation are responsible both for the work hardening of the material and for the very existence of the sources which produce the dislocations. Any physicist interested in plasticity should absolutely resist the temptation to forget either of these aspects. We have to deal both with out of equilibrium, long range interacting, non-conserved entities, and with short range reactions. The natural mode of dislocation motion is to glide in well defined crystallographic planes in response to a local stress which results from the superposition of the applied stress and of the stress arising from all of the surrounding dislocations.

Grain boundaries are somewhat more tractable. Like dislocations, they are non-equilibrium entities, but they have no long range stress field. Except for some very special grain boundaries which have a limited mobility and a low energy, their structure is such that they can have enough flexibility and ability to move so that the thermodynamic situation of having no boundary is, in principle, attainable.

The specific differences in these three basic defect entities will help explain the characteristics of the patterns they produce and the ease with which they can be explained.

12. GRAIN GROWTH

Grain growth is the simplest of the pattern forming systems.[110] There is a well defined thermodynamic state to be reached (infinite grain size) and the entities to be eliminated have no long range interaction. This apparent theoretical simplicity is less obvious from the experimental investigations. Grain growth is a 3D phenomenon, however very few 3D experimental studies have been reported in the literature.[111,112] Most experimental studies are made using 2D slices. In addition, the theoretical predictions are derived under the assumption of a linear relation between the boundary mobility and the local driving force (which is the

---

[110] A. P. Sutton and R. W. Balluffi, *Interfaces in Crystalline Solids*, Oxford University Press, Oxford (1996).
[111] F. N. Rhines and B. R. Patterson, *Metall. Trans. A* **13A**, 985 (1982).
[112] F. C. Hull, *Mat. Sci. and Tech.* **4**, 778 (1988).

reduction in grain boundary curvature, cf. Section II). Under these assumptions, the theories[113–115] predict that the length scale should behave as $t^{1/2}$ and that the size distribution of the grains, when scaled with the average grain size, should be invariant. Experimentally, the 1/2 exponent is almost never observed, although numerical simulations predict it, regardless of whether the boundary mobilities and energies depend on boundary misorientation. This suggests that the linear relationship between the velocity and the driving force is not valid. Natural reasons could be the impurity drag on moving boundaries[116] or the reduced mobility of triple junctions and quadruple points.[117] However, in spite of the disagreement between the predicted and experimentally observed exponents, the existence of an invariant scaled grain size distribution is observed both experimentally and in computer simulations. Patterning in grain growth is to be understood as the characteristics of this self-similar behavior:

(i) The existence and shape of an invariant scaled size distribution.
(ii) Topological relations such as the relation between the average size of the neighboring grains, or the number of neighbors of a grain of given size.

Since all simulations relying on a linear relation between local driving force and velocity must give a kinetic law in $t^{1/2}$, the issues on patterning are rather on these two characteristics. Extensive reviews of the topological characteristics of 'cellular structures' can be found in the articles by Weaire and McMurry[19] and Thompson.[20] Here we will focus on some of the open questions related both to theoretical pattern formation, and to some of the specifics of 'real systems'.

Two types of computer simulations dominate the literature on grain growth. On one side are those based on an adaptation of the Potts model,[118–120] and on the other are those based on a discretization of grain boundaries, called the 'vertex model'.[121–125] Both models have been run in 2D and 3D and give similar results,

---

[113] J. E. Burke and D. Turnbull, *Prog. Met. Phys.* **3**, 220 (1952).
[114] W. W. Mullins, *J. Appl. Phys.* **27**, 900 (1956).
[115] M. Hillert, *Acta metall.* **13**, 227 (1965).
[116] J. W. Cahn, *Acta metall.* **10**, 789 (1962).
[117] G. Gottstein and L. S. Shvindlerman, *Acta mater.* **50**, 703 (2002).
[118] M. P. Anderson, D. J. Srolovitz, G. S. Grest, and P. S. Sahni, *Acta metall.* **32**, 783 (1984).
[119] M. P. Anderson, G. S. Grest, and D. J. Srolovitz, *Phil. Mag. B* **59**, 293 (1989).
[120] A. D. Rollett, D. J. Srolovitz, and M. P. Anderson, *Acta metall.* **37**, 1227 (1989).
[121] K. Fuchizaki, T. Kusaba, and K. Kawasaki, *Phil. Mag. B* **71**, 333 (1995).
[122] F. J. Humphreys, *Scripta metall. mater.* **27**, 1557 (1992).
[123] C. Maurice and F. J. Humphreys, in *Proceedings of an International Conference on Thermomechanical Processing in Theory, Modelling and Practice*, ed. B. Hutchinson, The Society, Stockholm (1997), p. 201.
[124] D. Weygand, Y. Brechet, and J. Lepinoux, *Phil. Mag. B* **78**, 329 (1998).
[125] D. Weygand, Y. Brechet, J. Lepinoux, and W. Gust, *Phil. Mag. B* **79**, 703 (1999).

as far as the size distribution and topological relations are concerned. We will place emphasis on the vertex models since the input quantities are continuum ones (such as boundary energies or mobilities) which, in principle, can be derived from experimental measurements.

In both 2D and 3D, a scaled grain size distribution is obtained in the computer simulations as shown in Figure IV.1. However, this distribution does not corresponds to Hillert's distribution which is a sort of mean field distribution (i.e., it assumes that the average size around a given grain is the average grain size in the system). It seems that Louat's distribution, which relies on the unrealistic assumption of the totally random motion of a grain in the space of grain sizes[126] is much closer to the simulated distribution, whereas the experimental distributions are often observed to be Log Normal.

The ubiquity of Log Normal distributions (also observed in precipitation) is a puzzle in metallurgy. It seems that, although they do not correspond to steady state solutions of the Normalized Fokker–Planck equation, the transient to depart from them is very long. However, the reason for their appearance in the first place remains unknown.

The second item used to characterize the patterns, are local correlations. Two may be considered:

(i) Grains larger than average are more likely to be surrounded by grains smaller than average (Figure IV.2a).
(ii) Grains larger than average tend to have a larger number of neighbors (Figure IV.2b).

The correlations captured in item (i) are not included in Hillert's derivation and may be the reason for the discrepancies between computer simulations and Hillert's prediction. The correlations included in item (ii) are necessary in the Ryum–Hunderi scaling theory[127] which makes use of the Mullins–Von Neumann equation[114] to calculate the growth rate of an individual grain. The scaling theory of Ryum–Hunderi[127] describes very accurately the distribution observed in the simulation (Figure IV.3), but there is currently no analytical derivation of the correlation given in Figure IV.2b.

Even in this simplest case of pattern formation (scaling in grain growth), the distributions predicted both by the theory and the simulations are not in agreement, and they both disagree with the experimental data which appears to be a Log Normal distribution.

Besides the pure 'statistical physics' problem, which amounts to investigating domain sizes in a conserved order parameter system, one can also consider

---

[126] N. P. Louat, *Acta metall.* **22**, 721 (1974).
[127] O. Hunderi and N. Ryum, *J. Mat. Sci.* **15**, 1104 (1980).

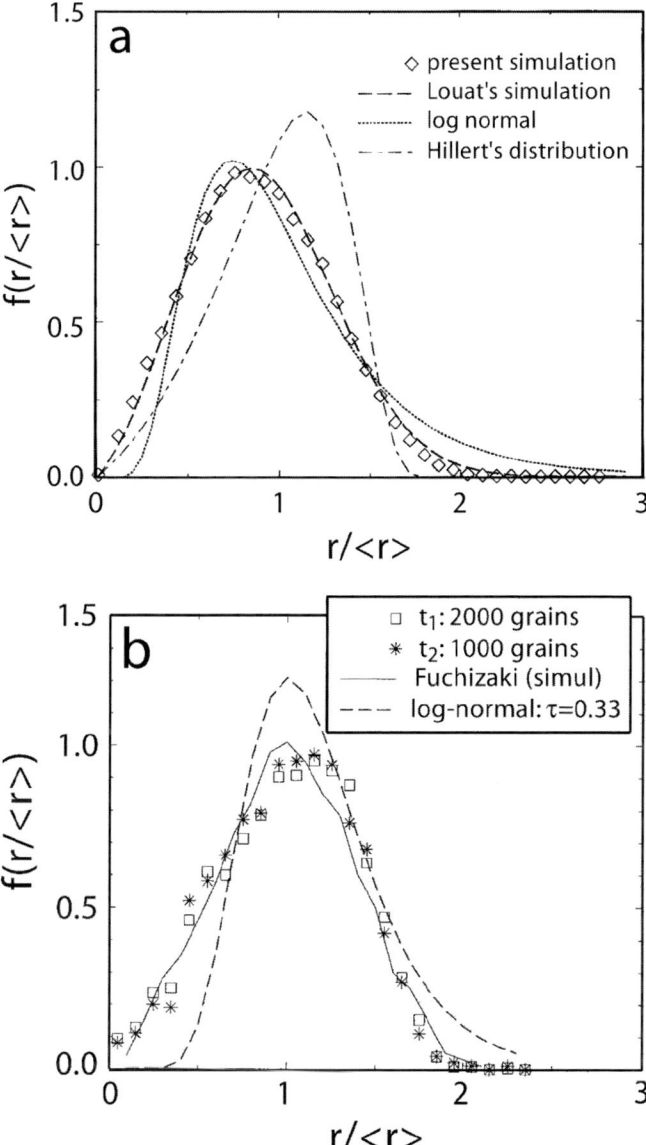

FIG. IV.1. (a) Scaled grain size distribution obtained from a 2D vertex dynamics simulation.[124] The log normal distribution as well as the distributions predicted by Hillert[115] and Louat[126] are shown for comparison. (b) Scaled grain size distribution obtained from a 3D vertex dynamics simulation[125] and the comparison with the log-normal distribution and a previous simulation by Fuchizaki et al.[121]

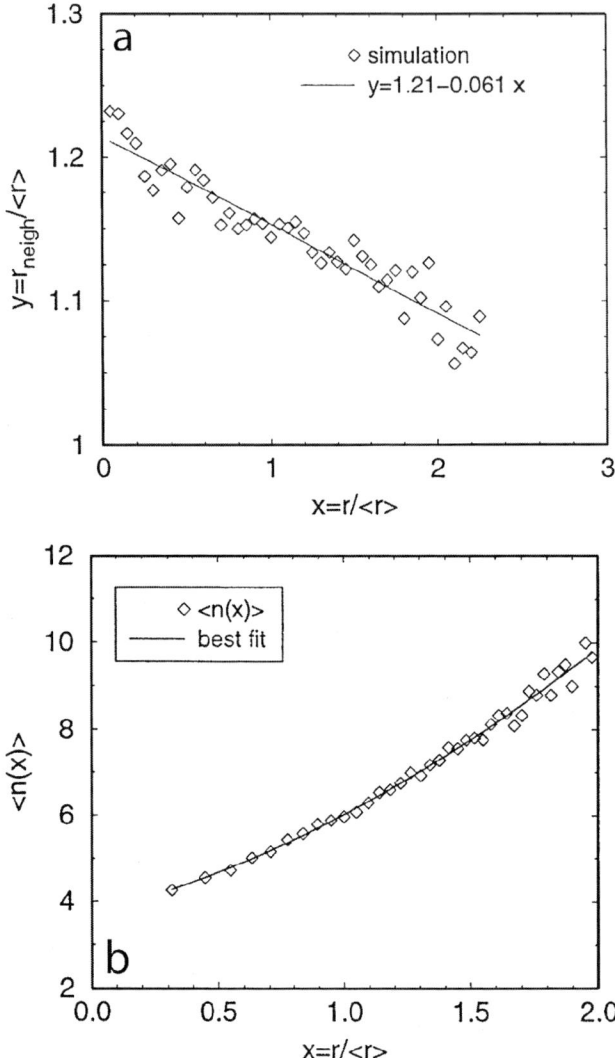

FIG. IV.2. 2D vertex dynamics simulations illustrating: (a) the average size of the neighbors of a grain of a given size[128] and (b) the average number of neighbors of a grain of given size.[124]

materials science specific issues: the anisotropy of grain boundary energies and mobilities, the finite mobility of triple junctions, pinning points, etc. For the sake

---

[128] D. Weygand, PhD, Institut National Polytechnique de Grenoble, Grenoble (1998).

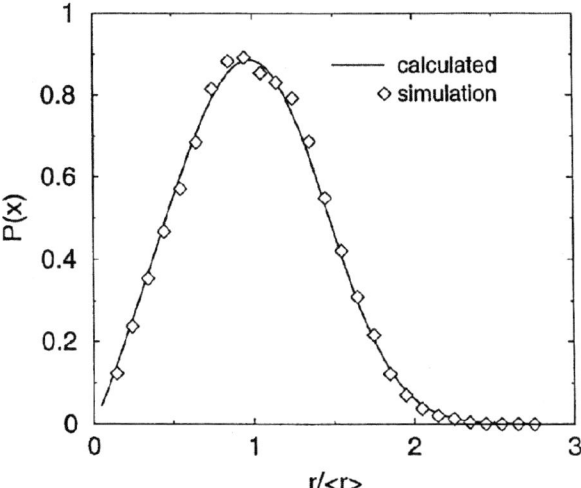

FIG. IV.3. Grain size distribution of the 2D vertex dynamics simulation compared to the distribution from the Ryum–Hunderi scaling theory.[128]

of illustration, let us consider the effect of particle pinning on the patterns formed by an evolving grain structure.[129] Of course, the pinning by particles (Zener Pinning) is known to lead to a stabilized grain size. The kinetics, as simulated by the vertex model, saturate toward this ultimate grain size according to a kinetic law corresponding to a subtraction from the capillary forces due to Zener pinning (which in itself is a surprising result since Zener pinning is in principle a threshold force). The simulated kinetics and the results of an analytical expression are shown in Figure IV.4.

The distributions are of course non-scale invariant, but the final distributions can be accurately fitted by a log normal distribution (Figure IV.5).

A perhaps more striking result is the behavior of the topological characteristics.[128,130] As shown in Figure IV.6, the correlation between the grain size of the neighbors and the average grain size is expected to be exactly opposite that predicted for grain growth in the absence of particles.

For this simple patterning system it seems that two ingredients are missing for a complete understanding: the analytical derivation of the local spatial correlations, and the reasons for the apparent stability of the Log Normal distribution for grain sizes. However, the tools developed for understanding the evolution of capillary driven structures does allow us to obtain a new insight into recrystallization and

---

[129] D. Weygand, Y. Brechet, and J. Lepinoux, *Acta mater.* **47**, 961 (1999).
[130] D. Weygand, J. Lepinoux, and Y. Brechet, in ed. J. Driver, Annecy, France, in press (2004).

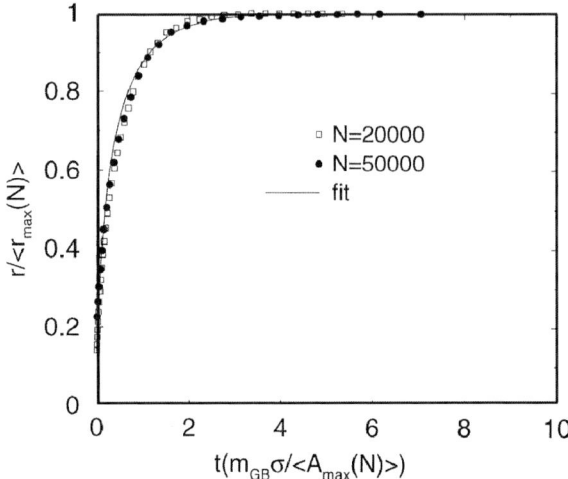

FIG. IV.4. Vertex dynamics simulation of the kinetics of grain growth in the presence of pinning particles.[129] The grain size has been normalized to its saturation value and the time has been rescaled accordingly.

recovery, which in turn will provide some important inputs for the understanding of pattern formation in plasticity.

13. Recovery and Recrystallization of Dislocation Structures

When heavily deformed, metals store large quantities of dislocations, in more or less organized structures (see Section V for the origin of this patterning). This certainly increases the free energy of the system, and provided some sufficient mobility is available to the defects, they will tend to move in order to reduce this energy. Accordingly, the dislocation density decreases, the substructures evolves and the hardness of the material decreases. This evolution of the structural defects during annealing can be followed by the evolution of the hardness of the material. Two situations are possible: either the evolution is slow in time (typically logarithmic) and occurs approximately everywhere at the same time and is referred to as 'recovery', or it is rapid in time, occurs heterogeneously, and corresponds to the nucleation of new grains free from defects (recrystallization). A detailed review of these mechanisms and their experimental studies can be found in Refs. [21,131].

---

[131] R. D. Doherty, D. A. Hughes, F. J. Humphreys, J. J. Jonas, D. J.Jensen, M. E. Kassner, W. E. King, T. R. McNelley, H. J. McQueen, and A. D. Rollett, *Mat. Sci. and Eng. A* **A238**, 219 (1997).

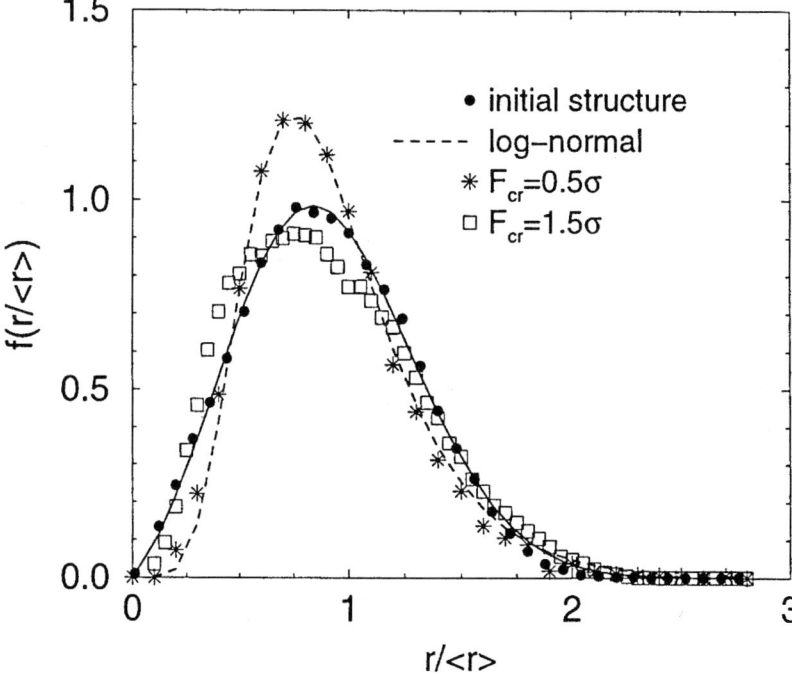

FIG. IV.5. Vertex dynamics simulations of the grain size distribution in the presence of particles, compared to the initial distribution.[129]

In the following, we will highlight some observations which may lead to a better insight into plastically induced patterning from their further thermal stability. The first remark concerns the role of fluctuations, whereas the second and third remarks concern the definition of a driving force for structural evolution.

The first set of comments concerns the nucleation of recrystallization. It is tempting to consider the nucleation of a dislocation free grain as a classical nucleation process where the gain in bulk energy pays for the cost in surface energy. The size of nuclei is such that the classical theory would predict a ridiculously low nucleation rate.[21] Recrystallization nucleation is often associated with pre-existing heterogeneities in the material: undeformable particles or initial grain boundaries. It seems natural to assume that gradients in stored energy are responsible for the motion of a grain boundary which delimits the new grain. This is perfectly understandable for particle nucleated grains, since geometrically necessary dislocations accumulate around the undeformable particles. However, it is more difficult to envisage in the vicinity of the grain boundary. Even if there is some crystalline anisotropy which may lead to different stored energies depend-

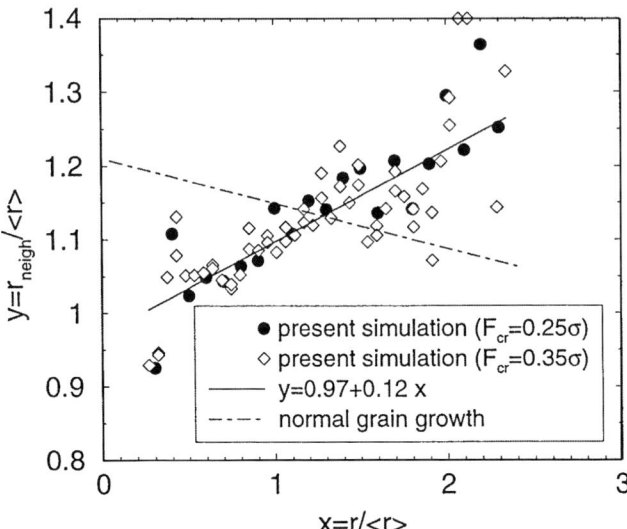

FIG. IV.6. The grain size correlation function for particle inhibited grain growth compared to the correlation for normal grain growth.[128] With increasing grain size, the average grain size of the neighboring grains increases in the presence of particles. The opposite behavior is observed for normal grain growth.

ing on the crystallographic orientation of the grains, these differences would tend to vanish when plastic deformation increases to the level attained in cold rolling. The paradox is that it is precisely when the strain increases that the nucleation of new grains at the boundary becomes easier. The dislocations stored in the material during cold rolling organize themselves into subgrains. The subgrain misorientations will increase, and their size will decrease when the strain in increased. An estimate of the stored energy can be obtained from the average subgrain size and the average misorientation. A vertex type simulation was performed in a situation where, across a grain boundary, a difference in stored energy was mimicked by a difference in subgrain size.[132] Sure enough, the bowing of the grain boundary was triggered by this somewhat artificial gradient. But another simulation can be made, with no macroscopic gradient between the two grains, but simply accounting for the fact that sub-boundaries have a lower energy and a lower mobility that a general grain boundary.[133] With these simple ingredients, it is found that, below a critical subgrain size or above a critical subgrain misorientation (which both can be understood as a critical strain) the nucleation of new grains by bowing out

---

[132] F. J. Humphreys, *Acta mater.* **45**, 4231 (1997).
[133] D. Weygand, Y. Brechet, and J. Lepinoux, *Phil. Mag. B* **80**, 1987 (2000).

occurs naturally (Figure IV.7). It is not the gradient in the average quantity, but the fluctuation in the quantity which drives the nucleation of a new pattern. The importance of the fluctuations, and the fact that they have been too neglected in the current treatments, is a remark we will come back to.

The second remark comes from measurements of the kinetics of recovery. If the sub-boundaries arising from deformation were low energy equilibrium structures, then the subsequent evolution of the microstructure would be simple subgrain growth. There would be no qualitative difference between subgrain growth and grain growth (it has been shown that the introduction of the dependence of the energy and mobility on grain boundary misorientation may change the prefactor, but the power law in $t^{1/2}$ should survive). Since the yield stress scales as the inverse subgrain size, the decrease in hardness should also follow a power law behavior, and not logarithmic kinetics. The experimentally observed logarithmic kinetics indicates that subgrain growth is slower, and suggests that the sub-boundary energy is decreasing during recovery. From this fact we can infer that

(i) the sub-boundaries formed during plastic deformation where not a minimum in energy, and
(ii) the sub-boundaries probably tend to a structure with minimum energy during recovery.

The third set of remarks relies on some experimental observations of recovery in Al-2.5%Mg alloys. These observations will also be useful to understand the nature of the patterns emerging from deformation. Recovery processes are in principle simple in the sense that the main parameter governing the evolution is time and not strain, and the system is out of equilibrium but not maintained far from equilibrium. The pertinent questions are: What is the evolution of dislocation density? How is it correlated to the evolution of internal stresses? What is the remaining stored energy in a system having undergone static recovery? Microdensity and resistivity[134] can be used to measure dislocation densities, the analysis of X-ray broadening allows us to evaluate the screening of dislocations in their substructure and calorimetry allows us to measure the stored dislocation energy which is released during recrystallization. Combining these measures on deformed samples, recovered for different times, provides information both on the pattern emerging from deformation, and its subsequent stability.[135–138]

---

[134] M. Verdier, PhD, Institut National Polytechnique de Grenoble, Grenoble (1992).
[135] M. Verdier, I. Groma, L. Flandin, J. Lendvai, Y. Brechet, and P. Guyot, *Scripta mater.* **37**, 449 (1997).
[136] M. Verdier, M. Janecek, Y. Brechet, and P. Guyot, *Mat. Sci. and Eng. A* **248**, 187 (1998).
[137] M. Verdier, Y. Brechet, and P. Guyot, *Acta mater.* **47**, 127 (1998).
[138] M. Verdier, F. Bley, M. Janecek, F. Livet, J. P. Simon, and Y. Brechet, *Mat. Sci. and Eng. A* **234**, 258 (1997).

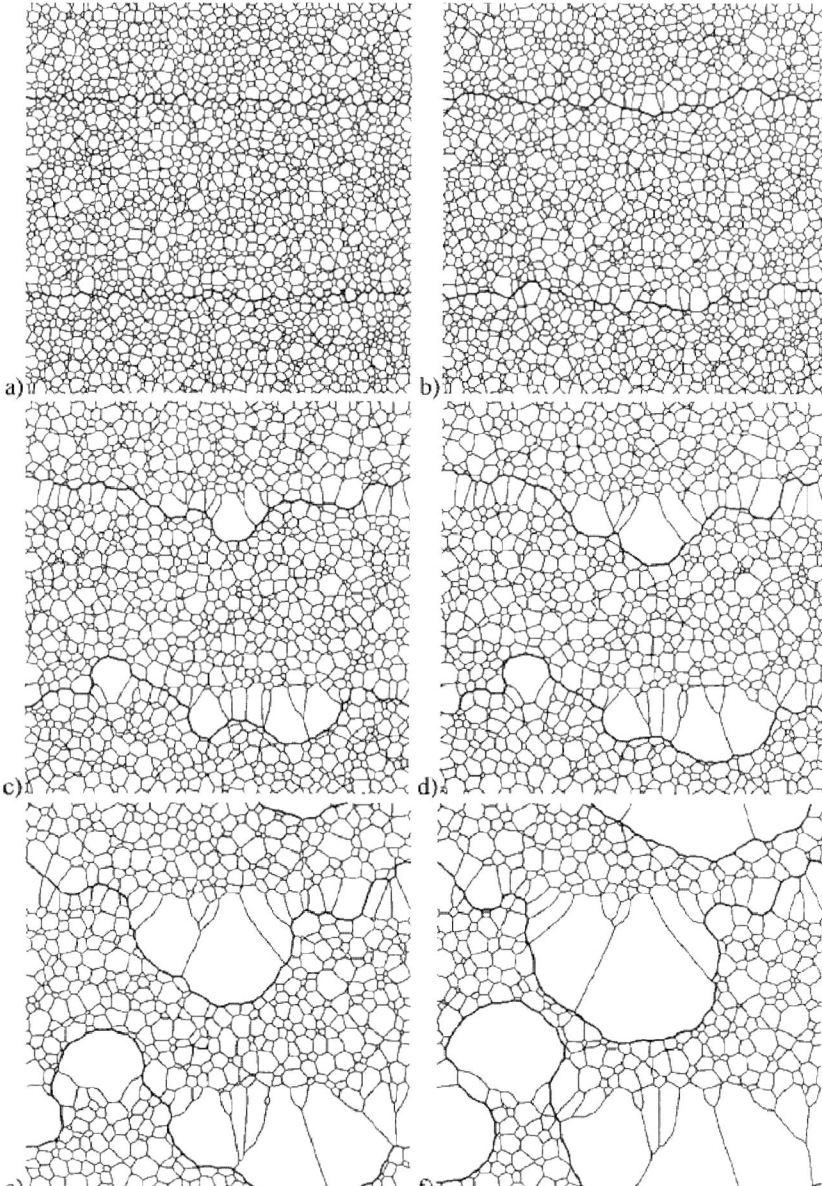

FIG. IV.7. The bulging mechanism for the nucleation of recrystallization calculated with 2D vertex dynamics simulations with no subgrain size misfit.[133] (a) Initial configuration and (b)–(f) evolution sequence.

The evolution of the dislocation density as a function of the cold rolling strain (ε) in an Al-Mg alloy is shown in Figure IV.8a. The dislocation storage is not simply proportional to the strain but involves also dynamic recovery taking place during deformation. As a consequence, the net dislocation storage rate decreases at large strains. Figure IV.8b shows the evolution of the dislocation density during recovery annealing. The behavior of the dislocation density with annealing time is characteristic of the recovery kinetics: a fast initial decrease followed by a slower decay at longer annealing times. The results from resistivity and from microdensity measurements are consistent. For each sample deformed and recovered, the density of dislocations is known.

In order to check that the broadening of the Bragg peak was mainly due to dislocations, the line width at half maximum has been measured for various peaks. After recovery annealing, the broadening of the peak decreases. In order to extract some useful information about the dislocation microstructure from the X-ray broadening, one has to postulate a model for the spatial arrangement of the dislocations. Instead, some information can be obtained directly from the behavior of the tail of the peak. Ideally, for an infinite straight dislocation the asymptotic decay of the intensity at large scattering vectors (q) would behave as q^{-3}, whereas for a dislocation dipole configuration, because of the screening of the strain fields, the intensity would decrease more slowly, as q^{-2}. The asymptotic behavior of the intensity for large q therefore gives information about the screening of the dislocation interactions. After static recovery, the tails at large q exhibit a slower decrease. This evolution can be interpreted as a consequence of the rearrangement of the dislocations into self screened structures of lower energy (Figure IV.9).

The density of dislocations can be estimated via the resistivity measurements which are directly related to the total length of dislocation core. The stored energy in a given state can be measured by the total energy released in the calorimeter when a sample is fully recrystallized. Figure IV.10 shows the stored energy as function of the dislocation density for increasing strains. Not only does the rough estimation of the stored energy given by $\rho\mu\mathbf{b}^2/2$ (ρ is the dislocation density, μ is the shear modulus and $\mathbf{b}$ is the burgers vector) tend to overestimate the true value, but including the logarithmic correction term ($\text{Ln}(\mathbf{b}\rho^{1/2})$) is also insufficient. This clearly shows that, while deformation increases, and as dislocation structures become better defined, the dislocation stress fields are more and more efficiently screened. This effect parallels the screening effect evidenced by the X-ray analysis.

During recovery, the dislocation density decreases, as expected. But the decrease in energy is much lower than would be expected from the simple estimates from the relation between dislocation density and energy in a cold rolled sample. Everything behaves as if there was a 'marginal cost' to incorporate and remove a dislocation from a given structure. This 'chemical potential' for dislocations governs the decrease of stored energy during recovery, as seen in Figure IV.11.

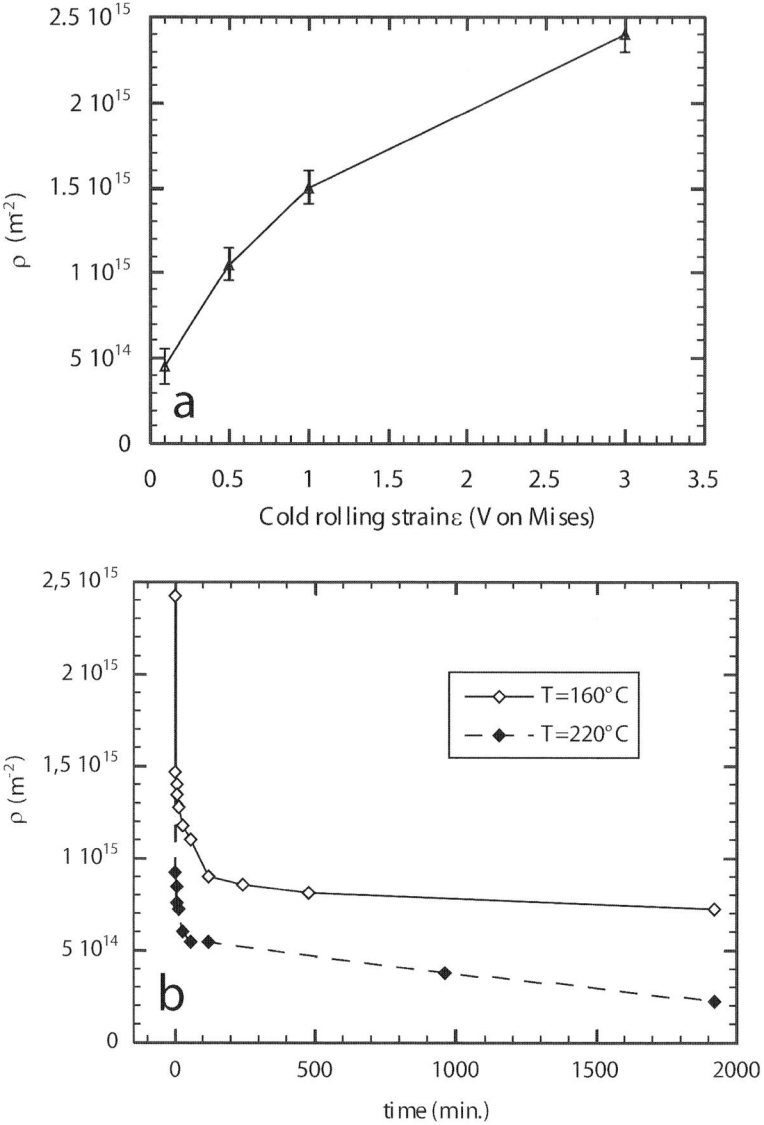

FIG. IV.8. The evolution of the dislocation density (a) as a function of the cold rolling strain, ε, in an Al-Mg alloy and (b) during annealing of deformed Al-Mg samples at 160 °C and 220 °C.

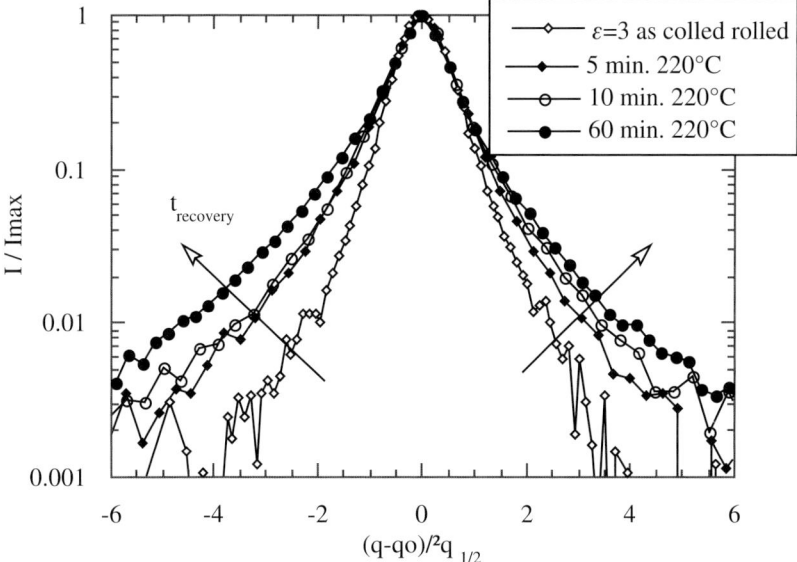

FIG. IV.9. Evolution of the X-ray Bragg peak during recovery of a deformed Al-Mg alloy after cold rolling to $\varepsilon = 3$.

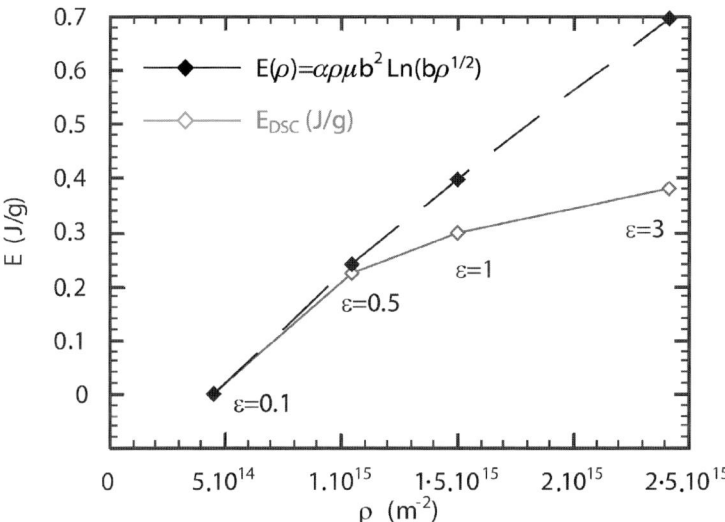

FIG. IV.10. Comparison between the measured stored energy (E_{DSC}) and the calculated $E(\rho)$ using the classical expression with a cut-off radius $R_e = 1/\nu\rho$, as a function of the dislocation density at different cold rolling strains in an Al-Mg alloy.

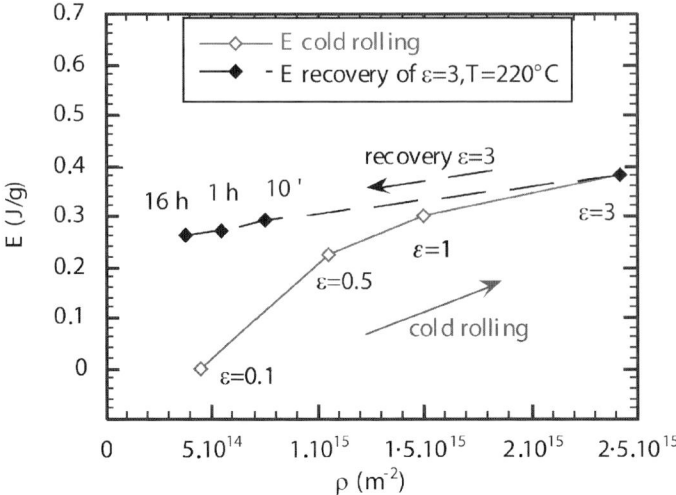

FIG. IV.11. Evolution of the stored energy during cold rolling and after recovery of $\varepsilon = 3$ at $T = 220\,°C$ as a function of the dislocation density in an Al-Mg alloy.

The main points to be retained from these results are the following:

(i) The dislocations structures emerging from cold deformation are *not* low energy structures: recovery leads to lower energy structures as well as to the reduction of their number per unit volume.
(ii) Dislocation density is not a sufficient variable to characterize a deformed state and the associated microstructure: as far a stored energy is concerned, the self screening of dislocations is an essential parameter.

These facts will provide guidelines to evaluate the existing models for dislocation patterning.

V. Structural Defect Patterning: Irradiation and Plastic Deformation

From the previous statements, one can infer that two factors will plague the understanding of patterns associated with dislocations. On one hand, the screening of the long range interactions between dislocations must be considered and this will be a key issue for understanding patterning. On the other hand, the motion of dislocations in response to an applied stress field must also be adequately addressed. Both of these aspects present current difficulties for the theoretician. In this sense, irradiation and plasticity are quite different. Irradiation induced dislocations are loops whose far field elastic stresses are much smaller than those associated with long dislocations generated by plasticity. Furthermore, whereas the point defects generated by irradiation are mobile entities, irradiation loops are effectively immobile. As a result of these two 'simplifications' our understanding of patterning in irradiation is more advanced than our understanding of patterns obtained during plastic deformation.

14. IRRADIATION INDUCED PATTERNING

The main effect of irradiation is the generation of point defects (both interstitials and vacancies[139]). These defects, far in excess of the equilibrium concentrations, tend to agglomerate[140] in the form of either irradiation loops or voids.[139,141] An example in an irradiated stainless steel[142] is shown in Figure V.1.

The defects visible in Figure V.1 are clearly not well organized. However, under particular conditions, these structural defects generated from the excess point defects do tend to self-organize into strikingly regular patterns as shown in Figure V.2.[143,144]

a. *The Experimental Facts: Dislocation Carpets and Void Lattices*

Detailed accounts of the experimental characteristics of irradiation patterns can be found in numerous reviews (Refs. [145–148] for voids and bubble lattices, Refs. [149–152] for dislocation patterning). We will focus our attention on the main features relevant to the modeling of irradiation patterns as dynamic instabilities.

The patterns shown in Figure V.2 have two seemingly contradictory characteristic features: their scale is very small, but still much larger than the interatomic

---

[139] M. T. Robinson, *J. Nucl. Mat.* **216**, 1 (1994).
[140] A. Barbu and G. Martin, *Solid State Phenomena* **30–31**, 179 (1993).
[141] A. Barbu, G. Martin, and L. Howe, in *Basic Mechanical Properties and Lattice Defects of Intermetallic Compounds*, eds. J. Westbrook and R. Fleisher, Wiley and Sons, New York (2000), p. 181.
[142] C. Pokor, Y. Brechet, P. Dubuisson, and J. P. Massoud, *J. Nucl. Mat.* **326**, 19 (2004).
[143] W. Jager, P. Ehrhart, and W. Schilling, *Solid State Phenomena* **3–4**, 279 (1988).
[144] G. L. Kulcinski and J. L. Brimhall, in *Effects of Irradiation on Substructure and Mechanical Properties of Metals and Alloys*, ed. J. Mateff, ASTM (1973), p. 258.
[145] J. H. Evans, *Nature* **229**, 403 (1971).
[146] J. H. Evans, *Solid State Phenomena* **3–4**, 303 (1988).
[147] J. H. Evans, *Rad. Eff.* **10**, 55 (1971).
[148] S. L. Sass and B. L. Eyre, *Phil. Mag.* **27**, 1447 (1973).
[149] D. J. Mazey and J. H. Evans, *J. Nucl. Mat.* **138**, 16 (1986).
[150] J. O. Stiegler and K. Farrell, *Scripta metall.* **8**, 651 (1974).
[151] W. Jager, P. Ehrhart, and W. Schilling, *Rad. Eff. Def. Solids* **113**, 201 (1990).
[152] R. Bullough, B. L. Eyre, and K. Krishan, *Proc. Roy. Soc. London A* **346**, 81 (1975).

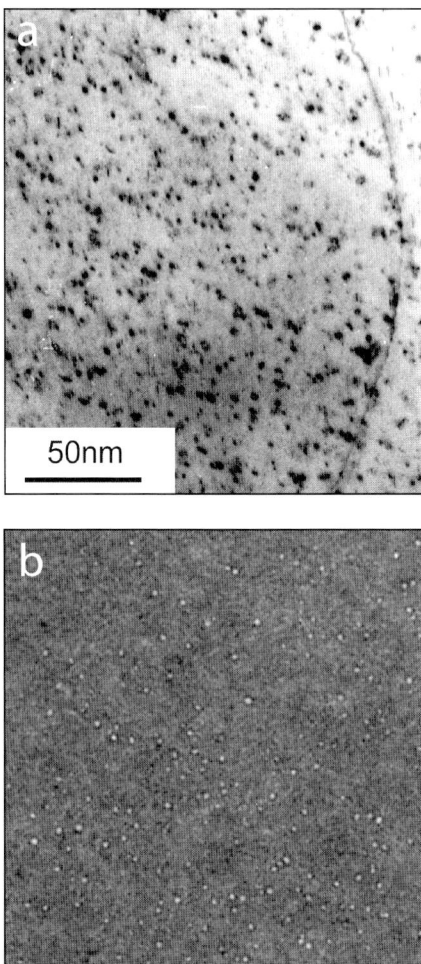

FIG. V.1. (a) Irradiation loops and (b) irradiation voids in an irradiated stainless steel.[142]

FIG. V.2. Self-organizing patterns under irradiation. (a) Dislocation carpets in Cu (periodic arrays of vacancy loops).[143] (b) Nb void lattices: electron micrograph showing a BCC void superlattice in Nb irradiated with Ta$^+$ ions at 800 °C.[144] The foil orientation is near $\langle 111 \rangle$.

DEFECT-INDUCED DYNAMIC PATTERN FORMATION IN METALS AND ALLOYS 251

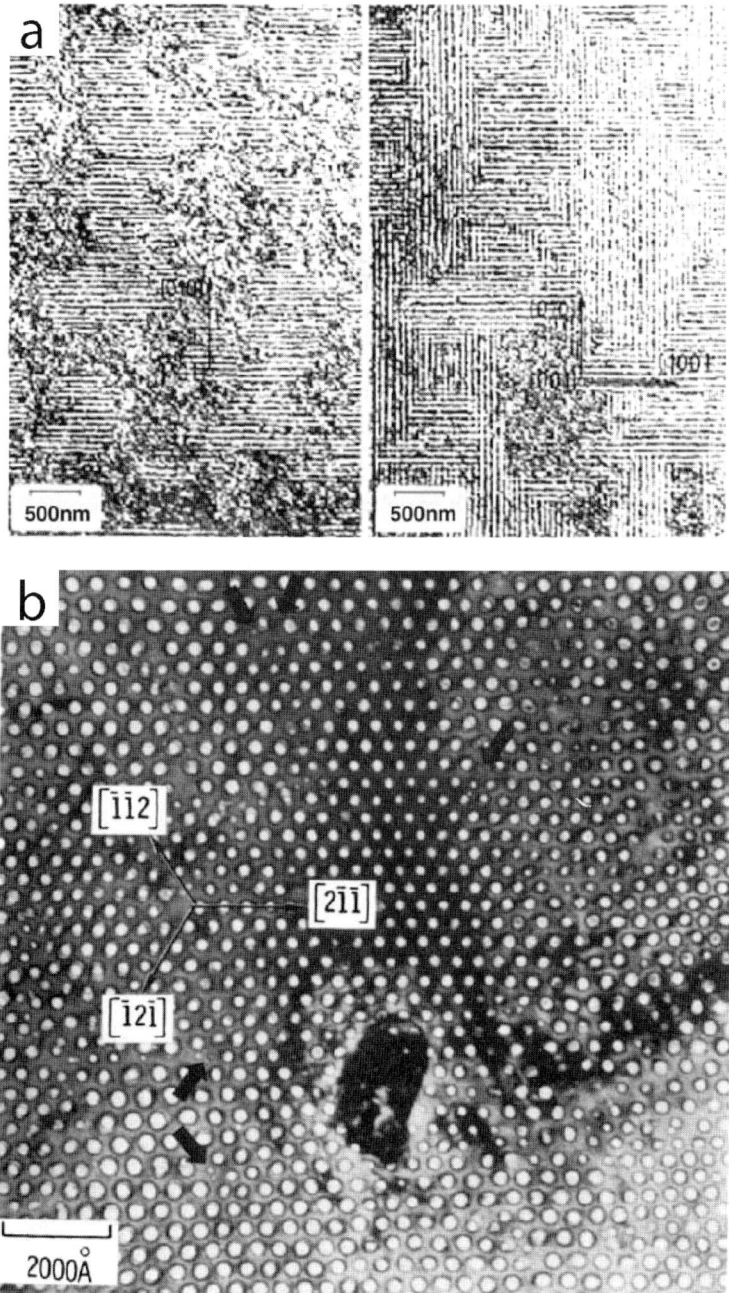

spacing (at least two orders of magnitude) and their alignment reflects the underlying crystallography of the lattice. In Figure V.2a, the dislocation vacancy loops generating the 'carpet pattern' in irradiated Cu are aligned along (001) planes. The 3D void lattice in irradiated Nb shown in Figure V.2b has the same symmetry and orientation as the underlying crystalline host lattice. The obvious question is what mechanism could transfer anisotropy at the atomistic level into patterns at a scale two orders of magnitude larger.

These well organized patterns are not observed under all irradiation conditions. The dislocation patterns become regular when the irradiation flux exceeds some critical value which is an increasing function of the irradiation temperature. However, the wave length developed seems to depend little on the irradiation conditions. If the flux exceeds a second critical value, the dislocation microstructure loses its regular patterning. The values of these two critical fluxes depend on the system and on the material purity.

The void lattice (Figure V.2b) seems to nucleate first at random, and then to order at a later stage when the irradiation dose has become sufficiently high. The conditions for void ordering also depend on the level of purity of the alloys and on their gas content. The void lattice parameter seems to depend on temperature and irradiation flux mainly through the total fraction of voids generated after a given irradiation dose.

b. The Framework of a 'Reaction Diffusion' Approach

The main effect of irradiation is to generate point defects (and possibly gases). These entities interact to create the mesoscopic defects whose patterning is of interest here: the point defects can generate dislocation loops and vacancies and gases can form bubbles. The point defects can also interact with the defects they have already formed. For example, the diffusion of vacancies or interstitials to the stationary dislocation loops and voids can give rise to absorption events. These mechanisms provide a framework for a 'chemical kinetics' approach to the evolution of the population of defects. A set of rate equations can be written to describe the evolution of the defect populations. The point defects (vacancies or interstitials) will be indexed by r, and the microstructural features (dislocation loops, vacancies clusters, voids) by q. Each defect in concentration, C_r, will be generated at a rate P_r by irradiation damage, and can be annihilated or created on either pre-existing sinks (such as grains boundaries) or those created by irradiation. Annihilation may also occur by pairs of vacancies/interstitials. A further feature of the rate equations is that the absorption of point defects by features such as dislocations presents a bias in favor of the interstitials. The rate of absorption is related to the mobility of point defects and therefore to their diffusion coefficients, whereas the relevant microstructural features are the defect densities and sizes.

The volume fraction Q_q occupied by a given type of microstructural feature (e.g., voids) may either grow or decay from the absorption or emission of point defects. The details of these equations will depend on the nature of the microstructural features under consideration, but the form of the equations is quite general. The form of the 'rate equations' for the point defects and the microstructural features can be written:

$$\dot{C}_r = P_r - AC_1C_2 + \Phi_r(C_r, Q_q)$$
$$\dot{Q}_q = P_q + \Psi(C_r, Q_q). \qquad (5.1)$$

These general 'rate equations', initially defined by Bullough et al.,[152–154] can be used to study the evolution of the specific features one is attempting to describe. For example, if the evolution of the distribution of dislocation loops is of concern, a detailed solution of the problem, using a Fokker–Plank approach, allows for a description of the number and size distribution.[142] If patterning (dislocation patterning or void patterning) is the phenomenon which has to be described, a spatial variable must be added. This appears as a diffusion term in the rate equation for the point defects. The microstructural features such as voids and dislocations are not supposed to move in space and their patterning reflects only the patterning in the field of the point defects. An example of such an approach was proposed by Murphy[154] for the patterning of vacancy loops. The technique used is always the same: one searches for a homogeneous steady-state solution to the rate equations, and then performs a linear stability analysis with respect to infinitesimal perturbations periodic in space. When at least one eigenvalue of the linearized equations in Fourier space has a positive real part, the corresponding eigenmode will grown and lead to a periodic pattern. This very simple linear stability analysis is able to provide a qualitative understanding of dislocation loop patterning, but is unable to provide a quantitative prediction for the selected wavelength. In order to go further, a non-linear analysis of the same set of rate equations must be performed. In a series of papers, Ghoniem and Walgraef[155–158] have thoroughly investigated this problem, both with weakly non-linear analysis (the next term in the expansion in terms of the amplitude of the perturbation) and with numerical simulations. They investigated the influence of defect densities (initial dislocation network) on the wavelength. The values of the parameters which lead to patterning have been derived and the ratio of the absorption bias to the emission bias

---

[153] K. Krishan, *Solid State Phenomena* **3–4**, 267 (1988).
[154] S. M. Murphy, *Solid State Phenomena* **3–4**, 295 (1988).
[155] N. M. Ghoniem, D. Walgraef, and S. J. Zinkle, *J. Comp. Aid. Mat. Des.* **8**, 1 (2002).
[156] D. Walgraef and N. M. Ghoniem, *Phys. Rev. B* **39**, 8867 (1989).
[157] N. M. Ghoniem and D. Walgraef, *Mod. Sim. Mat. Sci. and Eng.* **1**, 569 (1993).
[158] D. Walgraef, J. Lauzeral, and N. M. Ghoniem, *Phys. Rev. B* **53**, 14782 (1996).

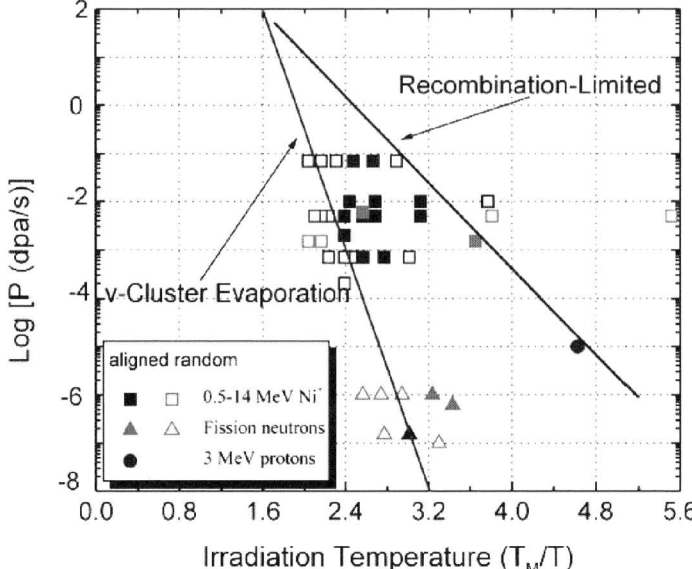

FIG. V.3. Comparison between experimental data and theoretical predictions for the temperature and dose rate dependence of defect cluster wall observations in irradiated Ni.[155] The filled symbols denote conditions where defect cluster wall formation was observed.

seems to play a key role. The anisotropy of diffusion of self interstitial atoms is held responsible for the crystallographic nature of the patterning of both the dislocation loops and the voids. It was also shown that this anisotropy need not to be larger than 1% to give a very well defined crystallographic organization. The conditions for patterning appear to be that the rate of point defect absorption exceeds their mutual annihilation, and that cascade induced vacancy cluster densities are above some critical value. These conditions lead to the prediction of the ranges in temperature and irradiation conditions for which dislocation patterning is expected (Figure V.3).[155] When the temperature is too high, the cluster evaporation decreases the net absorption rate by dislocation loops and when the temperature is too low the recombination between vacancies and interstitials dominates. Dislocation patterning is observed only within these two limiting temperatures.

The case of void formation is somewhat different: voids seem to nucleate 'at random', and progressively organize.[146] The crystallographic organization of the void lattice has also been interpreted as a consequence of the anisotropy of self interstitial diffusion. Furthermore, it has been shown that when a void is slightly displaced from its position in a regular lattice, it will tend to grow in an anisotropic manner, to bring it back to a central position.[159] This amounts to having a drift force on the void mediated by the diffusion field of point defects. In the case of the dislocation carpets discussed above it is assumed that spatial motion is only possible for the point defects. The dislocations themselves are assumed stationary. Such models are bound to fail in describing pattern formation for entities which are nucleated at random with a substantial size. For void patterning, it seems that the long range repulsive forces described above play a key role, more so than a diffusional instability. A similar situation will be suggested for dislocation patterning in plasticity.

15. Plasticity Induced Patterning

The first feature which arises from dislocation patterning in plasticity, especially when compared to the dislocation patterns issued from irradiation, is their relatively coarse scale (for an overview of experimental results for F.C.C. metals, see Ref. [160]). The scale is at least one order of magnitude larger than the 'carpet structures' developed under irradiation. This in itself can be seen as a sign of the long range elastic interactions between dislocations, and as a consequence of the dislocation glide transport mechanism which is far more efficient than diffusion.

a. The Experimental Facts

When plastic deformation takes place in a polycrystal, multiple slip systems are generally activated. The dislocations generated on these different slip systems interact and the material hardens. Eventually the flow stress reaches a saturation level where the annihilation of dislocations balances the accumulation. At large enough strains (or more accurately for large enough dislocation densities, i.e., when the interaction between dislocations exceeds the lattice friction) a cellular structure will form. The material is a composite structure of regions virtually free of dislocations (cell interiors) separated by regions of high dislocation densities (cell walls). The experimental conditions (T, ε) for the emergence of a cell structure in Fe are shown in Figure V.4.[161] At low temperatures the lattice friction is difficult to overcome and the dislocation pattern remains planar rather than cellular.

For F.C.C. materials, the lattice friction stress is generally negligible and cellular patterns emerge more readily than in BCC and HCP metals. When alloying

---

[159] P. Benoist and G. Martin, in *Proceedings of an Internations Conference on Fundamental Aspects of Radiation Damage in Metals*, Gatlinberg, Tennessee (1975), p. 1236.
[160] B. Bay, N. Hansen, D. A. Hughes, and D. Kuhlmann-Wilsdorf, *Acta metall mater.* **40**, 205 (1992).
[161] A. S. Keh and S. Weissmann, in *Electron Microscopy and Strength of Metals*, eds. G. Thomas and G. Washburn, Interscience, New York (1963), p. 231.

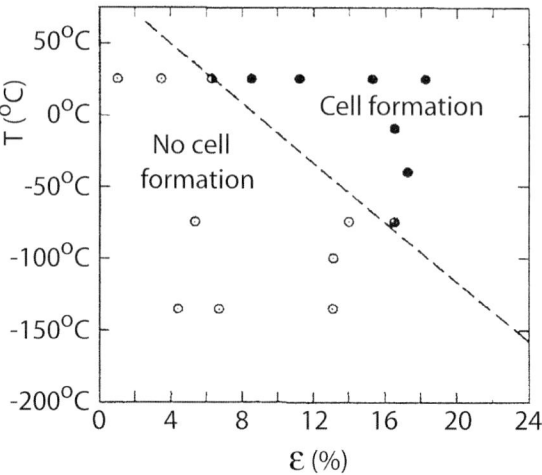

FIG. V.4. The transition between uniform and cellular dislocation structures in α-Fe as a function of strain and temperature.[161]

elements are added, any phenomenon which promotes the planarity of the slip (e.g., decreases the stacking fault energy, presence of shearable precipitates, etc.) will postpone to larger strains and larger dislocation densities the emergence of a cellular structure. The higher the deformation temperature, the better defined are the cells, i.e., the narrower are the cell walls.[162] Some examples taken from Ni, Cu and Al are shown in Figure V.5.[163]

The cell structure has a scale D which is inversely proportional to the flow stress, as shown in Figure V.6.[164] The relation between D (normalized by the Burgers vector **b**) and σ (normalized by the shear modulus μ) is given by Eq. (5.2). The constant K is close to 20.

$$\frac{\sigma}{\mu} = K \frac{b}{D} \qquad (5.2)$$

The 3D patterns developed during large strains have been extensively reviewed. They show the emergence of patterns elongated in the direction of cold rolling and the initiation of bands of intense localization of plastic strains, probably associated with crystalline rotations.[160]

In addition to this self imposed 'scale', more recent detailed examination of the apparent length scales at different magnifications have revealed the existence of a

---

[162] M. R. Staker and D. L. Holt, *Acta metall.* **20**, 569 (1972).

[163] I. Barker, N. Hansen, and B. Ralph, *Mat. Sci. and Eng. A* **A113**, 449 (1989).

[164] S. V. Raj and G. M. Pharr, *Mat. Sci. and Eng.* **81**, 217 (1986).

FIG. V.5. Bright field transmission electron microscope images of some FCC metals deformed to a true strain of 0.1.[163] (a) Ni, (b) Cu and (c) Al.

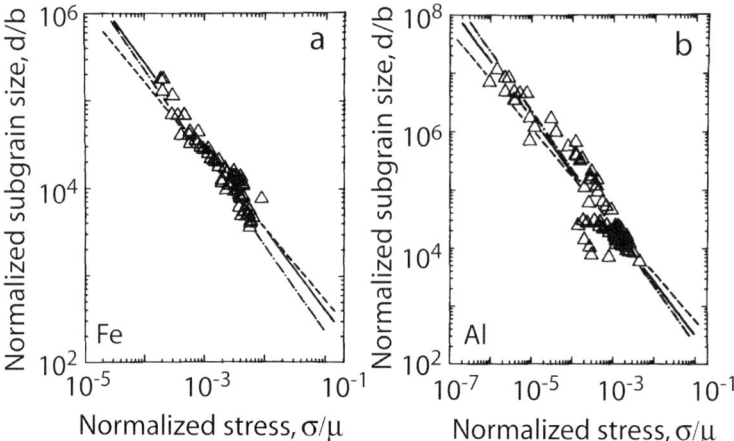

FIG. V.6. The relation between reduced cell size, D/b, and reduced flow stress, σ/μ, in (a) Fe and (b) Al.[164]

fractal structure in the dislocation cells.[165] The scaling works over two orders of magnitude for the cell sizes and the fractal dimension of the 2D slices of the cell structure ranges between 1.6 to 1.8 (with increasing flow stress), Figure V.7.

As pointed out by Kubin,[166] the scaling relation between the flow stress and the dislocation cell size (Eq. (5.2)) can be readily obtained either from the dependence in $1/r$ of the dislocation stress field, or from the dependence in $1/D$ of the activation stress for a source whose pinning points are separated by a distance D. Therefore, any acceptable theory for pattern formation should lead to this scaling law. The differentiation between the different theories should be made on the basis of both their predictions for the constant K and their ability to predict the conditions under which patterning is observed.

b. The Limits of Classical Approaches for Pattern Formation

Let us begin by summarizing a few key features of the formation of dislocation cells.

- Since dislocation patterning occurs during plastic deformation, and well before the saturation stress is reached, it should be considered as a dynamic instability, not an organization of a given number of dislocations.

---

[165] P. Hähner, K. Bay, and M. Zaiser, *Phys. Rev. Letts.* **81**, 2470 (1998).

[166] L. P. Kubin, in *Materials Science and Technology: Vol. 6 Plastic Deformation and Fracture of Materials*, ed. H. Mughrabi, VCH, Weinheim (1993), p. 138.

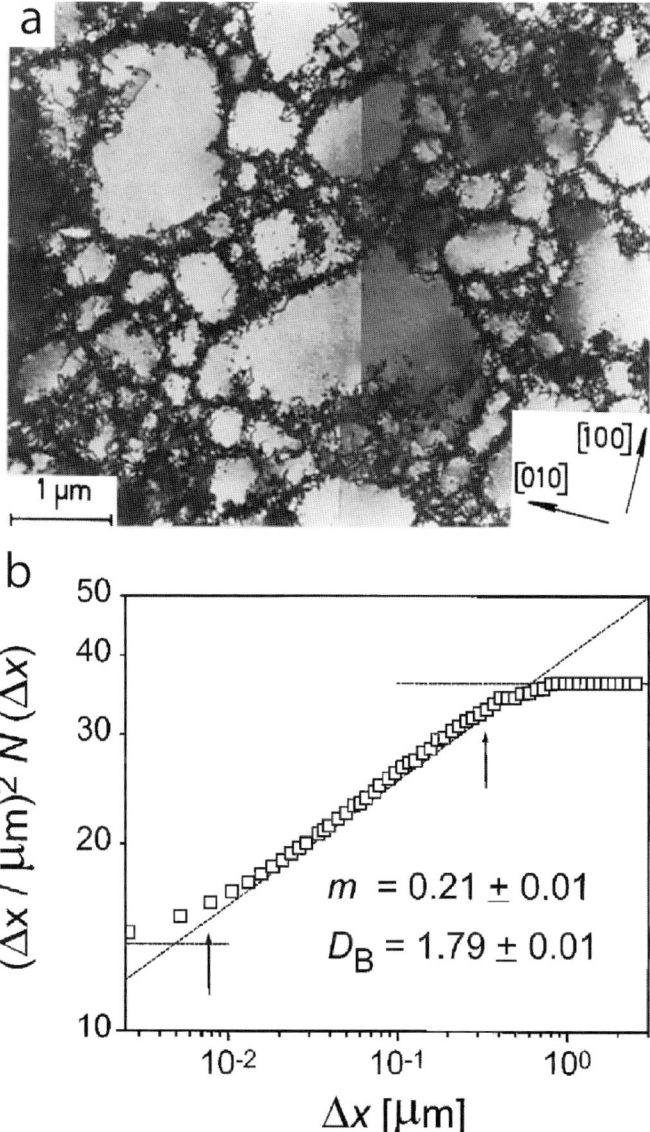

FIG. V.7. (a) Bright field transmission electron micrograph of the dislocation cell structure in a Cu single crystal after tensile deformation along a [100] direction at room temperature to a stress of 75.6 MPa. (b) Analysis of the cell structure shown in (a) by determining the 'box dimension' D_B.[165]

- Since cell formation is postponed by planarity of slip, and multiple slip seems necessary for the appearance of cells, the local reactions between different dislocation families must be a key ingredient for pattern formation.
- Since cellular patterns are observed only when the long range stresses between dislocations overcome the short range lattice friction stress, it is also necessary to include the stress field of a dislocation distribution as a crucial ingredient for pattern formation.

The classical approaches to dislocation patterning all fail in at least one of the above respects.

The first theory for dislocation patterning, due to Holt,[167] considers the formation of a cellular structure as an instability in a population of a given density of dislocations. He considers a two-dimensional distribution of supposedly infinite dislocation lines, and no short range reactions are allowed. The correlation function between dislocations is assumed to vanish on distances which are of the order of $\rho^{-1/2}$ where ρ is the dislocation density. Under these conditions, the system is expected to spontaneously 'un-mix' by a spinodal-like instability, the fastest growing wavelength being proportional to the cut-off radius. Since the flow stress is proportional to $\rho^{1/2}$, this gives the expected scaling law. However, besides the simplistic 2D realization, the model is fundamentally flawed in two respects: (a) the dislocations are assumed to respond to the internal stress only, whereas they should also be responding to the applied stress, and (b) the dislocation density is assumed to be constant, whereas this is obviously not generally the case during plastic deformation. Furthermore, in order to write the appropriate 'diffusion-like' equation for dislocation motion, the existence of a cut-off radius is crucial, and this gives the emergence of a characteristic length. The question of the spontaneous emergence of the screening length appears as a crucial issue in the problem of patterning and must be solved in a self consistent manner, not imposed '*a priori*'. The Holt model is, in essence, a model to minimize the energy of a closed system, whereas dislocation patterning is a dynamic instability in an open system. Along similar lines of a minimization of the total energy at constant dislocation density and in the absence of an external applied stress, is the family of models known as LEDS (Low Energy Dislocation Structures).[168–172] These

---

[167] D. L. Holt, *J. Appl. Phys.* **41**, 3197 (1970).
[168] D. Kuhlmann-Wilsdorf, *Metall. Trans. A* **16A**, 2091 (1985).
[169] D. Kuhlmann-Wilsdorf, *Trans. AIME* **224**, 1047 (1962).
[170] N. Hansen and D. Kuhlmann-Wilsdorf, *Mat. Sci. and Eng.* **81**, 141 (1986).
[171] D. Kuhlmann-Wilsdorf, *Mat. Sci. and Eng. A* **A113**, 1 (1989).
[172] U. F. Kocks, T. Hasegawa, and R. O. Scattergood, *Scripta metall.* **14**, 449 (1980).

models describe the scaling law, which only means that they are dimensionally correct, but they cannot describe the dynamic nature of the instability.

A completely different route was proposed by Kocks.[173,174] Whereas the previous models insisted on the forces acting on the dislocations, Kocks insisted on including the short range reactions between dislocations in their glide plane when they interact with the forest made by dislocations from other glide planes. The key idea is that the building up of a cell structure is intrinsically a statistical problem: regions where the forest trees are to narrowly spaced will not be penetrable by the gliding dislocations, which will leave around these regions loops which will make them even more impenetrable. Since the only length scale in the problem is $\rho^{-1/2}$, the distance between hard points will scale as this distance, which will indeed give the correct scaling law, since the flow stress scales as $\rho^{1/2}$. The simulations performed by Kocks confirm this finding. However the flaw in the approach is in two respects. On the one hand, dislocation patterning occurs in multiple slip and the forest dislocation density is certainly not constant, as is assumed in stage II work hardening. On the other hand, the hard points created by the mere fact that forest dislocations are too close together, are not really so hard. They may be too hard to be penetrated by the applied stress, but once a loop is left behind, they may be penetrated by the combined effect of the applied stress and the line tension. In order to stabilize these hard points, local reactions due to recovery are necessary to create sessile obstacles.

The dynamic nature of dislocation patterning and the non-conserved character of dislocations (i.e., their ability to annihilate by pairs or to multiply via various source mechanisms) has triggered interest in a 'chemical kinetics' approach. Various detailed sets of equations have been proposed,[174–182] but the framework always remains the same. The density of dislocations is divided into different 'species', ρ_i, each with a flux, J_i. The reaction terms (annihilation,

---

[173] U. F. Kocks, in *Dislocations and Properties of Real Materials*, The Institute of Metals, London (1985), p. 125.

[174] U. F. Kocks, A. S. Argon, and M. F. Ashby, *Prog. Mat. Sci.* **19**, 1 (1975).

[175] D. Walgraef and E. C. Aifantis, *J. Appl. Phys.* **58**, 688 (1985).

[176] L. P. Kubin and J. Lepinoux, in *Proceedings of an International Conference of the Strength of Metals and Alloys (ICSMA8)*, Vol. 1, eds. P. O. Kettunen, T. K. Lepisto, and M. E. Lehtonen, Pergamon, Tampere, Finland (1988) p. 35.

[177] U. Essmann and K. Differt, *Scripta metall.* **22**, 1337 (1988).

[178] K. Differt and U. Essmann, *Mat. Sci. and Eng. A* **A164**, 295 (1993).

[179] G. Malygin, *Soviet Physics State Solid* **31**, 96 (1989).

[180] G. Malygin, *Physics-Uspekhi* **42**, 887 (1999).

[181] Y. Estrin and L. P. Kubin, *Acta metall.* **34**, 2455 (1986).

[182] L. P. Kubin, in *NATO ASI Series: Stability of Materials*, eds. A. Gonis, P. E. A. Turchi, and J. Kudrnovsky, Plenum, New York (1995), p. 99.

multiplication...) are embedded into a function $R(\rho_i, \rho_j, \sigma)$ and the evolution equations can then be written as:

$$\frac{\partial \rho_i}{\partial t} + \mathrm{div}(J_i) = R(\rho_i, \rho_j, \sigma) \tag{5.3}$$

In some cases, when only dynamic processes are involved (i.e., when R is proportional to the velocity of dislocations), the differential equation can be written in terms of derivatives with respect to ε. The difficulty in the approach is in the expression of J_i. The flux is obviously related to the dislocation densities and velocities, and the velocities are related to the applied stress, σ_a, and to the internal stress, σ_{int}, due to the other dislocations by an 'effective' dislocation mobility, M.

$$J_i = \rho_i V = \rho_i M(\sigma_a - \sigma_{\mathrm{int}}) \tag{5.4}$$

All the difficulty lies in the evaluation of the internal stress. Since the dislocation field is long ranged, the expression for the internal stress is given by an integral, and only an expansion of this integral in terms of gradients in ρ allows us to reach an expression for σ_{int} which is local. Under this very questionable assumption, the basic equation can be reduced to a 'reaction diffusion' approach for which the standard techniques can be applied.

Another simplified condition under which a 'diffusion' term can be obtained is the case of spreading of the dislocation density via random events such as double cross-slip. But again, this procedure hides the fundamental fact that dislocations are gliding and not diffusing, and that the long range fields cannot be assumed to be screened without further justification. This is probably the reason why 'reaction diffusion' approaches have never been able to present for dislocation patterning the same success they have shown for irradiation patterning.

In spite of these pitfalls, the various 'classical approaches' presented above have brought to the forefront the key issues that need to be addressed by a future 'theory of dislocation patterning':

– local reactions must be accounted for, and in particular, the non-conservative nature of dislocation densities;
– long range stresses have to be accounted for, and the self-screening of dislocations, which can even be measured experimentally,[183,184] must be derived as part of the solution of the problem.

---

[183] T. Ungar, L. S. Toth, J. Illy, and I. Kovacs, *Acta metall.* **34**, 1257 (1986).
[184] T. Ungar, H. Mughrabi, D. Ronnpagel, and M. Wilkens, *Acta metall.* **32**, 333 (1984).

As a result of the difficulty in addressing these questions in an analytical manner, without making unreasonable assumptions, computer simulations of dislocation dynamics have been proposed to investigate the collective behavior of dislocation populations, and in particular, their tendency to patterning.[185]

c. Computer Simulations of Pattern Formation

The principle of all simulations described below is first to compute for each dislocation segment the local stress arising from the applied stress and from the interactions with other dislocations in the system. Under this local stress, each dislocation will move at a particular velocity. The simulations differ in the variety of local reactions allowed (multiplications, annihilation, junction formation), on the degree of naivety with which they are described and in the range of strains (or equivalently dislocation densities, or volume simulated) they can treat in a reasonable computing time.

The first simulation,[186,187] a cellular automaton carried out in two dimensions, included, for reasons of computational efficiency, a cut-off radius for dislocation interactions. Calculations were made with and without an externally applied stress. When no external stress was applied, but with the dislocation interactions artificially screened, the instabilities occur, and a length scale emerges which is related to the cut-off radius. When the long range interactions were accounted for without an imposed cut-off radius, as was done in another group of simulations using Molecular Dynamics techniques,[188,189] the patterning disappears. In any case, a partial self-screening is observed, in the sense that during relaxation, the net interaction stress with other dislocations is decreasing. When an external stress is applied, and no cut-off radius is artificially imposed, no clear patterning seems to occur. The definitions of the local rules for multiplication, which are randomly positioned in an artificial manner, are obviously too naïve to capture the spatial organization.

In spite of the numerical and topological difficulties associated with the 3D aspect of dislocation dynamics, it seemed necessary to go one step further and to explicitly include the long range dislocation interactions, the friction stress, the junction formation, dislocation multiplications and the various glide and cross

---

[185] L. P. Kubin, Y. Estrin, and G. Canova, in *NATO ASI: Patterns, Defects and Material Instabilities*, ed. D. Walgraef, Kluwer (1989).
[186] J. Lepinoux and L. P. Kubin, *Scripta metall.* **21**, 833 (1987).
[187] J. Lepinoux, *Solid State Phenomena* **3–4**, 389 (1988).
[188] N. M. Ghoniem and R. J. Amodeo, *Solid State Phenomena* **3–4**, 377 (1988).
[189] R. J. Amodeo and N. M. Ghoniem, *Phys. Rev. B* **41**, 6958 (1990).

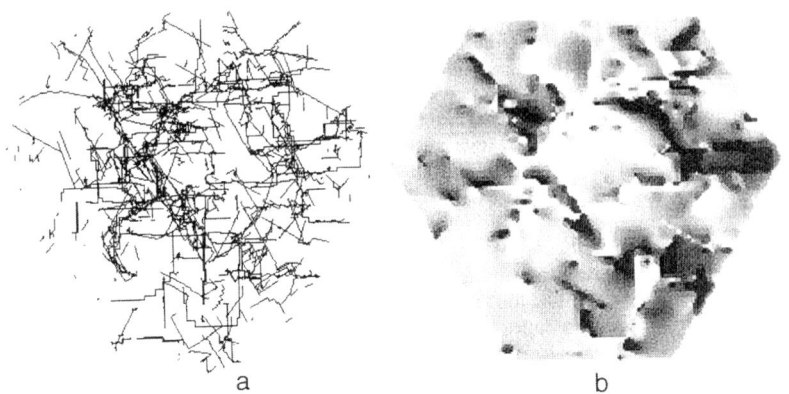

FIG. V.8. (a) Thin (111) slice (1 μm thickness) extracted from the simulated dislocation structure of a [100] aluminium crystal after deformation. (b) The corresponding mapping of the internal shear stress field resolved in the same plane.[196]

slip processes.[190–195] The details of the 3D dislocation simulation, known as DDD (Discrete Dislocation Dynamics) can be found in Ref. [196]. This ambitious program has been developed over the years and has finally led to the simulation of an emerging dislocation structure associated with a spatial patterning of the internal stress.[196,197] An example is shown in Figure V.8.

The limitations in terms of plastic strain attainable in a reasonable computational time, as well as the rather poorly defined cellular structure, prevents testing of the scaling law for cell size as function of the flow stress, although the length scale observed is reasonable. The lesson to be learnt from both the 2D and the 3D simulations is that both long range stresses and short distance reactions are necessary to obtain patterning, and that the short range reactions must be described with a sufficient spatial accuracy. The fact that a source is unlikely to operate

---

[190] L. P. Kubin and G. Canova, *Scripta metall. mater.* **27**, 957 (1992).

[191] L. P. Kubin and G. Canova, in *Electron Microscopy in Plasticity and Fracture Research of Materials*, eds. F. Appel and U. Messerschmidt, Akademie-Verlag, Berlin (1989), p. 23.

[192] L. P. Kubin, G. Canova, M. Condat, B. Devincre, V. Pontikis, and Y. Brechet, *Diff. and Def. Data B* **23–24**, 455 (1992).

[193] B. Devincre and L. P. Kubin, *Mod. Sim. Mat. Sci. and Eng.* **2**, 559 (1994).

[194] H. M. Zbib, M. Rhee, and J. P. Hirth, *Int. J. Mech. Sci.* **40**, 113 (1998).

[195] D. Weygand, L. H. Friedman, E. Van der Giessen, and A. Needleman, *Mod. Sim. Mat. Sci. and Eng.* **10**, 437 (2002).

[196] L. P. Kubin and B. Devincre, in *Deformation-Induced Microstructures: Analysis and Relation to Properties. Proceedings of 20th International Symposium on Materials Science, 6–10 Sept. 1999*, Riso Nat. Lab, Roskilde, Denmark (1999), p. 61.

[197] R. Madec, B. Devincre, and L. P. Kubin, *Scripta mater.* **47**, 689 (2002).

when dislocations in the surrounding region are closely spaced must be taken into account (e.g., distributing sources randomly will not be sufficient).

More recently, 2D simulations have been revised[198–201] with the idea of incorporating information from the 3D simulations, to simultaneously benefit from the much greater speed of the 2D calculations and the possibility of rigorously treating the mechanical boundary conditions, together with the necessity of relating the local rules to the local environment. Furthermore, lattice rotations can be naturally included.[202] Figure V.9 illustrates an emerging dislocation pattern in Cu

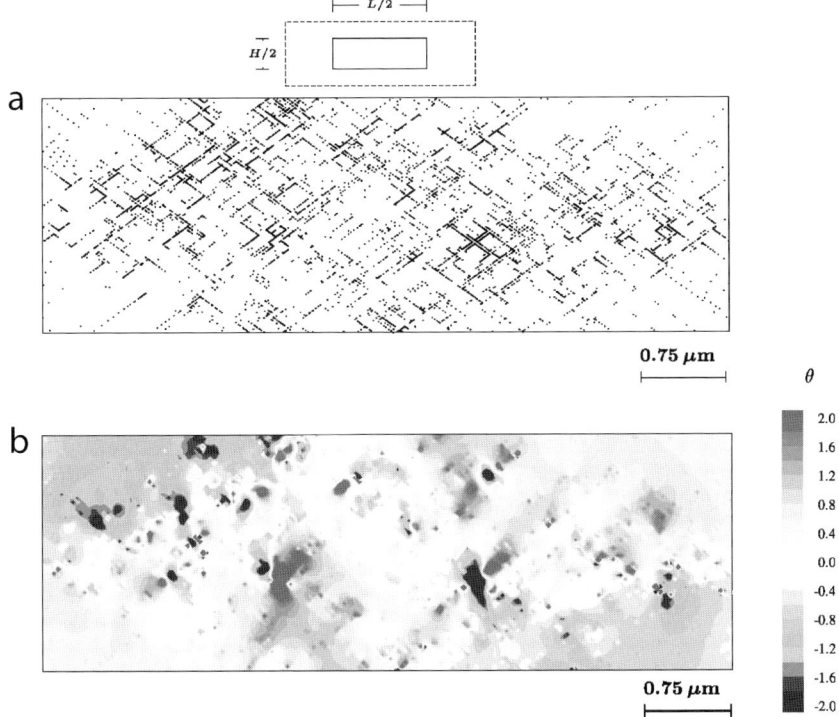

FIG. V.9. (a) Evolution of the bulk dislocation structure in [110] tension of crystal A at a strain of $\varepsilon = 0.03$. (b) The corresponding contours of lattice rotation, θ (in degrees).[202]

---

[198] H. H. M. Cleveringa, E. Van der Giessen, and A. Needleman, *Acta mater.* **45**, 3163 (1997).

[199] A. Needleman, *Acta mater.* **48**, 105 (2000).

[200] A. Needleman and E. Van der Giessen, *Mat. Sci. and Eng. A* **309**, 1 (2001).

[201] J. Y. Shu, N. A. Fleck, E. Van der Giessen, and A. Needleman, *J. Mech. Phys. Solids* **49**, 1361 (2001).

[202] A. A. Benzerga, Y. Brechet, A. Needleman, and E. Van der Giessen, *Mod. Sim. Mat. Sci. and Eng.* **12**, 159 (2004).

together with the local lattice rotations for a plastic strain of 3%. In this simulation, no artificial cut-off radius was imposed. When the rules for multiplication and annihilation are not defined according to the local environment, dislocation patterning does not emerge.

d. Necessary Ingredients for a Theory of Plastically Induced Pattern Formation

The situation concerning our understanding of dislocation patterning during plastic flow can be summarized in the following way. The energetic models and statistic models capture only one part of the problem and both fail to account for the long range interactions between dislocations which is a key ingredient. The local reactions must be explicitly accounted for, including their dependence on the local environment. The 'reaction diffusion' approaches oversimplify the dynamics of dislocation glide. In addition, one aspect which is not accounted for is the appearance, together with cell formation, of lattice rotations which may modify the dislocation dynamics on either side of a cell wall. Each of these ingredients are necessary for a future theory for pattern formation.

Recent work by Hähner and collaborators[203–211] has highlighted a completely new feature that may also need to be considered in the problem of dislocation patterning. Hähner emphasizes the importance of the stochastic aspect in dislocation dynamics. The local fluctuations in the shear rate, $\dot{\gamma}$, and the effective shear stress, τ_{eff}, are related to the internal shear stress, τ_{int}, and to the strain rate sensitivity, S, in the following way:

$$S = \langle \dot{\gamma} \rangle \frac{\partial \tau_a}{\partial \langle \dot{\gamma} \rangle} \tag{5.5}$$

Through an equivalent of the fluctuation dissipation theorem:

$$\langle (\delta \tau_{\text{eff}})^2 \rangle = S \langle \tau_{\text{int}} \rangle \quad \text{and} \quad \langle \delta \dot{\gamma}^2 \rangle / \langle \dot{\gamma} \rangle^2 = \langle \tau_{\text{int}} \rangle / S. \tag{5.6}$$

---

[203] P. Hähner, *Appl. Phys. A* **62**, 473 (1996).
[204] M. Zaiser, *Mat. Sci. and Eng. A* **309**, 304 (2001).
[205] S. Bross and P. Hähner, in *NATO ASI Series: Thermodynamics, Microstructures and Plasticity*, eds. A. Finel, D. Maziere, and M. Veron, Kluwer, Dordrecht (2002), p. 313.
[206] P. Hähner, *Acta mater.* **44**, 2345 (1996).
[207] M. Zaiser, M. Avlonitis, and E. C. Aifantis, *Acta mater.* **46**, 4143 (1998).
[208] M. Zaiser, K. Bay, and P. Hähner, *Acta mater.* **47**, 2463 (1999).
[209] M. Zaiser and P. Hähner, *Mat. Sci. and Eng. A* **270**, 299 (1999).
[210] P. Hähner and M. Zaiser, *Mat. Sci. and Eng. A* **272**, 443 (1999).
[211] P. Hähner, *Scripta Materialia* **47**, 415 (2002).

Zaiser, Bay and Hähner[208] then write an evolution equation for the dislocation density which is in fact a stochastic equation incorporating the noise in the local strain rate. The detailed form of the deterministic equation can be discussed, but the key feature is the following: the probability distribution of the dislocation density changes from a peaked distribution for small values of the noise (i.e., small internal stresses compared to the applied stress) to a power law distribution at larger values of the noise. In this approach, the cell size distribution is expected to show a fractal like behavior, and is to be understood as a noise induced phase transition, where the strain-rate sensitivity and the internal stress compared to the applied stress play a crucial role.[208,211]

In addition to the requirements concerning long range stresses and local reactions, the theory for dislocation cell formation still to be discovered will have to incorporate these stochastic aspects.

e. *The Special Case of Patterning in Fatigue*

Fatigue loading is quite special in plasticity. By a periodic reversal of the applied stress, dislocations are forced to move back and forth, and the resulting microstructure depends on the degree of irreversibility of this motion. As fatigue cycles accumulate, dislocation dipoles are formed, and these entities have elastic fields which are less extended in space. By the very construction of the fatigue dislocations, the long range interaction stresses tend to be naturally screened. The patterns observed in fatigue are very regular, much more so than the patterns obtained in monotonic plasticity. Since the phenomenon of plastic localization in fatigue was discovered,[212] the dislocations structures associated with intense localization have been extensively studied,[213–219] both by conventional TEM to characterize the structure, and by in-situ deformation, to observe the mode of operation.[220] We will summarize only the main features, since they have

---

[212] A. T. Winter, *Phil. Mag.* **30**, 719 (1974).
[213] J. G. Antonopoulos, L. M. Brown, and A. T. Winter, *Phil. Mag.* **34**, 549 (1976).
[214] U. Essmann and H. Mughrabi, *Phil. Mag. A* **40**, 731 (1979).
[215] U. Essmann, U. Gosele, and H. Mughrabi, *Phil. Mag. A* **44**, 405 (1981).
[216] F. Ackermann, L. P. Kubin, J. Lepinoux, and H. Mughrabi, *Acta metall.* **32**, 715 (1984).
[217] R. Wang, H. Mughrabi, S. McGovern, and M. Rapp, *Mat. Sci. and Eng.* **65**, 219 (1984).
[218] J. Lepinoux and L. P. Kubin, *Phil. Mag. A* **54**, 631 (1986).
[219] H. Mughrabi, *Acta metall.* **31**, 1367 (1983).
[220] J. Lepinoux and L. P. Kubin, *Phil. Mag. A* **51**, 675 (1985).

been extensively reviewed in Refs. [221–226]. Since the motion of dislocations is responsible for pattern formation, the most informative studies have been performed by applying a constant plastic strain amplitude to single crystals oriented for single slip. Below a threshold amplitude, the dislocation structure is formed of 'veins' rich in dislocations separated by 'channels' of equivalent volume fraction, poor in dislocations. This is referred to as the 'matrix structure', and the plastic deformation at the macroscopic level is homogeneous in space. When a critical threshold amplitude is attained, plastic deformation becomes strongly localized into so-called 'persistent slip bands' (PSB) which exhibit a very regular 'ladder structure'. This strikingly regular structure shown in Figure V.10^{214} has been thoroughly studied and is one of the nicest and simplest examples of dislocation patterning in plasticity. It is especially simple compared to the cellular patterns discussed in Section V.15.a–c since it is a structure developed while single crystals are fatigued so that a single burgers vector is activated.

This dislocation structure appears to be extremely efficient at carrying the important strains in the Persistent Slip Bands (PSB). Furthermore, the situation corresponds to a true steady state: in a test with constant applied plastic strain amplitude, it corresponds to the saturation of the applied stress amplitude, and to a steady state in the dislocation density. This implies that any creation of dislocation length has to be balanced by an equivalent annihilation process. The overall functioning of the structure is now well understood. The rungs in the ladder structure are of edge character, and the dislocations moving in the channels are of screw character (Figure V.10b). The rungs are formed by dislocation dipoles. Under the applied stress, the dipoles act as sources and emit dislocations in the channels. The screw dislocations moving in the channels leave new edge dislocations in the walls. The creation processes just described are counterbalanced by annihilation processes: dipoles in the walls can collapse by climb, annihilating edge dislocations, and two screw dislocations of opposite signs in the channels can annihilate by cross slip. The annihilation by climb in the walls requires the emission of point defects, which is the origin of the 'swelling' of the PSB as a whole, creating extrusions and intrusions. The operating stress of the structure is governed by the stress necessary to emit a dislocation from the wall and to propagate it in the channel.

---

[221] L. M. Brown, in *Proceedings of an International Conference on the Dislocation Modelling of Physical Systems*, eds. M. Ashby, R. Bullough, C. S. Hartley, and J. P. Hirth, Pergamon, Gainesville, Florida, USA (1980), p. 51.

[222] Z. S. Basinski and S. J. Basinski, *Prog. Mat. Sci.* **36**, 89 (1992).

[223] T. Magnin, J. Driver, J. Lepinoux, and L. P. Kubin, *Rev. de Phys.* **19**, 467 (1984).

[224] T. Magnin, J. Driver, J. Lepinoux, and L. P. Kubin, *Rev. de Phys.* **19**, 483 (1984).

[225] L. M. Brown, *Mat. Sci. and Eng. A* **A285**, 35 (2000).

[226] L. P. Kubin, C. Fressengeas, and G. Ananthakrishna, in *Dislocations in Solids*, Vol. 11, eds. F. R. N. Nabarro and M. S. Duesbery, Elsevier, Amsterdam (2002), p. 101.

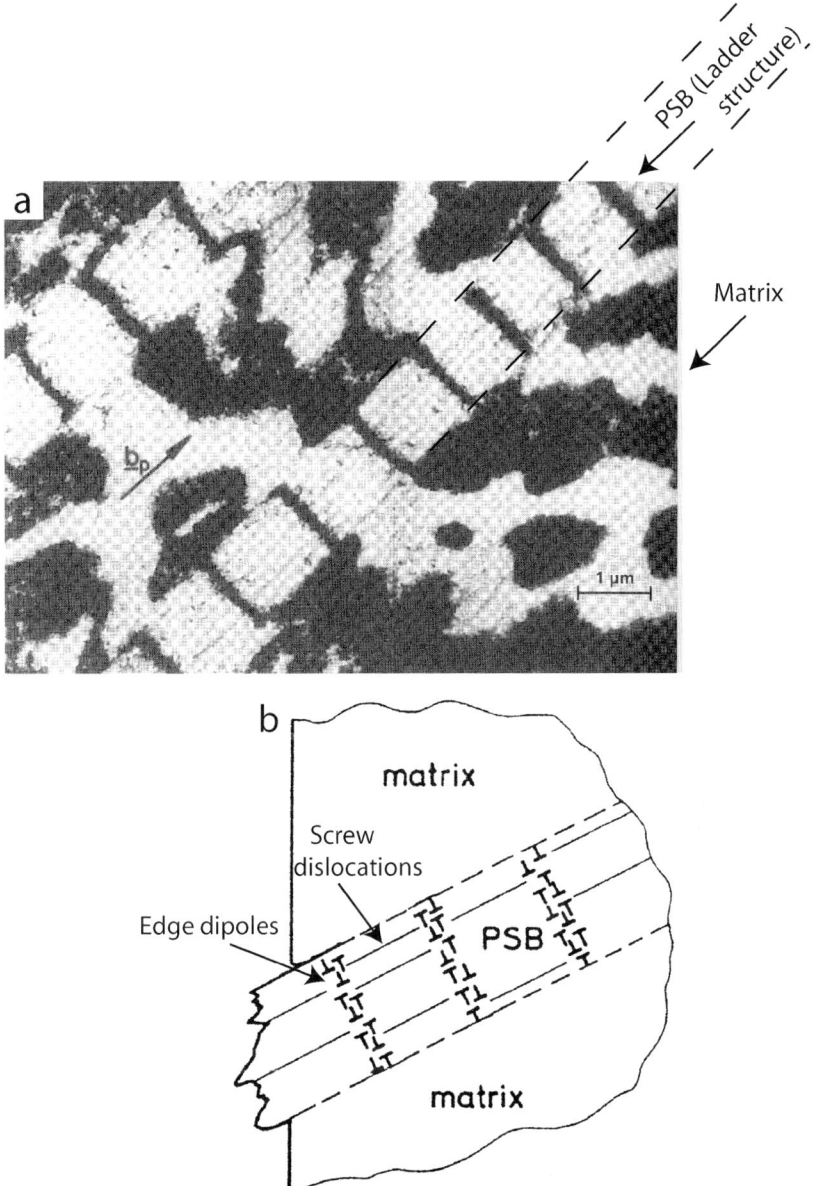

FIG. V.10. (a) Persistent slip bands (PSB) formed in Cu crystal subject to fatigue loading. The PSB and 'matrix' structures are clearly labeled.[214] (b) Schematic representation of the dislocation arrangement within a PSB.[215]

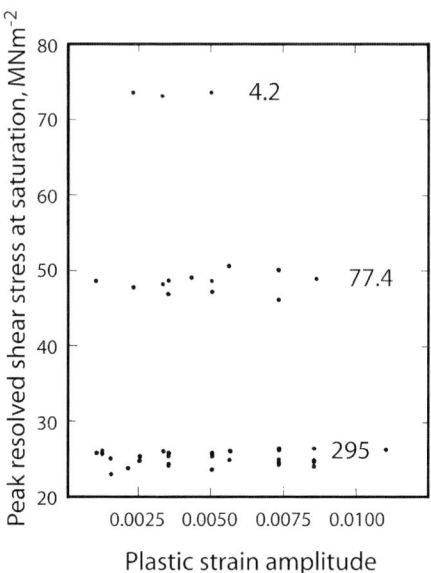

FIG. V.11. Saturation peak resolved shear stresses of Cu crystals fatigued at various constant plastic strain amplitudes and temperatures.[227]

Therefore the structure operates at a stress amplitude which scales as the inverse spacing of the ladder rungs.

In the domain of strain amplitude where the PSBs are observed, the applied saturation stress amplitude is constant and the volume fraction of PSBs is proportional to the strain[227] (Figure V.11). This qualitative description of the functioning of the structure has been cast into the format of a 'chemical kinetics approach' for the density of dislocations in the walls and in the channels by Differt and Essmann.[177,178]

Even though the functioning of the structure is well understood, its spontaneous emergence is still not clear. It has been studied in the framework of the 'reaction diffusion approach' in a series of papers by Walgraef and Aifantis, which can be seen as the most sophisticated examples of this type of modeling in plasticity.[175,228–232] Two dislocation populations, the fast and the slow,

---

[227] Z. S. Basinski, A. S. Korbel, and S. J. Basinski, *Acta metall.* **28**, 191 (1980).
[228] D. Walgraef and E. C. Aifantis, *Int. J. Eng. Sci.* **23**, 1351 (1985).
[229] D. Walgraef and E. C. Aifantis, *Int. J. Eng. Sci.* **23**, 1359 (1985).
[230] D. Walgraef and E. C. Aifantis, *Int. J. Eng. Sci.* **23**, 1365 (1985).
[231] D. Walgraef and E. C. Aifantis, *Int. J. Eng. Sci.* **24**, 1789 (1986).
[232] E. C. Aifantis, *Solid State Phenomena* **3–4**, 397 (1988).

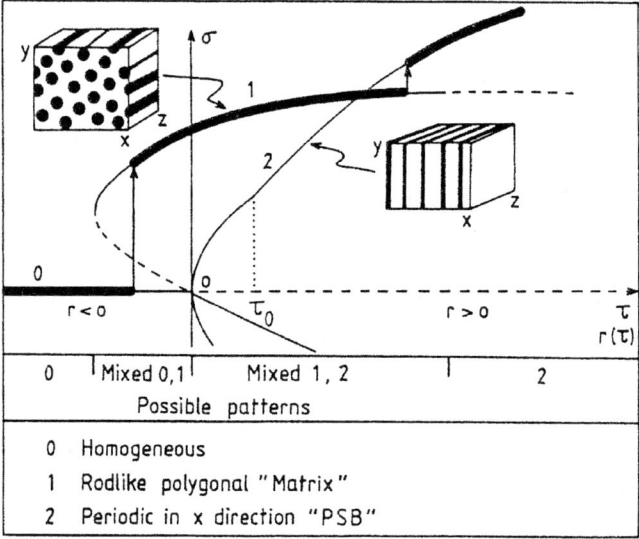

FIG. V.12. Bifurcation diagram in the WA model.[175] A is the amplitude of the modulation of the immobile dislocation density and is the stress at saturation during cycling. The heavy lines indicate the range of stability of each of the three possible solutions, a uniform structure (0), a 'vein' structure (1) and a 'ladder' structure (2).

with two different 'effective diffusion coefficients' evolve in a coupled manner. In fact, the exact form of the reaction terms is more inspired by the so called 'Brussellator'[5,6] than by any precise dislocation mechanism. The control parameter in this model is the applied stress and a sequence of instabilities is predicted via a 'Turing-type' scenario: from a homogeneous distribution, to a 'rod-like' distribution (identified as the so-called 'vein structure') to a 'ladder-like' distribution (Figure V.12).

Qualitatively, the sequence is indeed the observed one. However one may be surprised that in the set of equations studied by Walgraef and Aifantis, there are no terms specific to fatigue. It would predict the same kind of patterning in monotonic loading in single slip, which is known not to occur. In the current framework, the only difference in fatigue would be the ability to maintain the strain long enough so that saturation can indeed be reached and the control parameter increased sufficiently. However, very large flow stresses can be obtained in monotonic loading, and the PSB structure is never observed. This highlights the limitations of phenomenological approaches such as those developed in the reaction diffusion approach: it is necessary to incorporate some of the microstructural features which are known to be important.

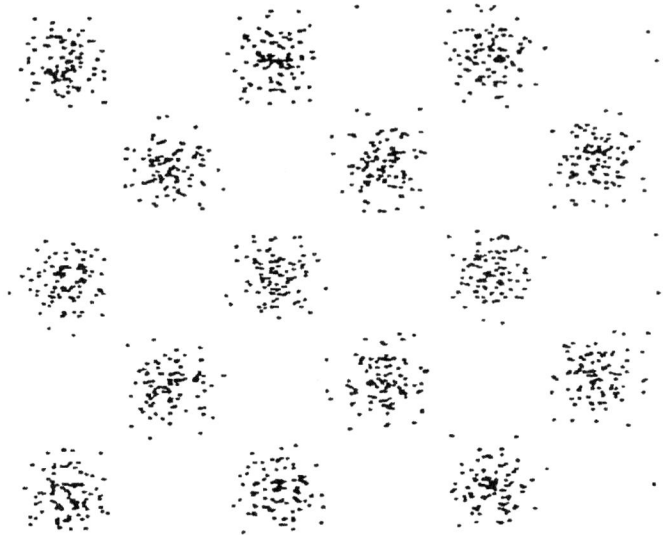

FIG. V.13. Simulated dipolar pattern viewed in a section perpendicular to the slip plane and parallel to the slip direction.[233]

One of these features specific to fatigue is the existence of a large density of dipoles. These dipoles are naturally created by the back and forth motion of dislocations during fatigue loading. An attempt to propose a working mechanism to generate dipole clusters, as incipient features of pattern formation, has been proposed by Kratochvil et al.[233–237] The basic idea is that a dipole drifts in a stress gradient generated by the moving screw dislocations. This drift tends to accumulate dipolar loops which in turn enhance the bowing of the dislocations and provoke further accumulation. This mechanism has been seen to operate in the 3D dislocation simulations,[233] and the simulation of dipole clustering shown in Figure V.13 shows that it may be considered as a possible mechanism for generating structures. However, we are still far from the nice regular structures we have described for the PSB.

---

[233] A. Franek, R. Kalus, and J. Kratochvil, *Phil. Mag. A* **64**, 497 (1991).

[234] J. Kratochvil, *Rev. de Phys.* **23**, 419 (1988).

[235] J. Kratochvil and M. Saxlova, *Scripta metall. mater.* **26**, 113 (1992).

[236] V. Gregor, J. Kratochvil, and M. Saxlova, *Mat. Sci. and Eng. A* **234**, 209 (1997).

[237] J. Kratochvil, *Mat. Sci. and Eng. A* **309**, 331 (2001).

VI. Spatio-Temporal Patterning in Plasticity: The Portevin–Le Chatelier Effect

After having examined the spatial organization of dislocations under various loading conditions, we finally consider a case of spatio-temporal instability in mechanical behavior. In this section, we consider patterning at scales which are much larger (∼mm) than typical dislocation spacings. At this scale, plastic deformation is normally homogeneous. Two stabilizing factors ensure this homogeneity: work hardening and a positive strain-rate sensitivity. When a material deforms, it hardens, and therefore a region which is more deformed tends to deform less easily. This is true for low strains (up to 10%) and is referred to as work hardening. When the strain is larger, and therefore when the stress has increased, the reduction in sample area associated with enhanced plastic deformation may overcome the work hardening rate (which decreases as deformation proceeds): this is the origin of the macroscopic instability called 'Necking' which ends any tensile test of a ductile material. Similarly, the textural evolutions associated with large strains may lead to a 'strain softening' and to the formation of deformation bands where deformation is much higher than in the rest of the sample. These are both instabilities in mechanical behavior. When a material is deformed at a faster rate, the stress necessary is in general higher and this is referred at as 'positive strain-rate sensitivity'. However, in some alloys, a negative strain-rate sensitivity can be observed, leading at the same time to a serrated stress-strain curve and to the spatial localization of plastic deformation. This phenomenon, called the 'Portevin–Le Chatelier' (PLC)[238] effect has been interpreted[239] as the interplay between the diffusion of impurity solutes toward dislocations, and de-pinning of the dislocations. The underlying physics is relatively clear. For a given applied strain rate, a given dislocation density spends part of the time waiting at obstacles (other dislocations), and part of the time flying towards another obstacle. While arrested at obstacles, dislocations attract mobile impurity solutes. The greater the concentration of impurities segregated to the dislocation core, the higher the stress necessary to unpin it for further flight. Therefore two characteristic times have to be considered: the diffusion time for impurities, and the waiting time for dislocations. When the diffusion time is very large compared to the waiting time, dislocations move with almost no solute interaction, with a high mobility, and a positive strain-rate sensitivity is observed. When the diffusion time is very small compared to the waiting time, dislocations are permanently loaded with solute, they will have a low mobility, but still a positive strain-rate sensitivity. When the two times are of the same order, the longer the waiting time (lower strain rate),

---

[238] H. Le Chatelier, *Rev. de Metall.* **6**, 914 (1909).
[239] A. Cottrell, *Dislocations and Plastic Flow in Crystals*, England, Carendon, Oxford (1953).

the greater the concentration of impurities that will segregate to the dislocation, and the lower will be its mobility, meaning a higher stress is needed to make it move at a given velocity: this results in a negative strain rate sensitivity. Since the waiting time depends on the dislocation density and the applied strain rate, and the diffusion time depends on both the temperature and dislocation density, this simple analysis allows for a basic understanding of why the PLC effect is observed only in a closed domain of strain-rates and temperatures, and only beyond a critical strain. This approach has been thoroughly explored, using rate equations for dislocation densities (mobile and immobile) and well established segregation kinetics for impurities at dislocations. The questions of conditions for serrated yielding (another name for PLC) in terms of strain rate, temperature and critical strain can be considered to be solved.[240,241] Similarly, a formal description of the range of negative strain-rate sensitivity,[242] together with spatial dependences in gradients of the stress or the strain rate[243,244] have allowed for a continuum description of the propagative plastic waves. The recent advances in the subject are related to the non-periodic serrated flow, and with the associated spatial aspects of strain localization.

In this section we will consider only the detailed features of the spatial organization of plasticity and the statistics of the serrated flow. After summarizing the variety of experimentally observed phenomena, we will outline two recent approaches: PLC as an example of self organized criticality (SOC), and PLC as an example of deterministic chaos.

16. Phenomenology of Portevin–Le Chatelier Effect

The PLC effect has two manifestations: a serrated stress-strain curve and a localized deformation pattern. Three types of stress strain curves are commonly distinguished, and they correspond to three different types of spatial organization of plasticity.[245–247] Type A serrations are associated with repetitive continuous propagation of deformation bands, nucleated at one end of the sample. Type B serrations correspond to a hopping propagation of localized bands in the axial direction of the sample. Type C serrations correspond to the nucleation of bands in a spatially non-correlated manner. The typical appearances of the various types of serrations are shown in Figure VI.1.

---

[240] L. P. Kubin and Y. Estrin, *Acta metall.* **33**, 397 (1985).
[241] L. P. Kubin and Y. Estrin, *Acta metall mater.* **38**, 697 (1990).
[242] P. Penning, *Acta metall.* **20**, 1169 (1972).
[243] V. Jeanclaude and C. Fressengeas, *Comp. Rendue Acad. des Sci.* **315**, 7 (1992).
[244] V. Jeanclaude and C. Fressengeas, *Comp. Rendue Acad. des Sci.* **316**, 867 (1993).
[245] K. Chihab, Y. Estrin, L. P. Kubin, and J. Vergnol, *Scripta metall.* **21**, 203 (1987).
[246] E. Pink and A. Grinberg, *Mat. Sci. and Eng.* **51**, 1 (1981).
[247] L. P. Kubin, K. Chihab, and Y. Estrin, *Acta metall.* **36**, 2707 (1988).

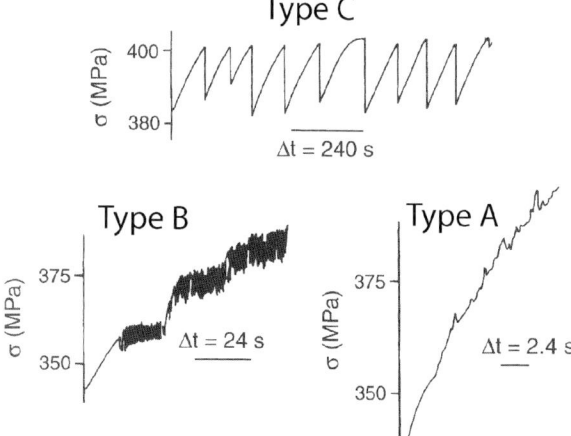

FIG. VI.1. Examples of the different types of serrated yielding seen in an Al-5Mg (wt.%) polycrystal deformed at room temperature with different imposed strain rates, $\dot{\varepsilon}$.[226] Type C serrations: $\dot{\varepsilon} = 5 \times 10^{-6}/s$. Type B serrations: $\dot{\varepsilon} = 5 \times 10^{-4}/s$. Type A serrations: $\dot{\varepsilon} = 5 \times 10^{-3}/s$.

The stress drops during serrated yielding have also been examined and the statistical analysis[248] in various situations can also lead to three types of histograms: the peak shaped distribution (type p), the asymmetric distribution (type as) and an intermediate distribution which presents both a peak and an asymmetry (type i) (Figure VI.2).

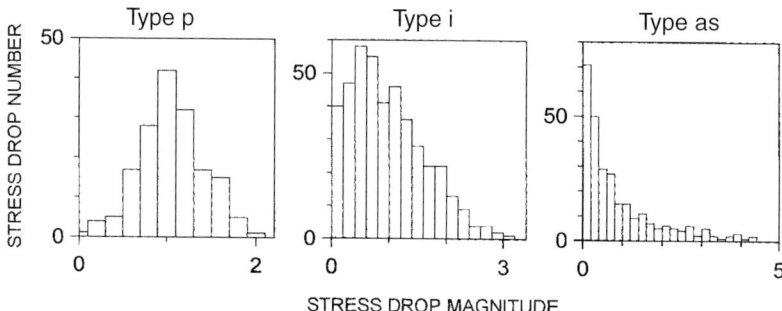

FIG. VI.2. Examples of 'stress-drop' histogram shapes for room temperature deformation:[249] (a) polycrystalline specimen, $\dot{\varepsilon} = 1.3 \times 10^{-5}/s$ (type p); (b) single crystal, $\dot{\varepsilon} = 1.3 \times 10^{-5}/s$ (type i); (c) polycrystalline specimen, $\dot{\varepsilon} = 6.1 \times 10^{-4}/s$ (type as).

---

[248] M. A. Lebyodkin, Y. Brechet, Y. Estrin, and L. P. Kubin, *Phys. Rev. Letts.* **74**, 4758 (1995).

[249] M. Lebyodkin, L. Dunin-Barkowskii, Y. Brechet, Y. Estrin, and L. P. Kubin, *Acta mater.* **48**, 2529 (2000).

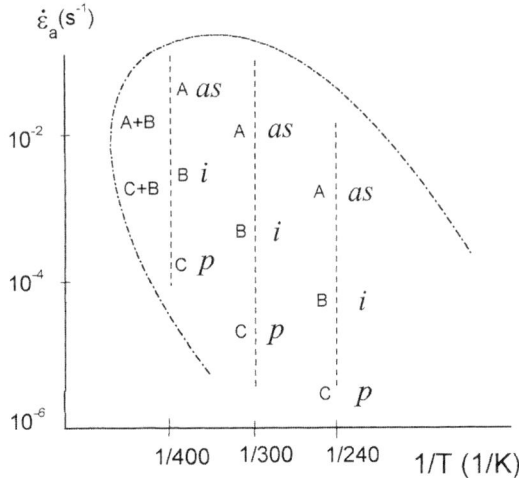

FIG. VI.3. Classification of the types of serrations and stress drop statistics in an ($\dot{\varepsilon}$, 1/T) map.[249] The boundaries between various regions are drawn in a semi-schematic way. The area where the PLC effect occurs is delineated by the horseshoe-shaped curve.

The parameters to be considered in a first approximation as controlling the characteristics of the PLC effect are the strain rate and temperature. It is convenient to map the domain in which serrated yielding occurs in terms of the type of instability and the type of stress-drop distribution. An example for the Al-Mg system is shown in Figure VI.3.

In the following sections we will consider two types of models developed to describe this rich phenomenology. The first (Section VI.17) relies on a phenomenological description, and considers as essential ingredients the negative strain-rate sensitivity and the spatial coupling between slices of material deforming at different rates. The second model (Section VI) delves deeper into the detailed dynamics of dislocations responsible for the phenomena.

17. STATISTICAL ANALYSIS OF SERRATED FLOW: AN EXAMPLE OF SELF ORGANIZED CRITICALITY (SOC)

The existence of a negative strain-rate sensitivity, which is the origin of the PLC instabilities, is reminiscent of the behavior of a dynamic friction coefficient which decreases when the sliding velocity increases. This behavior itself generates a similar type of instability during the sliding of two solids, the 'stick-slip'

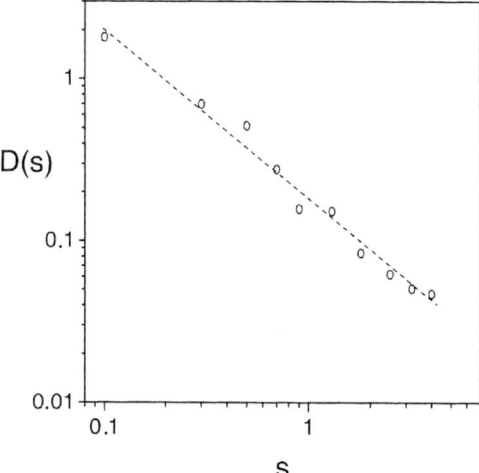

FIG. VI.4. The distribution density function $D(s)$ for an Al-Mg single crystal deformed at $120\,^\circ\text{C}$ and $\dot{\varepsilon} = 1.3 \times 10^{-5}/s$.[249] The dotted line corresponds to the exponent $\alpha \approx 1$ in the dependence $D(s) \approx s^{-\alpha}$.

instability.[250–252] This sliding instability is characterized by a power law distribution for the jump in the force/displacement curve. Similarly, the distribution of stress drops, when the distribution is asymmetrical, also exhibits a power law behavior as shown in Figure VI.4.

A simple model for the power law behavior in stick-slip has been proposed in the physics community: a set of blocks coupled by springs, with an appropriate friction law.[250–252] By analogy a simple model has been proposed for the PLC effect:[252,253] the material is 'sliced' into layers which may each deform at different rates. The relation between the local stress and the strain-rate is an N shaped curve which is obtained from the theory of dynamic strain aging (Figure VI.5). The local stress is the applied one corrected for 'incompatibility stresses' stemming from the difference in local strains. The equivalent of the position in Langer's 'stick-slip' model is the strain, and the equivalent of the friction law is the N-shaped stress-strain-rate curve. The equation governing the strain in each slice (ε_i) can be written:

$$\sigma = h\varepsilon_i + F(\dot{\varepsilon}_i) - K\big[(\varepsilon_{i-1} - \varepsilon_i) + (\varepsilon_{i+1} - \varepsilon_i)\big]. \tag{6.1}$$

---

[250] R. Burrige and L. Knopoff, *Bull. Seis. Soc. Amer.* **57**, 341 (1967).
[251] J. M. Carlson and J. S. Langer, *Phys. Rev. Letts.* **62**, 2632 (1989).
[252] J. M. Carlson and J. S. Langer, *Phys. Rev. A* **40**, 6470 (1989).
[253] M. Lebyodkin, Y. Brechet, Y. Estrin, and L. Kubin, *Acta mater.* **44**, 4531 (1996).

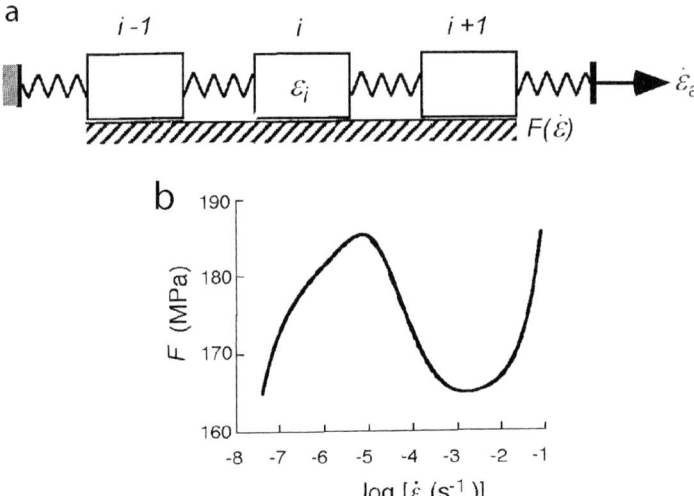

FIG. VI.5. (a) Schematic illustration of the coupled block model sliding with a strain-rate dependent friction ($F(\dot{\varepsilon})$) and deformed with a constant total imposed strain rate $\dot{\varepsilon}_a$. (b) The viscoplastic Penning's function $F(\dot{\varepsilon})$ for an Al-Mg alloy.[226] In the intermediate range of strain rates, the portion of the negative slope of F describes the negative strain-rate sensitivity associated with the PLC effect.

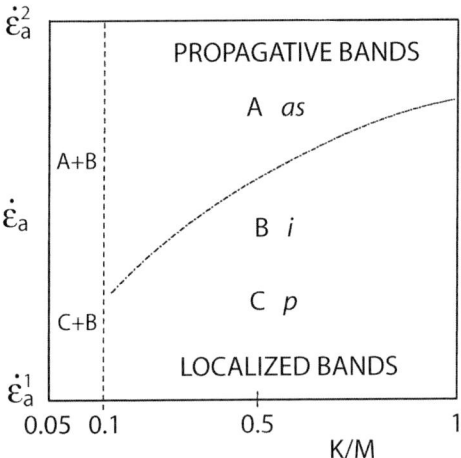

FIG. VI.6. Schematic phase diagram of the computer model behavior illustrating the types of plastic localization and stress drop statistics.[249] The dashed line separates domains of different spatial nature of strain localizations. $\dot{\varepsilon}_a^1$ and $\dot{\varepsilon}_a^2$ denote the lower and upper bounds of the applied strain rate interval where the PLC effect occurs.

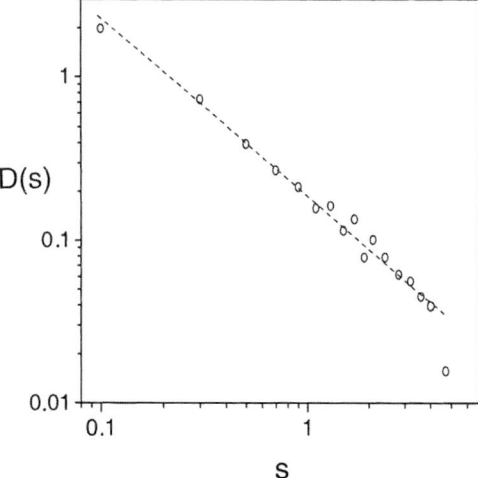

FIG. VI.7. The normalized distribution $D(s)$ of stress drops for a simulated deformation curve.[249] The dotted line corresponds to the exponent $\alpha \approx -1$ in the dependence $D(s) \approx s^{-\alpha}$.

The first term is the work hardening of the slice, the second describes the strain-rate sensitivity, and the third term describes the spatial coupling.

This model should be able to describe not only the power law behavior in the stress drops, but also the conditions under which the PLC effect is observed (in terms of applied strain rate and test temperature). It should also predict the type of spatial localization expected. The model is solved numerically, and indeed presents the same variety of behavior as observed experimentally. When the strain-rate is increased, the stress drop distribution changes from bell-shaped to a continuously decreasing one (cf. Figure VI.2). The asymmetric distribution is indeed a power law, and the exponent depends on the coupling constant between the blocks (slices of material). The value of this coupling constant is of the order of the elastic modulus and confirms the interpretation of the coupling as resulting mainly from elastic incompatibilities. The plastic localization which is observed in the numerical simulation also evolves with increasing applied strain-rate, from spatially uncorrelated (Type C) to spatially correlated, hopping or propagating (Type B and A).

The band structure predicted in each of the three cases, in relation with the stress strain curve is shown in Figure VI.8.

The velocity of the propagating band has also been computed with this simple model and it is found to exhibit a non-monotonic behavior.

Within this framework, the statistical behavior of the PLC instability, namely the existence of a power law distribution of stress drops can be seen, in certain

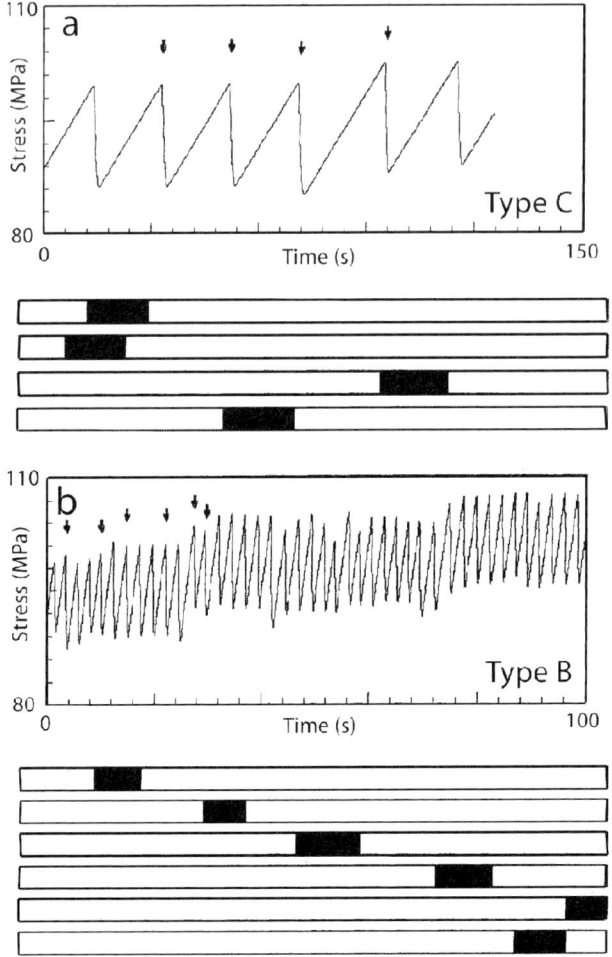

FIG. VI.8. Stress versus time curve and the attendant deformation band patterns for (a) $\dot{\varepsilon}_a = 10^{-5} s^{-1}$, (b) $\dot{\varepsilon}_a = 5 \cdot 10^{-5} s^{-1}$ and (c) $\dot{\varepsilon}_a = 4 \cdot 10^{-4} s^{-1}$.[253] (*Continued on the next page.*)

domains, as an example of Self Organized Criticality.[22,254–256] However, some other experimental findings suggest another type of analysis.

---

[254] P. Bak, C. Tang, and K. Wiesenfeld, *Phys. Rev. Letts.* **59**, 381 (1987).
[255] P. Bak, C. Tang, and K. Wiesenfeld, *Phys. Rev. A* **38**, 364 (1988).
[256] P. Bak, *How Nature Works: The Science of Self-Organized Criticality*, Oxford University Press, Oxford, England (1997).

DEFECT-INDUCED DYNAMIC PATTERN FORMATION IN METALS AND ALLOYS 281

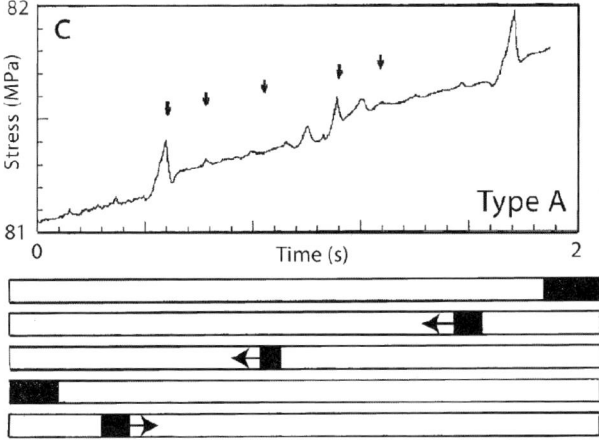

FIG. VI.8. (*Continued.*)

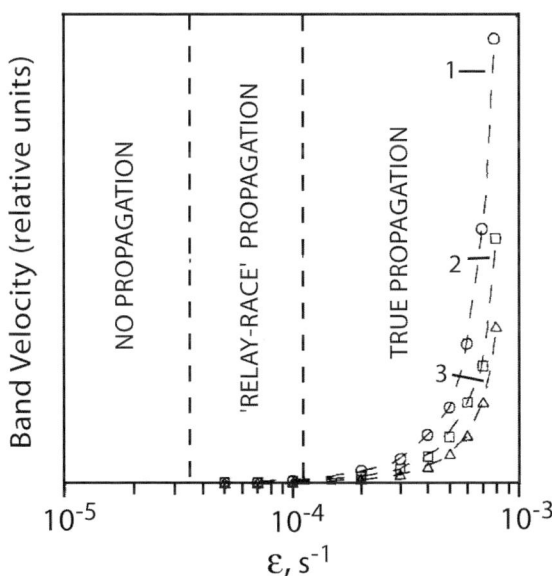

FIG. VI.9. Dependence of the band propagation velocity on the imposed strain rate for three values of the coupling constant for $h = 10^3$ MPa. Curve 1: $k = M$, Curve 2: $k = 0.5M$ and Curve 3: $k = 0.27M$. A demarcation line between the regions of the true and apparent (relay-race) propagation is shown.[253]

18. PORTEVIN–LE CHATELIER EFFECT: AN EXAMPLE OF DETERMINISTIC CHAOS

The details of the stress strain curve appear to be 'noisy' but what is the structure of this noise? As detailed in Ref. [226], this curve is considered as a scalar time series on which the standard techniques of chaos analysis are applied. From a series of N values of stress taken at regular intervals of time Δt, $\{\sigma(k\Delta t), k = 1$ to $N\}$, one constructs a vector $X(k)$ of M values corresponding to a sampling every $L\Delta t$, starting from the point $\sigma(k\Delta t)$, $X(k) = \{\sigma(k + nL)\Delta t, n = 0$ to $M − 1\}$. The mapping $X(k + 1) = F(X(k))$ in the M-dimensional space is known as the 'reconstructed attractor'. The embedding dimension M is *a priori* unknown. If M is smaller than the space in which the 'true attractor' lives, two points which have different orbits in the 'true attractor' will seem to be on the same orbit on the 'reconstructed one'. This will influence the correlation function, $C(r)$ namely the fraction of vectors $X(k)$ in the M dimensional space which are closer than a distance r. When we have a fractal attractor, $C(r)$ shows a power law behavior $C(r) = r^\nu$. The correlation dimension ν depends on the dimension M of the space in which the reconstructed attractor is built. When this space is made larger and larger (M increasing), ν tends towards a limiting value. The value of M above which the correlation dimension ceases to evolve indicates the number of variables necessary to describe the deterministic chaos.

This method[257] has been applied to experimental data on Portevin–Le Chatelier serrated flow. The dimension of the attractor, as obtained from the correlation function in the phase space, is shown to converge toward a value of 3.2 when the dimension M is increased beyond 5 (Ref. [258], Figure VI.10).

This indicates that indeed the behavior of the time series studied is an example of deterministic chaos, and that no more than 5 variables are required to analyze this behavior. Well before this experimental proof of deterministic chaos in PLC (serrated yielding) was given, a model was proposed by Ananthakrishna,[259–264] with evolution equations for the densities of mobile dislocations, of immobile dislocations and of dislocations interacting with impurity clouds, to which an equation for the stress was added. These four variables indeed show chaotic behavior.

---

[257] F. Takens, *Dynamical Systems and Turbulence*, Springer-Verlag (1981).
[258] S. J. Noronha, G. Ananthakrishna, L. Quaouire, C. Fressengeas, and L. P. Kubin, *Int. J. Bifur. Chaos* **7**, 2577 (1997).
[259] G. Ananthakrishna and M. C. Valsakumar, *J. Physics D* **15**, 171 (1982).
[260] G. Ananthakrishna and M. C. Valsakumar, *Phys. Letts. A* **95A**, 69 (1983).
[261] G. Ananthakrishna, *Solid State Phenomena* **3–4**, 357 (1988).
[262] G. Ananthakrishna, *Solid State Phenomena* **23–24**, 417 (1992).
[263] G. Ananthakrishna, *Scripta metall. mater.* **29**, 1183 (1993).
[264] M. S. Bharathi, S. Rajesh, and G. Ananthakrishna, *Scripta mater.* **48**, 1355 (2003).

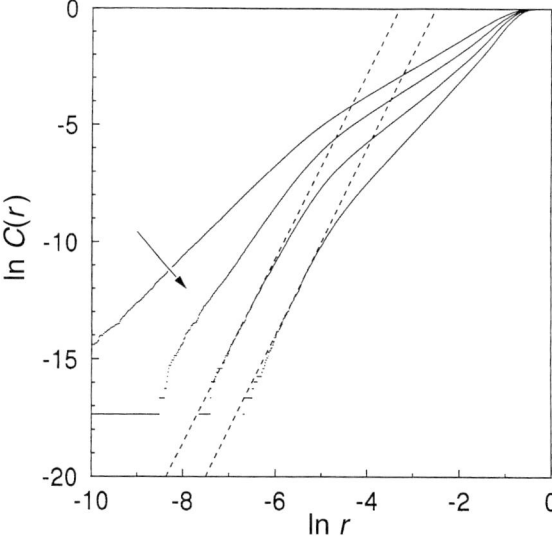

FIG. VI.10. Log-log plot of the correlation integral $C(r)$ as a function of distance r in the phase space.[258] Note the convergence of the slope to 3.2 (dotted line) as the embedding dimension is increased (arrow) from two to five.

A systematic statistical analysis of experimental data has shown that, depending on the loading conditions, the stress drops showed either deterministic chaos, or power law behavior, the transition being the two regimes being characterized by a multi-fractal behavior.[265–269]

The attractor reconstructed from experimental data on Al-Mg polycrystals, and the prediction of the model proposed by Ananthakrishna, shown in Figure VI.11 shows striking similarities, indicating that the non-linearities in this set of equations are relevant to capture the physics of the problem.[270]

---

[265] G. Ananthakrishna, C. Fressengeas, M. Grosbras, J. Vergnol, C. Engelke, J. Plessing, H. Neuhauser, E. Bouchaud, J. Planes, and L. P. Kubin, *Scripta metall. mater.* **32**, 1731 (1995).

[266] G. Ananthakrishna, S. J. Noronha, C. Fressengeas, and L. P. Kubin, *Phys. Rev. E* **60**, 5455 (1999).

[267] G. Ananthakrishna, S. J. Noronha, C. Fressengeas, and L. P. Kubin, *Mat. Sci. and Eng. A* **309**, 316 (2001).

[268] M. Lebyodkin, C. Fressengeas, G. Ananthakrishna, and L. P. Kubin, *Mat. Sci. and Eng. A* **319**, 170 (2001).

[269] M. S. Bharathi, M. Lebyodkin, G. Ananthakrishna, C. Fressengeas, and L. P. Kubin, *Phys. Rev. Letts.* **8716**, (2001).

[270] S. Kok, M. S. Bharathi, A. J. Beaudoin, C. Fressengeas, G. Ananthakrishna, L. P. Kubin, and M. Lebyodkin, *Acta mater.* **51**, 3651 (2003).

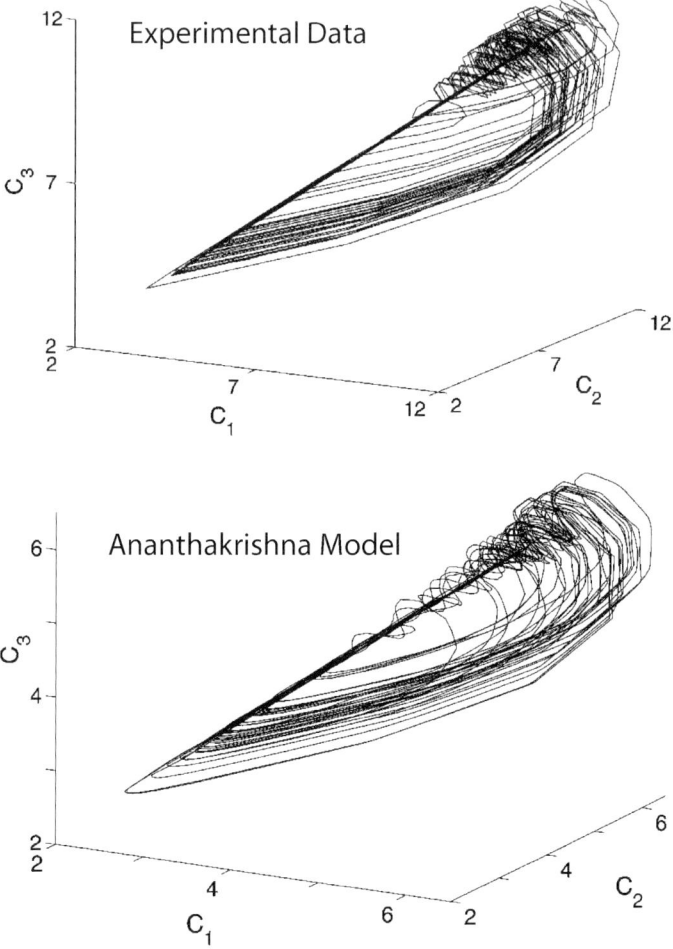

FIG. VI.11. Type B bands. Strange attractor from experiment (a) and from model (b) in a 3D space.[270] Note the striking similarity.

The spatial features of the chaotic regime need further investigation to be fully understood, as well as the mechanisms for the cross over between chaos and SOC with increasing strain rate.[270–272]

---

[271] G. Ananthakrishna, M. S. Bharathi, et al., *Phys. Rev. E*, in press (2004).
[272] P. Hähner, A. Ziegenbein, E. Rizzi, and H. Neuhauser, *Phys. Rev. B* **65** (2002).

VII. Concluding Remarks

As we bring this chapter to a close, we wish to stress some features of the patterning problems we have examined which would certainly benefit from the input of the physics community. However, before these 'open problems' are outlined, we want to insist on the necessity to incorporate in any modelling those specific features which are experimentally well established. Non-linear physics has justly shown that some universal behavior can be derived without knowledge of the precise details of the equations involved and of the microscopic mechanisms. Bifurcation theory, in both its linearized and non-linear (amplitude equations) versions and Chaos theory have shown that it is possible to generate a large variety of behavior with a limited number of ingredients. Similarly, Self Organized Criticality has been claimed to provide a very general framework. However, as outlined in the previous examples, some specific features of the problems have to be incorporated: the N shape of the stress/strain rate curve in Portevin–Le Chatelier modelling brings some new features to the richness of the observed behavior and the fact that the lamellar structure in discontinuous precipitation involves two different driving forces depending on the nature of the interface has important consequences on the removal of the velocity/spacing degeneracy. The analogies between problems (PLC and solid friction, discontinuous precipitation and eutectic solidification, dislocation self-screening and ion screening in a 2D coulomb gas) have both their strengths and weaknesses. Part of the challenge in applying the methods of non-linear physics is to identify the necessary ingredients provided by a detailed knowledge of the underlying mechanisms, without being overflowed by too many details.

In the problems of patterning of lamellar structures in the solid state, a key issue is to identify precisely the differences (and implication) between the solid state structures and those resulting from solidification. One of the aspects certainly lies in the difference between solids and liquids, both in their ability to sustain shear, and in the requirements imposed by crystallography. Misfit stresses may modify the driving forces, but may also drive the details of the geometry of the pattern. The crystallography of the interface determines both its energy and its mobility. In a crystalline solid, depending on the driving force, an interface can move either by lateral propagation of steps, or by normal growth. This behavior is probably associated with the 'stop and go' motion which has been described in Section III.7.a, but the macroscopic consequences on the patterning are still to be explored. Another feature is of course the fact that, not a single wavelength but a range of wavelengths is experimentally observed. The limited data available in the literature indicates a scaling behavior for the population of spacings observed. These results are still waiting to be modelled. Another feature, also related to the spacing selection problem, is the transition between the patterns which are nucleated and those which are growing, or the adjustment mechanisms when a change in the transformation conditions requires a modification of the interlamellar spacing (pattern repeat distance). Are the mechanisms operating in eutectic spacing selection still valid in systems where the anisotropy of the interface properties would make difficult any 'tilt wave'? On the whole, patterning in phase transformations in the solid state is nevertheless on relatively solid grounds, in the sense that the relevant equations are available, and agreed upon by the community.

Comparatively less advanced is the spatial patterning of structural defects. In spite of decades of experimental work and many attempts to developed models for these problems, the state of 'irradiation patterning of dislocations' outlined in Section V.14, where there is a relative consensus on the equations to be solved, is far from general. The role of long range stresses seems to be central to many of these 'structural patterning' problems. It seems crucial to explain how random nucleation of voids can lead, above a critical density, to a well organized lattice. The case of dislocation patterning in plasticity is far from being solved. The 're-action diffusion' approaches have shown their limitations, due to the difficulty in mimicking a glide process by a diffusion-like description. The energetic approaches, as long as they ignore the fact that the system is maintained far from equilibrium, are unlikely to provide further contributions. Both the short range interactions and the long range interactions must be accounted for. A key issue seems to be the self-screening of dislocations. Like in ionic solutions, one could envisage a population of dislocations as a set of positive and negative entities, in principle able to screen each others field. Due to the specific forces between dislocations, it would be a sort of 'Coulomb Gas' with logarithmic interaction energies. But unlike ionic solution, this screening cannot be obtained from standard statistical physics: the motion of dislocations is mainly restricted to glide in certain directions, their number is not conserved since they can annihilate by pairs, and since they are generated, not spontaneously, but by source mechanisms. A proper theory of screening under such restrictions is still to be developed. Another issue put forward recently, which brings a new light to these problems, is a systematic method to account for fluctuations in plasticity. Already these approaches have led to interesting results (e.g., Refs. [133,206]). The incorporation of these ingredients together with the local events which control both the pinning and the annihilation and multiplication of dislocations is a promising route. It is likely that the intensive computer simulations developed in the last decade for dislocation dynamics will have to incorporate these stochastic features to overcome their current limitations.

In the cases we have reported, spatio-temporal patterning in plasticity known as the Portevin–Le Chatelier phenomena is specially encouraging and interesting. The set of equations at the level of dislocation dynamics can be seen as sufficiently consensual to provide a realistic description of macroscopic data such as the critical strain to observe the phenomenon of serrated flow. A continuum description of band propagation has been proposed, so the 'simplified regular PLC effect' can be said to be reasonably well understood. When one delves deeper into the details of the statistics of stress jumps, two regimes have been identified, corresponding respectively to a type of 'Self Organized Criticality' with well defined scaling laws, and to a regime of deterministic chaos, corresponding to a correctly reproduced 'strange attractor'. However, the transition from a regime in which an infinite number of variables is required to describe the statistics, to a regime where a finite and small number of variables is sufficient, is still not understood. The phenomenon being well studied experimentally and the models to describe it at a microscopic level being reasonably safe, it could be very promising to use this problem as a prototype for understanding the transition from deterministic chaos to SOC.

In any case, the variety of the patterning phenomena observable in the solid state, and associated with structural defects is certainly a driving force for new modelling approaches which could be both respectful of the underlying mechanisms but not obsessed by them.

Author Index

Numbers in parentheses are reference numbers and indicate that an author's name is not cited in the text.

A

Aaronson, H. I., 196(42), 198(53)
Aarts, J., 2(2)
Abdou, S., 219(88)
Abrahams, E., 13(46), 73(195), 160(374)
Abrikosov, A. A., 9(28)
Acker, F., 34(106)
Ackermann, F., 267(216)
Adams, G., 67(184)
Adelmann, P., 12(35), 174(410)
Adrian, H., 60(168)
Adroja, D. T., 119(274)
Aeppli, G., 62(177), 91(218), 119(274)
Aifantis, E. C., 261(175), 266(207), 270, 271
Albessard, A. K., 17(64), 20(72, 73)
Alexander, M., 174(410)
Allen, J. W., 17(66, 67), 50(138)
Amici, A., 61(174)
Amir, Q. M., 212
Amodeo, R. J., 263(188, 189)
Ananthakrishna, G., 268(226), 282, 283, 284(271)
Anderson, M. P., 235(118–120)
Anderson, P. W., 59(160), 108(257), 125(286)
Andersson, P. H., 50(141)
Andres, K., 2
Anisimov, V. I., 48(132), 125(290), 131(303)
Antonopoulos, J. G., 267(213)
Antonov, V. N., 13(45), 97(239), 125(290)
Aoki, H., 17(64), 20(72, 73), 74(197), 79(202), 85(210, 211), 86(213), 87(215), 90(217)
Appel, J., 59(152)
Argon, A. S., 261(174)
Arko, A. J., 50(141)
Armstrong, P. E., 17(66, 67)

Asamitsu, A., 114(266)
Asamitsu, Z., 108(255)
Ashby, M. F., 261(174)
Aso, N., 49(136, 137), 60(169, 170)
Assmus, W., 46(126)
Asundi, M. K., 193(28)
Avella, A., 6(23)
Avlonitis, M., 266(207)

B

Baas, J., 94(232)
Bak, P., 280(254–256)
Balatsky, A. V., 59(163)
Balents, L., 146(327, 330)
Balluffi, R. W., 234(110)
Barbu, A., 249(140, 141)
Bareiter, S., 135(311)
Barker, I., 256(163)
Basinski, S. J., 268(222), 270(227)
Basinski, Z. S., 268(222), 270(227)
Baumann, S. F., 196(41)
Bay, B., 255(160)
Bay, K., 258(165), 266(208), 267
Beaudoin, A. J., 283(270)
Beck, H., 91(219)
Bedell, K. S., 160(375)
Bednorz, J. G., 2(5)
Bellon, P., 184(10, 11)
Beni, G., 92(222)
Benner, H., 87(215)
Benoist, P., 255(159)
Benzerga, A. A., 265(202)
Berge, P., 182(2)
Berk, N. F., 33(97)
Bernardini, J., 193(34)
Bernert, A., 98(241), 99(242, 244)
Bernhoeft, N., 49(136), 60(170), 62(178), 66
Bernhoeft, N. R., 34(108)

Bertrand, G., 183(9)
Bethe, H., 8, 85
Betouras, J., 140(319)
Bharathi, M. S., 282(264), 283(269, 270), 284(271)
Billinge, S. J. L., 109(258)
Binder, K., 185(14)
Biroli, G., 159(368)
Blaha, P., 127(295)
Bley, F., 243(138)
Bogdanov, P. V., 154(357)
Bogel, A., 198(65), 219, 220
Bolze, G., 198(62)
Bonitz, M., 67(185)
Bonner, J. C., 91(219), 93, 130(302)
Bonsall, L., 70(189)
Borsa, F., 126(291)
Bouchaud, E., 283(265)
Boucher, J. P., 103(247)
Bougourzi, A. H., 79(204)
Bozorth, R. M., 34(107)
Brandt, W. H., 198(45), 203
Brechet, Y., 184(12), 212(82), 215(85), 232(105–107), 235(124, 125), 239(129, 130), 242(133), 243(135–138), 249(142), 264(192), 265(202), 275(248, 249), 277(253)
Brechet, Y. J. M., 186(18), 198(66)
Bredl, C. D., 2(2)
Brener, E. A., 182(4)
Brimhall, J. L., 249(144)
Brodsky, M. B., 34(105)
Broholm, C., 91(218)
Bross, S., 266(205)
Brown, D., 209(79)
Brown, J. D., 95(234)
Brown, L. M., 267(213), 268(221, 225)
Brugger, T., 12(35)
Bruls, G., 24(84)
Bucher, E., 59(156, 157)
Budnick, J. I., 34(104)
Bullough, R., 249(152), 253
Bulut, N., 159(370)
Bünemann, J., 164(395)
Burke, J. E., 235(113)
Burrige, R., 277(250)
Butler, E. P., 196(38)
Butterfield, M. T., 50(141)

C

Cahn, J. W., 198, 203–206, 210, 212, 213, 215, 218, 220, 222, 230, 231, 235(116)
Caliebe, W., 113(265)
Callen, E., 125(288)
Canals, B., 131(305)
Canfield, P. C., 17(66)
Canova, G., 263(185), 264(190–192)
Carbotte, J. P., 155(365)
Carlson, J. M., 277(251, 252)
Carpy, A., 92(227)
Cavadini, N., 38(115)
Ceperley, D. M., 71(191), 73(193)
Chaboussant, G., 119(274)
Chaffron, L., 184(11)
Chandrasekhar, S., 182(1)
Chatterji, T., 94(229), 99(244), 103(247), 109(259), 113(265), 119(272, 273, 275, 276), 120(278)
Chaudhari, P., 161(376)
Cheetham, D., 193(29, 31), 207
Cheynet, M. C., 212(82)
Chi, E. O., 113(264)
Chihab, K., 274(245, 247)
Choi, H. S., 113(264)
Choi, S., 225(92, 94)
Christensen, N. E., 20(71), 172(405)
Chubukov, A. V., 163(381)
Chuk, T., 154(358)
Cichorek, T., 22(79), 39(119), 74(197), 86(213)
Claessen, R., 164(395)
Cleveringa, H. H. M., 265(198)
Cliff, G., 225(93)
Cline, H. E., 229(102)
Coldea, R., 38(117)
Coleman, P., 39(120)
Collins, R. T., 161(376)
Colten, R. A., 209(77)
Condat, M., 264(192)
Continentino, M. A., 32(94), 45(125)
Cooper, B. R., 59, 62(175), 66, 70
Cottrell, A., 273(239)
Courtney, T. H., 227(99)
Cox, D. L., 15(52)
Craco, L., 48(135), 131(304)
Cross, J. O., 138(315)
Cross, M. C., 92(224), 102

D

Cullen, J. R., 125(288)
Custers, J., 39(120), 74(197), 86(213)
Czjzek, G., 12(35)
Czycholl, G., 162(378)

D

Dagotto, E., 122(281)
Daliachaouch, Y., 17(66)
d'Ambrumenil, N., 75(198), 85(212)
Darken, L. S., 193(26)
Das, A., 193(32, 33)
de Boer, P. K., 110(260)
de Groot, R. A., 110(260)
de Visser, A., 24(84), 47(130)
deBoer, J. L., 94(232)
DeGroot, S. R., 188(24)
Delin, A., 50(141)
Dender, D. C., 91(218)
Denlinger, J. D., 17(66, 67), 21
des Cloizeaux, J., 79
Devereaux, T. P., 154(358)
Devincre, B., 264(192, 193, 196, 197)
Dhalenne, G., 113(265), 119(272, 273, 275), 120(278)
Dho, J., 113(264)
Differt, K., 261(177, 178), 270
Dischner, M., 97(240), 105(252)
Dmitriev, D. V., 91(221)
Doherty, R. D., 240(131)
Doniach, S., 14(49), 33(98), 39, 40(121)
Donnevert, L., 23(82)
Dönni, A., 60(172)
Dooryhee, E., 138(315)
Driver, J., 268(223, 224)
Dubuisson, P., 249(142)
Ducastelle, F., 164(382, 383)
Duly, D., 212, 215, 220
Dunin-Barkowskii, L., 275(249)
Durakiewicz, T., 50(141)
Dzyaloshinski, I. E., 9(28)

E

Eberhardt, W., 147(332)
Ebihara, T., 20(72)
Eder, R., 159(372)
Edington, J. W., 205(69)
Edwards, D. M., 164(390, 391)
Efremov, D. V., 57(147)
Ehm, D., 16(55), 164(395)
Ehrenreich, H., 153(340, 343)
Ehrhart, P., 249(143, 151)
ElBoragy, M., 219(88)
Elgazzar, S., 51(142)
Elliott, R. J., 153(342)
Eloirdi, R., 171(403)
Endoh, Y., 60(169, 170, 172), 113(263), 117(271), 121(279)
Engelke, C., 283(265)
Engelsberg, S., 33(98)
Enz, C. P., 33(99)
Eriksson, O., 50(141)
Erkelens, W. A. C., 20(70)
Eschrig, H., 13(45), 97(239)
Eschrig, M., 155(362, 364)
Essmann, U., 261(177, 178), 267(214, 215), 270
Estrin, Y., 261(181), 263(185), 274(240, 241, 245, 247), 275(248, 249), 277(253)
Evans, J. H., 249(145–147, 149)
Eyert, V., 127(296)
Eyre, B. L., 249(148, 152)

F

Fagot-Revurat, Y., 95(236)
Fak, B., 90(217)
Falakshahi, H., 71(192)
Falicov, L. M., 164(393)
Farrell, K., 249(150)
Faulhaber, E., 22(79)
Fauth, F., 84(209), 138(315)
Fay, D., 59(152)
Fedorov, A. V., 155(363)
Feild, C., 161(376)
Fertig, W. A., 59(157)
Filinov, A. V., 67(185)
Fink, J., 174(410)
Firsov, Y. A., 115
Fischer, M., 96(237)
Fischer, Ø., 61(173)
Fischer, P., 60(171)
Fisher, D. S., 92(224)
Fisher, M. E., 93, 102, 130(302)
Fisher, M. P. A., 146(327, 330)
Fisher, R. M., 193(26)

Fisk, Z., 2(3, 4), 17(66, 67), 34(106)
Flandin, L., 243(135)
Fleck, N. A., 265(201)
Flouquet, J., 20(70), 47(130)
Fournelle, R. A., 227, 231
Fradkin, E., 140(320)
Franek, A., 272(233)
Franz, W., 2(2)
Fratzl, P., 185(14, 15)
Freeman, A. J., 34(105)
Freericks, J., 154(352)
Fressengeas, C., 268(226), 274(243, 244), 282(258), 283(265–270)
Freye, D. M., 39(118)
Fridberg, J., 222(89, 90)
Friedel, J., 6, 81
Friedman, L. H., 264(195)
Fröhlich, H., 59
Frye, J. H., 209(78)
Fuchizaki, K., 235(121), 237
Fuji, Y., 94(233), 103(248, 249)
Fujimori, A., 6(22), 108(254), 127(294), 135(312)
Fujimori, S., 50(139), 171
Fujimoto, S., 132(309)
Fujioka, H., 117(271), 121(279)
Fujiwara, K., 127(297, 298)
Fukuyama, H., 70(188)
Fulde, P., 5(14), 9(29), 11(31, 33), 12(36), 13(41, 42, 45), 17(62), 27(88, 89), 35(109), 46(127, 128), 49(137), 53(144), 57(147), 59(153–155, 157), 61(174), 63(180), 74(196), 75(198), 81(205), 85(212), 87(216), 97(239), 99(242–244), 129(299), 136(313), 137(314), 140(318, 319), 145(323), 146(324), 154(346, 353–356), 164(396, 397), 165(401), 174(407, 408)
Furrer, A., 38(115)
Furukawa, N., 119(277), 139(316)

G

Gajewski, D. A., 126(291)
Galy, J., 92(227)
Garst, M., 46(129)
Gaskell, T., 149
Gebhard, F., 164(395)

Geerk, J., 61
Gegenwart, P., 23(82, 83), 39(119, 120), 74(197), 86(213), 90(217)
Geibel, C., 16(55), 22(79), 23(82, 83), 39(119, 120), 49(137), 60(171), 94(230), 95(235), 96(237), 97(238, 240), 105(252)
Georges, A., 69(186)
Gerhardt, W., 34(103)
Ghoniem, N. M., 253, 263(188, 189)
Girvin, S. M., 146(327)
Gogolin, A. O., 12(37)
Goldman, A. I., 126(291)
Goremychkin, E. A., 22(77)
Gorkov, L. P., 9(28)
Gosele, U., 267(215)
Goto, T., 76(199), 78(201)
Gottstein, G., 235(117)
Gottwick, U., 20(69)
Gouder, T., 171(403)
Graebner, J. E., 2
Graf, J., 162(380)
Grauel, A., 60(171)
Greer, A. L., 232(108, 109)
Gregor, V., 272(236)
Grenier, B., 103(247)
Grest, G. S., 235(118, 119)
Grewe, N., 20(69)
Grin, Y., 94(229), 129(299)
Grinberg, A., 274(246)
Gröber, C., 159(372)
Groma, I., 243(135)
Gros, C., 97(238), 105(250)
Grosbras, M., 283(265)
Grosche, F. M., 39(118)
Grüner, G., 69(187)
Gu, G., 154(357)
Gu, G. D., 155(363)
Gu, Q., 40(122)
Gu, T., 126(291)
Güdel, H. U., 38(115)
Güntherod, G., 96(237)
Gupta, S. P., 212
Gust, W., 193(32–35), 196(36), 198(65), 202(68), 205(70–72), 211(80), 215, 219, 220, 235(125)
Gutzwiller, M. C., 6, 164
Guyot, P., 243(135–137)
Gweon, G. -H., 17(66, 67), 162(380)

H

Haase, R., 188(25)
Haayman, P. W., 3(8), 124
Habicht, K., 38(115)
Hackenberg, R. E., 197(44), 211(81)
Haen, P., 47(130)
Haga, Y., 49(136), 52(143), 65(181), 66(183), 77(200)
Hagel, W. C., 206(75), 222
Hähner, P., 258(165), 266, 267, 284(272)
Haken, H., 183(8)
Haldane, F. D. M., 129
Halperin, B. I., 71(190)
Hammar, P. H., 91(218)
Hanke, W., 159(371, 372)
Hansen, N., 255(160), 256(163), 260(170)
Haque, M., 73(194)
Harasawa, A., 164(395)
Harima, H., 127(297)
Harrison, N., 86(213)
Hase, M., 92(225)
Hasegawa, H., 108(257)
Hasegawa, T., 260(172)
Hasselmann, N., 57(147)
Hauptmann, R., 95(235), 96(237)
Hayden, S. M., 34(108)
Heeger, A. J., 12(38)
Heiniger, F., 59(156)
Heitler, W., 3
Held, K., 17(59)
Helfrich, R., 23(82), 85(211), 87(215)
Hermele, M., 146(330)
Hertz, J. A., 34, 164(390, 391)
Herzig, C., 193(34)
Hettler, M. H., 159(367, 369)
Hewson, A. C., 15(50)
Hidaka, H., 127(297)
Hiess, A., 49(136), 60(170), 103(247), 119(275)
Hillert, M., 198, 202, 203, 205–207, 209, 210, 212, 215, 218, 222, 225, 235(115), 236, 237
Hinkel, C., 135(311)
Hinma, T., 52(143)
Hino, O., 115(268)
Hinze, P., 23(83)
Hirai, K., 164(397)
Hirooka, S., 154(347)
Hirota, K., 113(263), 117(271), 121(279)
Hirsch, J. E., 59(158)
Hirst, L. L., 59(153)
Hirth, J. P., 264(194)
Hisazaki, Y., 77(200)
Hoang, A. T., 110(261)
Höck, K. -H., 127(296)
Hoffmann, J. U., 120(278)
Hohlwein, D., 120(278)
Holt, D. L., 256(162), 260
Holtzberg, F., 161(376)
Honig, J. M., 135(311)
Horibe, Y., 139(316)
Horn, R., 23(82)
Horn, S., 127(296)
Hornbogen, E., 198(56)
Horsch, P., 93(228)
Howe, L., 249(141)
Huang, C. Y., 34(106)
Hubbard, J., 2, 6–8, 14, 27, 32, 68, 69, 74, 85, 86, 97, 98, 100, 105, 108, 110, 111, 125, 130, 132, 144, 148, 151–155, 157–161, 163–165, 167, 169, 172, 174–177, 179
Huber, F., 171(403)
Hückel, E., 3
Hüfner, S., 16(55), 147(331)
Hughes, D. A., 240(131), 255(160)
Hull, F. C., 209(77), 234(112)
Humphreys, F. J., 186(21), 235(122, 123), 240(131), 242(132)
Hunderi, O., 236, 239
Hunt, J. D., 198(51)
Hur, K. L., 131(303)
Hur, N. H., 113(264)
Hüser, D., 34(103)
Hutchinson, C. R., 211(81)
Huth, M., 60(168)
Hybertsen, M. S., 172(405)

I

Igarashi, J., 164(387, 388, 396, 397)
Ihle, D., 125(289)
Ikeda, H., 154(357)
Ikeda, N., 139(316)
Illy, J., 262(183)
Imada, M., 6(22)
Inada, Y., 52(143)

Ishibashi, H., 139(316)
Ishida, K., 66(183)
Ishihara, S., 117(271), 154(359, 361)
Ishikawa, M., 61(173)
Isobe, M., 92(226), 94(233), 103(248, 249), 105(251)
Isoda, M., 124(285)
Iwasa, K., 11(32), 77(200), 79(202), 90(217)
Izawa, K., 66(182)

J

Jackeli, G., 109(259), 121(280)
Jackson, K. A., 198(51)
Jaclic, G., 42(123)
Jager, W., 249(143, 151)
Jaime, M., 86(213)
Janecek, M., 243(136, 138)
Jarrell, M., 154(352), 159(367, 369), 160(373)
Jayprakash, C., 16(56, 57)
Jeanclaude, V., 274(243, 244)
Jedrak, J., 56(145)
Jeevan, H., 22(79)
Jegoudez, J., 103(247)
Jensen, D. J., 240(131)
Jensen, H. J., 187(22)
Jensen, J., 35(110, 111), 62(176)
Jepsen, O., 94(229)
Jobst, A., 94(230)
Johnson, J. D., 87(214)
Johnson, P. D., 155(363)
Johnston, D. C., 123(282), 124(283, 284), 126(291)
Jonas, J. J., 240(131)
Jones, B., 16(58)
Jourdan, M., 60(168)
Joyce, J. J., 50(141)
Julian, S. R., 17(65), 20(74), 39(118)

K

Kadowaki, K., 76, 126
Kagami, C., 103(248)
Kaganov, M. I., 23(81)
Kaiser, A. B., 46(128)
Kajihara, M., 225(92, 94)
Kakehashi, Y., 154(344–346, 348, 349, 353, 355, 356), 165(401), 179

Kakizaki, A., 164(395)
Kakol, Z., 135(311)
Kakurai, K., 60(172), 103(248)
Kalmeyer, V., 146
Kalus, R., 272(233)
Kampe, J. C. M., 227(99)
Kampf, A. P., 152(336, 337)
Kampmann, R., 185(13)
Kanamori, J., 6, 164
Kanao, R., 115(268)
Karbach, M., 79(204)
Kasahara, Y., 65(181)
Kassner, M. E., 240(131)
Kastrinakis, G., 162(379)
Kasuya, T., 13(43), 17(60), 180
Kato, K., 139(316)
Katsnelson, M. I., 164(394)
Katsufuji, T., 133(310), 139(316)
Kaur, I., 193(35), 211
Kawabata, A., 14(48), 34(100)
Kawakami, K., 135(312)
Kawakami, N., 48(133)
Kawasaki, K., 235(121)
Kawatra, M. P., 34(104)
Keh, A. S., 255(161)
Keller, J., 9(29), 59(154, 155, 157)
Keller, S. A., 154(357)
Kendziora, C., 155(363)
Khachaturian, A. G., 185(17)
Khomskii, D. I., 100(245)
Kikuchi, M., 225
Kim, W. S., 113(264)
Kimura, N., 66(183)
Kimura, T., 113(262, 263), 119(274)
King, C. A., 20(75)
King, W. E., 240(131)
Kirkaldy, J. S., 198(55, 61, 62), 206(74), 208, 215
Kirkpatrick, S., 153(340)
Kishio, K., 154(357)
Kita, H., 60(172)
Kitaoka, Y., 15(51), 66(183)
Kleinmann, L., 165(398)
Klinger, L., 198(66), 215(85), 216, 218, 220, 232
Kmety, C. R., 109(258)
Knaupp, M., 94(229)
Knigavko, A., 154(360)
Knoester, J., 100(245)

Knopoff, L., 277(250)
Kobayashi, T., 127(298)
Kobayashi, T. C., 127(297)
Kocks, U. F., 260(172), 261
Koepernik, K., 51(142)
Koga, A., 48(133)
Kohgi, M., 11(32), 74(197), 77(200), 79(202), 80, 90(217)
Koike, Y., 49(136)
Koitzsch, A., 21(76)
Kok, S., 283(270)
Komatsubara, T., 49(136, 137), 60(169, 170, 172)
Kondakov, D. E., 48(132)
Kondo, S., 11, 14–17, 21, 32, 38–43, 45, 46, 74, 76, 79, 117, 120, 123(282), 126(291, 293), 132, 179
Kondoh, Y., 225(92)
Konno, R., 28(91)
Köppen, M., 87(215), 95(235)
Korbel, A. S., 270(227)
Korotin, M. A., 125(290), 131(303)
Koshizuka, N., 155(363)
Koslowski, A., 135(311)
Kotegawa, H., 127(297)
Kotliar, G., 69(186), 159(368), 164(394)
Kovacs, I., 262(183)
Kovrizhin, D., 38(117)
Koyama, I., 113(263)
Krämer, K., 38(115)
Krane, H. G., 94(230)
Kratochvil, J., 272
Krauth, W., 69(186)
Kremer, R. K., 94(229), 95(236), 97(240)
Krimmel, A., 60(171)
Krishan, K., 249(152), 253(153)
Krishna-murthy, H. -R., 16(56, 57)
Krishnamurthy, H. R., 159(367)
Krivnov, V. Y., 91(221)
Kroha, J., 16(55)
Krumhansl, J. A., 153(342)
Kubin, L., 277(253)
Kubin, L. P., 258, 261(176, 181, 182), 263(185, 186), 264(190–193, 196, 197), 267(216, 218, 220), 268(223, 224, 226), 274(240, 241, 245, 247), 275(248, 249), 282(258), 283(265–270)
Kubota, M., 117(271), 121(279)
Küchler, R., 39(119)

Kudasov, Y., 81(205)
Kuhlmann-Wilsdorf, D., 255(160), 260(168–171)
Kulcinski, G. L., 249(144)
Kumai, R., 113(262)
Kuramoto, Y., 15(51)
Kurushima, K., 139(316)
Kurz, W., 182(4)
Kusaba, T., 235(121)
Kusunose, H., 132(307)
Kuwahara, H., 114(266)

L

Laad, M. S., 48(135), 131(304)
Labbé, J., 81(208)
Lacerda, A., 47(130)
Lacroix, C., 11(34), 131(305), 132(306)
Landau, L. D., 7, 9–11, 13, 18, 32, 45, 47, 76, 87, 148
Lander, G. H., 49(136), 60(169, 170)
Lang, I. G., 115
Lang, M., 23(82, 83), 74(197), 86(213), 87(215), 90(217), 95(235)
Langari, A., 101(246)
Langer, J. S., 182, 277
Langhammer, C., 23(82, 83), 74(197)
Lanzara, A., 154(357, 359), 162(380)
Larche, F., 185(16)
Laughlin, R. B., 13(40), 146
Lauzeral, J., 253(158)
Le Chatelier, H., 184, 187, 273, 274, 282, 286
Leath, P. L., 153(342)
Lebowitz, J., 185(15)
Lebyodkin, M., 275(249), 277(253), 283(268–270)
Lebyodkin, M. A., 275(248)
Lee, B. -W., 17(66)
Lee, D. -H., 162(380)
Lejay, P., 20(70), 47(130)
Lemarchand, H., 182(7)
Lemmens, P., 96(237)
Lendvai, J., 243(135)
Leonov, I., 125(290), 136(313)
Lepinoux, J., 235(124, 125), 239(129, 130), 242(133), 261(176), 263(186, 187), 267(216, 218, 220), 268(223, 224)
Levin, M., 140(321)

Lhuillier, C., 146(329)
Li, J. Q., 113(262)
Li, Q., 155(363)
Li, Y. M., 75(198), 85(212)
Lichtenstein, A. I., 164(394)
Lieb, E., 36(112), 85
Liebsch, A., 48(134), 164(385, 386)
Liecke, W., 2(2)
Lifshits, I. M., 23(81)
Link, A., 23(82)
Littlewood, P. B., 13(46, 47), 160(374)
Liu, Y. C., 198(53)
Livet, F., 243(138)
Livingston, J. D., 230, 231
Loewenhaupt, R., 22(79)
Loh, E., 59(158)
Löhneysen, H. v., 34(108)
Loidl, A., 60(171), 127(296)
London, F., 3
Longinotti, L. D., 59(156)
Lonzarich, G. G., 17(63, 65), 20(74, 75), 34(108), 39(118), 59(164–166)
Lorenz, B., 125(289)
Lorenzo, J. E., 103(247)
Lorimer, G. W., 225(93)
Louat, N. P., 236, 237
Louchet, F., 184(12)
Lozovik, Y. E., 67(185)
Lu, D. H., 154(358)
Lu, E. D., 154(357)
Lüdecke, J., 94(230, 231)
Luengo, C. A., 59(157), 124(284)
Lühmann, T., 23(83), 38(117)
Luther, A., 13(44), 59(153)
Lüthi, B., 24(84), 46(126), 78(201), 135(311)
Luttinger, J. M., 10, 12, 17, 160

M

Machida, A., 133(310)
Machida, K., 28(92)
Mack, F., 93(228)
Mackintosh, A. R., 35(111), 62(176)
Madec, R., 264(197)
Maekawa, S., 117(271), 174(409)
Maezawa, K., 66(183)
Magnin, T., 268(223, 224)
Mahajan, A. V., 126(291)

Maier, T., 159(369)
Maier, T. A., 160(373)
Maitra, J. P., 59(156, 157)
Maki, K., 65(181)
Maletta, H., 60(171)
Malygin, G., 261(179, 180)
Mancini, F., 6(23)
Manna, I., 193(32), 196, 230
Maple, M. B., 17(66, 67), 59(157), 124(284), 126(291)
Maradudin, A. A., 70(189)
Marchionni, C., 184(12)
Marsiglio, F., 154(360)
Martin, G., 184(10, 11), 249(140, 141), 255(159)
Martin-Delgado, M. A., 101(246)
Mason, T. E., 62(177)
Massoud, J. P., 249(142)
Mathur, N. D., 39(118)
Matsuda, Y., 65(181), 66(182)
Matsui, Y., 113(262)
Matsumoto, H., 164(392)
Matsumoto, M., 38(116)
Matsuno, J., 127(294), 133(310)
Mattheiss, L. F., 127(294)
Matthias, B. T., 34(107)
Mattis, D., 36(112)
Mattis, D. C., 81(207)
Mattix, L., 59(157)
Maurice, C., 235(123)
Maxwell, E., 59(149)
Mayer, H. M., 20(69)
Mazey, D. J., 249(149)
Mazin, I. I., 127(295)
Mazur, P., 188(24)
McCallum, R. W., 59(157), 124(284)
McElroy, D. L., 209(78)
McEwen, K. A., 35(110)
McGovern, S., 267(217)
McHale, P., 63(180)
McIntyre, G. J., 113(265), 119(272)
McMahan, A. K., 17(59)
McMullen, G. J., 17(65), 20(74)
McMurry, S., 186(19), 235
McNelley, T. R., 240(131)
McQueen, H. J., 240(131)
Medrick, K., 165(398)
Meetsma, S. A., 94(232)
Mehl, R. F., 206(75), 209(77)

Mehring, M., 95(236)
Menovsky, A., 24(84)
Meschede, D., 2(2)
Metoki, N., 49(136)
Metzner, W., 154(350)
Meyrick, G., 196(40)
Michael, J., 196(41)
Mignot, J. -M., 11(32), 79(202), 90(217)
Miller, L. L., 123(282), 126(291)
Millis, A. J., 48(131)
Misguish, G., 146(329)
Mishin, Y., 193(34)
Mitchell, J. F., 109(258)
Mittemeijer, E. J., 193(33)
Miyake, K., 59(159), 132(307)
Miyake, K. ., 49(137)
Miyoshi, K., 127(297, 298)
Mizuta, M., 50(139)
Moessner, R., 146(326)
Monthoux, P., 59(163–166)
Moodenbaugh, A. R., 155(363)
Moore, D. P., 50(141)
Morales, L. A., 50(141)
Moreo, A., 122(281)
Mori, H., 150(334)
Mori, S., 124(285), 133(310), 139(316)
Moritomo, Y., 108(255), 113(263),
    114(266), 117(271), 121(279), 133(310)
Moriya, T., 14, 34, 59(162), 88, 179
Morré, E., 94(230), 105(252)
Mostovoy, M. V., 100(245)
Motome, Y., 139(316)
Motrunich, O. I., 146(328)
Mott, N. F., 2, 68, 69, 74, 98, 108, 111
Mughrabi, H., 262(184), 267(214–217, 219)
Mukherjee, M., 159(367)
Müller, G., 79(204), 91(219)
Müller, K. A., 2(5)
Müller-Hartmann, E., 48(135), 131(304),
    154(351)
Muller-Krumbhaar, H., 182(4)
Mullins, W. W., 218, 227, 228, 235(114),
    236
Murakami, Y., 94(233), 103(249), 113(263)
Muraoka, K. Y. J., 103(249)
Murata, K. K., 14(49)
Murphy, S. M., 253
Mutka, H., 38(115)
Mydosh, J. A., 34(103, 104), 39(119)

N

Nagaosa, N., 154(358, 359, 361)
Nakajima, K., 103(248)
Nakajima, N., 77(200)
Nakao, H., 94(233), 103(249)
Needleman, A., 264(195), 265(198–202)
Neef, M., 26(86, 87)
Nekrasov, I. A., 48(132)
Nelson, D. R., 71(190)
Nemoto, Y., 76(199)
Nersesyan, A. A., 12(37)
Nesbitt, L. B., 59(150)
Neuhauser, H., 283(265), 284(272)
Neumaier, K., 39(119, 120)
Nichols, F. A., 227(97)
Nicolay, G., 16(55)
Nicolis, G., 182(5)
Niksch, M., 46(126)
Ninomiya, E., 94(233), 103(249)
Nishi, M., 103(248)
Nishibori, E., 133(310)
Noda, Y., 94(233)
Nohara, M., 126(293)
Nohdo, S., 115(268)
Norman, M. R., 155(362, 364), 163(381)
Normand, B., 38(116)
Noronha, S. J., 282(258), 283(266, 267)
Nücker, N., 174(410)

O

Ochiai, A., 11(32), 13(43), 74, 76(199),
    77(200), 79(202), 84(209), 85(210, 211),
    86(213), 90(217), 138(315)
Oeschler, N., 39(119)
Ohama, T., 94(233)
Ohishi, Y., 127(297)
Ohm, T., 164(395)
Ohoyama, K., 121(279)
Ohwada, K., 94(233), 103(249)
Okane, T., 50(139)
Okazaki, K., 108(254)
Okubo, T., 126(293)
Okumura, T., 225(92)
Oleś, A., 5(14), 165(399)
Olson, C. G., 17(66, 67)

Onuki, Y., 17(64), 20(72, 73), 52(143), 65(181), 66(183)
Opahle, I., 51(142)
Oppeneer, P. M., 13(45), 51(142)
Osborn, R., 22(77)
Osumi, H., 77(200)
Ott, H. R., 2
Ouchni, F., 119(276)
Ovchinnikov, A., 81(205)
Ovchinnikov, A. A., 91(221)
Ozaki, M., 28(92)

P

Pabi, S. K., 193(32), 196(36)
Pallson, G., 159(368)
Palstra, T. T. M., 94(232)
Pande, C. S., 196(42)
Pankert, D., 87(215), 95(235)
Pankov, S., 73(194)
Paschen, S., 74(197)
Patterson, B. D., 84(209), 138(315)
Patterson, B. R., 234(111)
Paul, I., 73(194)
Paulus, B., 154(354)
Pawlowski, A., 196
Pearson, J. J., 79
Penc, K., 140(318)
Penn, D. R., 164(384)
Penning, P., 274(242), 278
Penrose, O., 185(15)
Pepin, C., 39(120)
Perkins, N. B., 121(280)
Perlov, A. Y., 13(45)
Perovic, A., 197(43)
Perring, T. G., 119(274)
Peschel, I., 13(44)
Petermann, J., 198(56)
Petit, L., 50(140)
Peuchert, U., 97(238)
Pfeuty, P., 36(113)
Pfleiderer, C., 34(108)
Pharr, G. M., 256(164)
Pincus, P., 92(222)
Pines, D., 59(163)
Pinettes, C., 11(34)
Pink, E., 274(246)
Plakida, N. M., 59(167)
Planes, J., 283(265)

Platzmann, P. M., 70(188)
Plessing, J., 283(265)
Plummer, E. W., 147(332)
Pochet, P., 184(11)
Pokor, C., 249(142)
Pollmann, F., 56(146), 57(148), 140, 145(323), 146(324)
Polyakov, A. M., 140(317)
Pomeau, Y., 182(2)
Pontikis, V., 264(192)
Poppe, U., 20(69)
Porter, D. A., 205(69)
Pott, R., 46(126)
Predel, B., 215(84), 219(88)
Prelovsek, N. B., 42(123)
Presura, C., 97(240)
Preuss, R., 159(371)
Prigogine, I., 182(5)
Pruschke, T., 131(303), 154(352), 159(369), 160(373)
Puls, M. P., 198(61, 62)
Pulst, U., 22(78), 23(80)
Purdy, G., 232(105–107)
Purdy, G. R., 186(18), 196(39), 197(43), 198(66), 215
Pyka, N., 79(202)
Pytte, E., 92(223)

Q

Qin, S., 91(220)
Qiu, X., 109(258)
Quaouire, L., 282(258)

R

Radu, T., 38(117)
Raj, S. V., 256(164)
Rajesh, S., 282(264)
Ralph, B., 256(163)
Ranzetta, G. V. T., 193(27)
Rapp, M., 267(217)
Rauchschwalbe, U., 20(69)
Raw, G., 188(23)
Rayleigh, Lord, 227
Razavi, F., 34(103)
Rebizant, J., 171(403)
Reese, M., 26(87), 57(148)

Regnault, L. P., 20(70), 103(247), 119(273, 275)
Reich, D. N., 91(218)
Reinert, F., 16(55)
Revcolevschi, A., 103(247), 113(265), 119(272, 273, 275), 120(278)
Reynolds, C. A., 59(150)
Rhee, M., 264(194)
Rhines, F. N., 234(111)
Rice, T. M., 38(116), 48(132, 133), 108, 131(303), 173–176
Ridley, N., 193(29, 31), 207, 209(79)
Riseborough, P. S., 127(296)
Rizzi, E., 284(272)
Robinson, M. T., 249(139)
Roessli, B., 49(136), 60(169–171)
Roll, U., 215(84)
Rollett, A. D., 235(120), 240(131)
Romberg, H., 174(410)
Ronnpagel, D., 262(184)
Rosch, A., 46(129)
Rosenberg, M. J., 69(186)
Rossat-Mignot, J., 20(70)
Roth, G., 12(35), 97(238)
Roth, L., 164(389)
Ruckenstein, A., 13(46), 160(374)
Rudigier, H., 2(3)
Ruegg, C., 38(115)
Runge, E., 13(42), 20(71), 57(147), 146(324)
Ruvalds, J., 162(377)
Ryum, N., 236, 239

S

Sachdev, S., 32(93)
Sahni, P. S., 235(118)
Saito, Y., 50(139)
Sakai, F., 126(293)
Sakai, O., 81(206), 91(220)
Sakata, M., 133(310)
Sakurai, K., 52(143)
Sarrao, J. L., 17(67)
Sasagawa, T., 162(380)
Sass, S. L., 249(148)
Sato, K., 50(139)
Sato, N., 23(83), 49(137), 50(139), 60(169, 170, 172)
Sato, N. K., 49(136)
Satoh, K., 20(72)

Savrasov, S., 159(368)
Sawa, H., 94(233), 103(249)
Sawatzky, G., 166(402)
Saxlova, M., 272(235, 236)
Scalapino, D. J., 59(158), 159(370)
Scalettar, R. T., 17(59)
Scattergood, R. O., 260(172)
Schachinger, E., 155(365)
Schäfer, H., 2(2)
Schank, C., 60(171)
Schefzyk, R., 46(126)
Scheil, E., 198(46), 203
Schilling, J. S., 34(103)
Schilling, W., 249(143, 151)
Schlesinger, Z., 161(376)
Schlüter, M., 172(405)
Schmidt, B., 11(31), 42(124), 74(197), 87(215, 216)
Schmidt, S., 16(55)
Schmidt, W., 119(275)
Schmidt-Rink, S., 59(159)
Schmitt-Rink, S., 13(46), 160(374)
Schneider, R., 120(278)
Schreiner, T., 12(35)
Schrieffer, J. R., 12(38, 39), 33(97), 152(336)
Schultz, T., 36(112)
Schulze-Briese, C., 84(209), 138(315)
Schwartz, L. M., 153(343)
Schwarz, K., 127(295)
Schwedler, A. R., 59(157)
Schweitzer, H., 162(378)
Schwenk, H., 135(311)
Seki, M., 50(139)
Seki, S., 164(392)
Senthil, T., 146(328)
Seo, Y., 225(92)
Serin, B., 59(150)
Shannon, N., 109(259), 119(276), 140(318)
Shapiro, J. M., 198(55)
Shen, Z. -X., 154(358, 359)
Shi, M., 84(209), 138(315)
Shiba, H., 81(206), 91(220)
Shiflet, G. J., 197(44), 211(81)
Shiina, R., 49(137)
Shimaoka, Y., 127(298)
Shimizu, K., 127(297)
Shimizu, M., 154(347)
Shimizu, T., 20(72)

Shingu, M., 139(316)
Shiraki, T., 126(293)
Shu, J. Y., 265(201)
Shvindlerman, L. S., 235(117)
Si, Q., 39(119), 46(129)
Sigrist, M., 28(90), 38(116), 48(132, 133)
Simon, J. P., 243(138)
Simoyama, J. -I., 154(357)
Singh, D. J., 127(295)
Skalski, S., 34(104)
Slater, J. C., 6, 72
Smith, J. L., 2(3, 4), 34(106)
Smolinsky, H., 97(238)
Soderholm, L., 84(209), 138(315)
Solorzano, G., 219(88)
Solorzano, I. G., 196(39), 215
Somer, M., 94(229)
Sommerfeld, A., 8, 76, 81, 87, 126, 130, 132
Sondhi, S. L., 146(326)
Soven, P., 153(338)
Spalek, J., 59(161)
Spanjaard, D., 164(382, 383)
Sparn, G., 16(54), 20(69), 23(82, 83), 105(252)
Srolovitz, D. J., 235(118–120)
Staker, M. R., 256(162)
Stansbury, E. E., 209(78)
Staub, U., 84(209), 138(315)
Steglich, F., 2(2), 16(54), 20(69), 22(79), 23(82, 83), 38(117), 39(119, 120), 49(137), 60(171), 85(211), 86(213), 87(215), 95(235), 96(237)
Steiner, M. J., 39(118)
Stewart, G. R., 2(4)
Stiegler, J. O., 249(150)
Stirling, W. G., 35(110)
Stockert, O., 16(54), 22(79)
Stoll, H., 154(354), 159(366)
Stollhoff, G., 165(399–401)
Stoner, E., 32–34
Stuesser, N., 22(79)
Su, W. P., 12(38, 39)
Sundquist, B. E., 198(54, 58, 63), 212
Suryanarayanan, R., 113(265), 119(272, 273, 275), 120(278)
Sutton, A. P., 234(110)
Suzuki, T., 11(32), 13(43), 79(202), 85(210, 211), 90(217)
Svane, A., 50(140)

Swenson, C. A., 126(291)
Szotek, Z., 50(140)

T

Tachiki, M., 164(392)
Tajima, K., 77(200)
Takagi, H., 126(293), 133(310), 162(380)
Takahashi, Y., 59(162)
Takata, M., 133(310)
Takeda, K., 127(297)
Takens, F., 282(257)
Takeuchi, J., 127(297, 298)
Takke, R., 46(126)
Tamura, K., 50(139)
Tanaka, A., 50(139)
Tanaka, R., 225(92)
Tanatar, B., 71(191)
Tang, C., 280(254, 255)
Tautz, F. S., 17(65), 20(74)
Tayama, T., 23(83)
Taylor, D. W., 153(339)
Tegus, O., 39(119)
Temmerman, W. M., 50(140)
Terasaki, I., 92(225)
Thalmeier, P., 11(31), 13(45), 16(53, 54), 24(84), 37(114), 42(124), 46(127), 49(137), 61(174), 62(179), 63(180), 65(181), 66(182), 87(215, 216), 97(239), 99(242–244), 101(246), 106(253), 110(261), 114(267), 119(272, 273, 275, 276), 137(314), 144(322), 165(400)
Thomas, H., 91(219)
Thompson, C. V., 186(20), 235
Tohyama, T., 174(409)
Tokiwa, Y., 39(120), 52(143)
Tokura, Y., 6(22), 108(255), 113(262, 263), 114(266), 119(274)
Tomioka, Y., 114(266)
Toth, L. S., 262(183)
Tou, H., 66(183)
Tran, M. -T., 144(322)
Treaftis, H. N., 201(67)
Treglia, G., 164(382, 383)
Trovarelli, O., 16(55), 39(120)
Tsakiropoulos, P., 193(30)
Tsuda, N., 135(312)
Tsvelik, A. N., 12(37)
Tu, J. J., 155(365)

AUTHOR INDEX

Tu, K. -N., 198(52)
Turnbull, D., 198(48, 52), 201–203, 235(113)
Tyer, R., 50(140)

U

Uchinokura, K., 92(225)
Ueda, K., 28(90, 91), 59(162), 81(206), 91(220), 132(308)
Ueda, Y., 92(226), 94(233), 103(248, 249), 105(251), 108(254)
Uhlarz, M., 34(108)
Uimin, G., 81(205)
Uji, S., 17(64), 20(72, 73)
Umehara, I., 20(72)
Umezawa, L., 164(392)
Umstätter, R., 164(395)
Ungar, T., 262(183, 184)
Unger, P., 164(396, 397), 174(407, 408)
Urabe, T., 225(93)
Urano, C., 126(293)

V

Valenti, R., 105(250)
Valla, T., 155(363)
Valsakumar, M. C., 282(259, 260)
van de Kamp, R., 119(272, 275)
Van der Giessen, E., 264(195), 265(198, 200–202)
van der Marel, D., 97(240), 166(402)
van der Merwe, A., 188(23)
van Smaalen, S., 94(230, 231)
Varelogiannis, G., 49(137)
Varma, C. M., 13(46, 47), 16(58), 59(159), 160(374)
Vaughan, G. B. M., 84(209), 138(315)
Velický, B., 153(340)
Verdier, M., 243(134–138)
Verga, S., 154(360)
Vergergaugry, J. L., 184(12)
Vergnol, J., 274(245), 283(265)
Verwey, E. J. W., 3, 124, 125
Victora, R. H., 164(393)
Vidal, C., 182(2, 7)
Virosztek, A., 162(377)
Vleck, J. H. V., 6
Vollhardt, D., 154(350)

Vollmer, R., 34(108)
von der Linden, W., 159(371)
von Schnering, H. G., 94(229)
Voorhees, P. W., 185(13), 229(103)
Vorderwisch, P., 38(115), 119(275)

W

Wagner, R., 185(13)
Waintal, X., 71(192)
Wakabayashi, N., 77(200)
Wakabayashi, Y., 113(263)
Walgraef, D., 182(6), 253, 261(175), 270, 271
Walker, I. R., 39(118)
Walz, F., 125(287)
Wand, B., 87(215)
Wang, R., 267(217)
Ward, J. C., 10(30)
Wastin, F., 171(403)
Watanabe, K. I. T., 65(181)
Watanabe, T., 66(182)
Weaire, D., 186(19), 235
Weatherly, G. C., 196(39)
Weber, D., 24(84)
Weber, W., 97(238), 164(395)
Weickert, F., 24(85)
Weiden, M., 94(229), 95(235), 96(237), 97(238)
Weiser, S., 164(395)
Weissmann, S., 255(161)
Wells, B. O., 155(363)
Wen, X. -G., 140(321)
Werner, E., 227
Wertheim, G. K., 147(331)
West, D. F. R., 193(27, 28)
Weygand, D., 235(124, 125), 238(128), 239(129, 130), 242(133), 264(195)
White, R., 53(144)
White, S. R., 159(370)
Whiting, M. J., 193(30)
Wiesenfeld, K., 280(254, 255)
Wigner, E., 3, 7, 40, 66–74, 91, 108, 177, 179
Wildes, A., 38(115)
Wilhelm, H., 38(117), 39(120)
Wilkens, M., 262(184)
Wilkins, J. W., 16(56, 57)
Williams, D. B., 196(38, 41)

Willis, J. O., 2(4)
Wills, J. M., 50(141)
Winter, A. T., 267(212, 213)
Wohlleben, D. K., 46(126)
Wolf, B., 24(84)
Woods, S. B., 76, 126
Wright, W. H., 59(150)

Y

Yamada, Y., 115(268)
Yamagami, H., 17(67)
Yamagani, H., 52(143)
Yamaki, K., 50(139)
Yamamoto, E., 52(143), 66(183)
Yamamoto, N., 133(310)
Yamashita, Y., 132(308)
Yamauchi, T., 108(254)
Yanagisawa, T., 52(143)
Yaresko, A., 136(313), 137(314)
Yaresko, A. N., 13(45), 97(239), 125(290), 129(299)
Yonezawa, F., 153(341)
Yoshitake, R., 77(200)
Yoshizawa, H., 117(271), 121(279)
Yosihama, T., 103(248)
Yotsuhashi, S., 132(307)
Yourgrau, W., 188(23)
Yu, L., 40(122)
Yuan, Q., 114(267)
Yunoki, S., 122(281)
Yushankhai, V., 37(114), 38(117), 144(322)
Yusof, Z., 155(363)

Z

Zaanen, J., 5(14)
Zaiser, M., 258(165), 266(204, 207–210), 267
Zawadowski, A., 15(52)
Zbib, H. M., 264(194)
Zener, C., 108(256), 198–203, 208–210, 215, 239
Zerec, I., 42(124)
Zevin, V., 12(36)
Zhang, F. C., 108, 173–176
Zhang, G. M., 40(122)
Zhang, Y. Z., 137(314), 144(322)
Zheng, H., 115(270)
Zheng, W. J., 154(357)
Zhou, S. Y., 162(380)
Zhou, X. J., 154(357, 358)
Zhu, L., 39(119), 46(129)
Zieba, P., 196, 202(68), 205(70–73)
Ziegenbein, A., 284(272)
Ziman, T., 103(247)
Zimanyi, G. T., 160(375)
Zinkle, S. J., 253(155)
Zölfl, M., 131(303)
Zvyagin, A. A., 129(299)
Zwanzig, R., 150(335)
Zwicknagl, G., 9(29), 11(33), 12(36), 16(53, 54), 17(61), 18(68), 20(71), 22(78, 79), 26(87), 27(88, 89), 56(145, 146), 57(147, 148)

Subject Index

A

AF structures, 60–61
AlV$_2$O$_4$
  characterization, 133–134
  charge disproportionation, 136–139
  structural transitions, 134–136
  V ions, 134
α'-NaV$_2$O$_5$
  coupled CO, 98–107
  crystal structure, 94–95
  electronic structure, 97–98
  exchange dimerization in, 98–107
  ground state, 91
  Jordan–Winger mapping, 91–92
  structural phase transition, 92–93
  temperature structure, 93–97

B

Bands
  persistent slip, 268–272
  shadow, 151–153
  theory, 18–19
BEC. See Bose–Einstein condensation
Bilayer manganites. See Double exchange bilayer manganites
Bose–Einstein condensation, 38

C

Cahn's treatment, 204–206
CeCu$_2$Si$_2$
  heavy fermions, 19–23
  phase diagram, 23–30
CEF. See Crystalline electric fields
CeRu$_2$Si$_2$, 19–23
Charge ordering
  $4f$ compounds, 67–68
  concept of, 66
  density in metals, 68–69
  double exchange bilayer manganites
    as function of e_g, 110–112
    model, 117–122

    phase diagram, 112–114
    physics of, 108–110
  in 3D-valence elections, 67, 68
  in homogeneous 2D electron systems, 69–74
  original proposal, 66–67
  quasi-1D spin excitations
    α'-NaV$_2$O$_5$, 91–107
    electronic structure, 97–98
    Hubbard model, 98–99
    Ising-spin Peierls model, 101–107
    ITF model, 99–101
    Jordan–Wigner mapping, 91–92
    β-vanadium bronzes, 107–108
    structural phase transition, 92–93
    temperature structure, 93–97
  Wigner crystallization
    $4f$ model Hamiltonian, 81–87
    Dzyaloshinsky–Moriya interaction, 88–91
    experimental summary, 74–76
    Labbé–Friedel model, 81
    underlying physics, 77–81
Chemical patterning
  characterization, 185–186, 192–193
  coarsening
    discontinuous, 230–231
    overview, 225, 227
    spheroidization, 227–230
  discontinuous precipitation
    characterization, 194–196
    divergent lamellar growth, 224–225
    interface motion, 219–220
    multilayer stability, 232
    spacing selection, 216, 218
  morphological evolution, 225, 227–230
  non-steady state growth
    concerns, 218–219
    divergent lamellar, 220–222, 224–225
    interface motion, 219–220
  spacing selection, 213, 215–218
  steady state growth
    Cahn's treatment, 204–206
    discontinuous precipitation, 211–213

Chemical pattering (*Continued*)
    Hillbert's treatment, 202–203
    pearlite observations, 206–211
    theoretical history, 197–199
    Turnbull's treatment, 202–203
    Zener treatment, 199–202
  transformations
    -discontinuous precipitation, comparison, 198–199
    characterization, 193–195
    divergent lamellar growth, 220–222, 224
    experimental observations, 206–207
    Fe-C steels, 208–208
    identification, 196
    multilayer stability, 232
    regularity of, 193, 197
    steady state growth, 206–211
CMR. *See* Colossal magnetoresistance effect
Co-Si system, 230–231
Coarsening
  discontinuous, 230–231
  overview, 225, 227
  spheroidization, 227–230
Coherent potential approximation, 153–154
Colossal magnetoresistance effect, 108
Condensation, Bose–Einstein, 38
Cooper pairs, 59
Copper-oxide planes, 171–176
Correlated systems with orbital degeneracies, 48–51
CPA. *See* Coherent potential approximation
Creation operators, 26–27
Crystalline electric fields
  characterization, 15
  doublet-quartet scheme, 23, 35
  in rare earth ions, 59
  singlet states, 36–38
Crystallization. *See* Wigner crystallization
Cu-Al systems, 206–207
Cuprates perovskite structures, 13–14

D

Discontinuous precipitation
  -transformation, comparison, 198–199
  chemical patterning
    characterization, 194–195
    identification, 196

    interface motion, 219–220
    Klinger's alternative, 216, 218
    theory/experiment comparisons, 211–213
  divergent lamellar growth, 224–225
Dislocations
  characterization, 233–234
  patterning
    cell formation, 258
    dynamic nature of, 261–262
    first theory, 260
    Hähner model, 266–267
    Holt model, 260–261
    Kocks model, 261
  structures
    carpets, 249, 252
    density of, 245
    in PSB, 268–272
    recovery
      density decreases, 245
      kinetics, 243
      processes, 243
      study results, 248
    recrystallization
      evolution of, 240–241
      nucleation of, 241–243
      study results, 248
    void lattices, 249, 252
Divergent lamellar growth, 220–222, 224
Double exchange bilayer manganites
  reentrant charge ordering
    as function of e_g, 110–112
    electronic structure calculations, 110
    model, 117–122
    phase diagram, 112–114
    physics of, 108–110
    polaron formation, 114–117
    structural properties, 109–110
Dynamical mean field theory, 17
Dzyaloshinsky–Moriya interaction, 88–91

E

Elastic strains, 190
Energy
  chemical, 188–189
  free, 77–78
  free energy, 189–191

high (*See* High-energy excitations)
interfacial, 190–191
low, scales, 10–12
madelung, 87
strain, 190
Equilibrium
systems maintained far from, 191–192
systems prepared far from
chemical energy, 188–189
interfacial energy, 190–191
strain energy, 190
structural patterning in, 186–187
Eutectoid decomposition, 220–222, 224–225

F

Fatigue loading, 267–268, 270–272
Fault migration, 229
Fe-C steels, 208–208
Fe-C-Mn systems, 222, 224
Fe-C-Mo systems, 222, 224
Fe-C-Si systems, 222, 224
Fe_3/O_4, 124–125, 130
Fermi-liquid state
$CeCu_2Si_2$, 19–23
$CeRu_2Si_2$, 19–23
high-energy excitations
combined models, 153–154
findings, 159–163
methodology, 154–159
mean-field theory, 16–17
renormalized band theory, 18–19
strongly correlated electrons deviations, 12–14
Wigner crystallization, 69, 71
Fermions, heavy, 19–23
Five f systems, 168–171
Free energy
at phase/grain boundary interface, 190–191
elastic strain-induced, 190
strain-free regions, 189
Free energy changes
systems continuously maintained, 191–192
systems preparation
characterization, 188–189
chemical energy, 189

interfacial energy, 190–191
strain energy, 190

G

Geometrically frustrated lattices
characterization, 123
charge disproportionation, 133–139
fractional charges, 139–147
function, 122–123
metallic spinels, 124–133
structural transition, 133–139
Grain growth
characterization, 234
computer models, 235–236
experiments, 234–235
log normal distributions, 236–240
Green function, 7–8

H

Hähner model, 266–267
Heavy quasiparticles
in U compounds
dual model application, 51–54
dual model theory, 50–51
superconductivity, 59–66
High-energy excitations
copper-oxide planes, 171–176
energies, 147–148
Hubbard model, 151–153
in $5f$ systems, multiplet effects, 168–171
kink structure, 153–163, 154–159
marginal Fermi liquid behavior
combined models, 153–154
findings, 159–163
methodology, 154–159
Ni satellite structure, 163–168
projection operators, 148, 149–150
research summary, 176–178
zero-point fluctuations, 149
Hillbert's treatment, 202–203
Holt model, 260–261
Hubbard model
for high-energy excitations, 151–153
quasi-1D spin excitations, 98–99
Hückel theory, 3

I

Interatomic correlations, 3–4
Interface motion, 219–220
Interfaces at phases/grain boundaries, 190–191
Intra-atomic correlations
  function, 4
  in superconductivity, 59–66
  strength measure, 5–6
Irradiation
  main effect of, 252
  rate equations, 253
  structural defect patterning
    experimental characteristics, 249, 252
    rate equations, 253
    reaction diffusion approach, 252–255
    void formation, 254–255, 258
Ising model in transverse field, 99–101
ITF. See Ising model in transverse field
Itinerant ferromagnetism models, 32–37

J

Jordan–Wigner transformation, 91–92

K

Kink excitations, 12–13
Kink structure, 153–163
Klinger's alternative, 216, 218
Kocks model, 261
Kondo lattice systems
  Ce-based compounds, 15–16
  $CeCu_2Si_2$, 19–23
  $CeRu_2Si_2$, 19–23
  Fermi-liquid state, 18–19
  function, 14–15
  heavy fermions, 19–23
  heavy quasiparticles, 18–19
  magnetic ions, 16–17
  quantum criticality in, 38–45
  renormalized band theory, 18–19
  scaling theory, 45–48
Kondo necklace model, 40

L

Labbé–Friedel model, 81
Lamellar growth, divergent, 220–222, 224

Lamellar patterning
  characterization, 225, 227
  discontinuous precipitation, 224–225
  pearlite, 220–222, 224
  spheroidization, 227–230
$LaMnO_3$
  electronic structure calculations, 110
  physics, 108
  structural properties, 109–110
Landau theory, 9–10
$LaSr_2Mn_2O_2$
  electronic structure calculations, 110
  physics of, 108
  structural properties, 109–110
LDA. See Local density approximations
Line defects. See Dislocations
LiV_3/O_4, 125–133
Local density approximations, 51
Log normal distributions, 236–240
Low-energy scales, 10–12

M

Madelung energy, 87
Magnetic ion lattice, 16–17
Magnets, itinerant, 32–38
Magnets, localized, 32–38
Manganites. See Double exchange bilayer manganites
Metallic spinels
  Fe_3/O_4, 124–125
  LiV_3/O_4, 125–133
Metals. See also specific metals
  density in, charge ordering, 68–69
  Landau theory, 9–10
  quasiparticle energies, 8–9
  transitions, 6

N

Nickel satellite structure, 163–168
Non-steady state growth
  chemical patterning
    concerns, 218–219
    divergent lamellar, 220–222, 224–225
    interface motion, 219–220

O

Orbital degeneracies
  in correlated systems
    dual model, 49–51
    examples, 48–49

P

Partial localization
  concept of, 48–49
  in U compounds
    heavy quasiparticles in, 51–54
    LDA, 51
    microscopic model calculations, 54–58
Patterns
  Belgian school, 182–183
  chemical (See Chemical patterning)
  concept of, 182
  energetic costs, 183
  equilibrium departure, 184
  in metallic alloys, 183–184
  spatio-temporal (See Spatio-temporal patterning)
  structural (See Structural defect patterning)
Pearlite transformations
  -discontinuous precipitation, comparison, 198–199
  characterization, 193–195
  divergent lamellar growth, 220–222, 224
  experimental observations, 206–207
  Fe-C steels, 208–208
  identification, 196
  regularity of, 193, 197
  steady state growth, 206–211
Persistent slip bands, 268–272
Phase diagrams
  $CeCu_2Si_2$, 23–30
  charge ordering, 112–114
Phase transitions
  $CeCu_2Si_2$, 23–30
  free energy, 190–191
  quantum
    broken symmetry state, 30–32
    in itinerant magnets, 32–38
    in localized magnets, 32–38
  structural, 92–93
Plasticity
  in structural defect patterning

classical approaches, 258, 260–263
computer simulations, 263–266
experimental facts, 255–256
fatigue loading, 267–268, 270–272
spatio-temporal patterning in
  chaotic behavior of, 282–284
  characterization, 273–274
  phenomenology of, 274–276
  serrated flow, statistical analysis, 276–280
stress-strain curve, 187
PLC. See Portevin–Chatelier effect
Polarons
  formation in double exchange bilayer manganites, 114–117
  hopping element, 116
  JT type, 115
Polyacethylene, 12–13
Portevin–Chatelier effect
  chaotic behavior of, 282–284
  characterization, 273–274
  phenomenology of, 274–276
  serrated flow, statistical analysis, 276–280
PSB. See Persistent slip bands
Pyrochlore lattice
  band structure, 127–128
  characterization, 123
  spin degrees of freedom, 129

Q

QCP. See Quantum critical point
QCT. See Quantum phase transitions
QPC. See Quantum critical point
Quantum critical point
  non-Fermi behavior in, 14
  QPT in, 31
  scaling theory, 45–48
Quantum criticality, 38–45
Quantum phase transitions
  broken symmetry state, 30–32, 31–32
  in itinerant magnets, 32–38
  in localized magnets, 32–38
  onset of, 35–36
Quasi-1D spin excitations
  charge ordering
    α'-NaV_2O_5, 91–107
    electronic structure, 97–98
    Hubbard model, 98–99

Quasi-1D spin excitations (*Continued*)
    ITF model, 99–101
    β-vanadium bronzes, 107–108
    structural phase transition, 92–93
    temperature structure, 93–97
  Ising-spin Peierls model, 101–107
  Jordan–Wigner mapping, 91–92
Quasiparticles. *See* Heavy quasiparticles

R

Raleigh instability, 227–228
Rate equations, 253
Reaction diffusion approach, 252–255
Recrystallization, 240–243

S

β-Vanadium bronzes, 107–108
SDW. *See* Spin-density wave
Self organized criticality, 276–280
Serrated flow, 276–280
Shadow bands, 151–153
SOC. *See* Self organized criticality
Spacing selection
  chemical patterning
    discontinuous precipitation, 216, 218
    problems, 213, 215–218
Spatio-temporal patterning
  PLC effect
    chaotic behavior of, 282–284
    characterization, 273–274
    phenomenology of, 274–276
    serrated flow, statistical analysis, 276–280
    stress-strain curve, 187
Spheroidization, 227–230
Spin excitations. *See* Quasi-1D spin excitations
Spin-change separation, 12–13
Spin-density wave, 24
Steady state growth. *See also* Non-steady state growth
  chemical patterning
    Cahn's treatment, 204–206
    discontinuous precipitation, 211–213
    Hillbert's treatment, 202–203
    pearlite observations, 206–211
    theoretical history, 197–199

    Turnbull's treatment, 202–203
    Zener treatment, 199–202
Stick-slip model, 277, 279
Stoner–Wolfarth theory, 32
Strain energy, 190
Strain-free regions, 189
Strong correlations, 139–147
Strongly correlated electrons
  charge ordering
    α'-NaV$_2$O$_5$, 91–108
    concept of, 66–69
    generalized Wigner lattice, 74–91
    homogeneous 2D systems, 69–74
    ID spin excitations, 91–108
    reentrant, 108–122
    Wigner crystallization, 69–74
    Yb$_4$As$_3$, 74–91
  Fermi-liquid behavior deviations, 12–14
  geometrically frustrated lattices
    characterization, 122–124
    charge disproportionation, 133–139
    fractional changes, 139–147
    metallic spinels, 124–133
    structural transitions, 133–139
  high-energy excitations
    $5f$ systems, 168–171
    characterizations, 147–149
    copper-oxide planes, 171–176
    future research, 176–178
    Hubbard model, 151–153
    kink structure, 153–163
    marginal Fermi liquid behavior, 153–163
    Ni satellites, 163–168
    projection operators, 149–150
    shadow bands, 151–153
  Kondo lattice systems
    band theory, 18–19
    CeCu$_2$Si$_2$, 19–30
    CeRu$_2$Si$_2$, 19–23
    characterization, 14–18
    Fermi-liquid state, 18–19
    heavy fermions, 19–23
    heavy quasiparticles, 18–19
    low-temperature phase, 23–30
    quantum criticality, 38–45
  low-energy scales, 10–12
  overview, 2–8
  partial localization

SUBJECT INDEX

characterization, 48–51
intra-atomic excitations, 59–66
microscopic model calculation, 54–59
quasiparticles, 51–54
superconductivity, 59–66
UPd_2Al_3, 51–54
quantum phase transitions
characterization, 30–32
critical point, 45–48
criticality, 38–45
itinerant magnets, 32–38
localized magnets, 32–38
scaling theory, 45–48
theories, 8–10
Structural defect patterning
characterization, 233–234
dislocation structures
cold rolling stain, 245, 248
density of, 245, 248
evolution, 240–241
nucleation of, 241–243
recovery kinetics, 243, 248
recovery observations, 243
grain growth
characterization, 234
computer models, 235–236
experiments, 234
Log Normal distributions, 236–240
irradiation-induced
experimental characteristics, 249, 252
rate equations, 253
reaction diffusion approach, 252–255
void formation, 254–255
plasticity-induced
classical approaches, 258, 260–263
computer simulations, 263–266
experimental facts, 255–256, 258
fatigue loading, 267–268, 270–272
Hähner model, 266–267
Structural patterning, 186–187
Structural transition, 133–139
Superconductivity
A phase interplay, 24–26
$CeCu_2S_2$, 24–29
intra-atomic excitations, 59–66

T

Transformations. *See* Pearlite transformations
Transitions metals, 6
Turnbull's treatment, 202–203

U

U compounds
AF structures, 60
heavy quasiparticles
dual model application, 51–54
dual model theory, 50–51
microscopic model calculations, 54–58
superconductivity, 59–66

V

Void formation, 254–255

W

Wigner crystallization
charge ordering
$4f$ model Hamiltonian, 81–87
Dzyaloshinsky–Moriya interaction, 88–91
experimental summary, 74–76
Labbé–Friedel model, 81
underlying physics, 77–81
in homogeneous 2D electrons, 69–74

Y

Yb_4, change ordering
$4f$ model Hamiltonian, 81–87
elastic constant, 78–80
free energy, 77–78
interacting chains, 81
lattice structure, 74–76
Madelung energy, 87

Z

Zener treatment, 199–202